D1573994

Drug Testing In Vitro

Edited by
Uwe Marx and Volker Sandig

1807–2007 Knowledge for Generations

Each generation has its unique needs and aspirations. When Charles Wiley first opened his small printing shop in lower Manhattan in 1807, it was a generation of boundless potential searching for an identity. And we were there, helping to define a new American literary tradition. Over half a century later, in the midst of the Second Industrial Revolution, it was a generation focused on building the future. Once again, we were there, supplying the critical scientific, technical, and engineering knowledge that helped frame the world. Throughout the 20th Century, and into the new millennium, nations began to reach out beyond their own borders and a new international community was born. Wiley was there, expanding its operations around the world to enable a global exchange of ideas, opinions, and know-how.

For 200 years, Wiley has been an integral part of each generation's journey, enabling the flow of information and understanding necessary to meet their needs and fulfill their aspirations. Today, bold new technologies are changing the way we live and learn. Wiley will be there, providing you the must-have knowledge you need to imagine new worlds, new possibilities, and new opportunities.

Generations come and go, but you can always count on Wiley to provide you the knowledge you need, when and where you need it!

William J. Pesce
President and Chief Executive Officer

Peter Booth Wiley
Chairman of the Board

Drug Testing In Vitro

Breakthroughs and Trends in Cell Culture Technology

Edited by
Uwe Marx and Volker Sandig

WILEY-VCH Verlag GmbH & Co. KGaA

The Editors

Dr. Uwe Marx
Dr. Volker Sandig

ProBioGen AG
Goethestrasse 54
13086 Berlin
Germany

Library of Congress Card No.: applied for

British Library Cataloguing-in-Publication Data:
A catalogue record for this book is available from the British Library.

Bibliographic information published by the Deutsche Nationalbibliothek
The Deutsche Nationalbibliothek lists this publication in the Deutsche Nationalbibliografie; detailed bibliographic data are available in the Internet at http://dnb.d-nb.de.

Typesetting Manuela Treindl, Laaber
Printing betz-druck GmbH, Darmstadt
Binding Litges & Dopf Buchbinderei GmbH, Heppenheim

Printed in the Federal Republic of Germany
Printed on acid-free paper

ISBN: 978-3-527-31488-1

Foreword

In recent years, few methods have changed so dramatically as those used *in vitro* for drug development. The main areas of progress involve target finding and validation with molecular biology, and *in-vitro* testing of drug safety to the production of biological molecules. The application of these methods has made all steps of drug development not only faster, but also less costly.

As research in all areas is continuing apace with major efforts, we can expect major breakthroughs with the maturation of human micro-organoid *in-vitro* cultures in the near future. Clearly, however, because of the increased sophistication and specialization of these investigations, an even greater need for team work is indicated.

Hence, it is mandatory for those of us involved in drug development to keep pace with the continuous progress in methodology, by receiving the experts' overview, as presented in this book.

Frankfurt a. M., July 2006

Prof. Dr. Rolf Krebs
former Chairman of the
Board of Managing Directors
of Boehringer Ingelheim GmbH

Drug Testing In Vitro: Breakthroughs and Trends in Cell Culture Technology
Edited by Uwe Marx and Volker Sandig

ISBN: 978-3-527-31488-1

Preface

At the beginning of the 21st century, the development of medicine is suffering from two major obstacles.

First, new drug candidates directed at *pivotal human receptors* can have unprecedented positive or negative biological effects involving systemic interactive networks specific to humans. None of the animal species or human cell lines can properly imitate the biological effects on these networks. Consequently, few relevant data on the efficacy and safety of new drugs can be obtained for evaluation prior to human testing. A prime example is the super-agonist antibody TGN1412, which was developed to direct the immune system to fight cancer cells or to reduce arthritis pain, and has triggered multiple organ failure in healthy volunteers undergoing experimental testing. In binding the CD28-receptor, the antibody overrides the basic control mechanism of the whole immune system. Yet whilst adhering to standard clinical rcsearch guidelines, the drug showed absolutely no adverse effects in studies with animals.

Second, significant drawbacks – such as severe adverse side effects – often occur after drugs have entered the market. Today, there are increasing indications that *specific genetic predisposition* is one of the key reasons for these high-profile recalls. This human genetic diversity is rarely addressed in preclinical and clinical safety studies at the present time. A sound hypothesis on the correlation of morbidity of patients treated with roferoxib (Vioxx) with the genotype for 5-LOX and 5-LOX activating protein polymorphisms, is one of many examples describing this obstacle.

The breakthrough might be to develop high-throughput, human micro-organoid *in-vitro* test systems. In mammals, organs and systems are built up by multiple identical functionally self-reliant structural units, with easily remembered examples *in vivo* being the liver acinus, β-cell islets in the pancreas, alveoli in the lung, or germinal centers in lymph nodes. When science and industry succeed in designing human micro-organoids *in vitro* that fully emulate these *in-vivo* counterparts, the dream of drug testing predictive to individual human exposure might became reality. For at least 30 years the vision of proper modeling of these human micro-organoids *in vitro* to gain knowledge about their performance and function in man – and consequently to use them for highly predictive drug screening and testing purposes – has been set back by prohibitive scientific and technological bottlenecks. However, achievements made over the past seven

Drug Testing In Vitro: Breakthroughs and Trends in Cell Culture Technology
Edited by Uwe Marx and Volker Sandig

ISBN: 978-3-527-31488-1

years have substantially changed this starting position, and the multidisciplinary contributions in this book introduce different aspects leading towards anticipated short-term progress in that area.

- Part I brings together an overview of new and forthcoming tissue models, the challenges to be met by the development of bioreactors, and the biosensoric microstructures for control and measurement. An illustration of complexity is provided by the biomonitoring of airborne contaminants *in vitro*.
- Part II combines overviews of state-of-the-art *in-vitro* techniques in conventional monolayer and suspension culture systems, with the potential of two relatively new technological platforms – the creation of human designer cell lines and stem cell technologies. The latter provides basic guidelines of how to overcome the chronic bottleneck of sustainable, human genotyped cell and tissue supply.
- Part III emphasizes the tension between ethical, regulatory and commercial aspects of drug testing and screening on human micro-organoids *in vitro* as a viable alternative to animal testing.
- Part IV concludes with the tremendous potential of the anticipated emerging *in-vitro* drug evaluation platform technology, including a road map enforcing them.

The book is introduced by a personal statement of Rolf Krebs, former chairman of the Board of Managing Directors of Boehringer Ingelheim.

Progress anticipated in the emerging platform technology can have significant impact beyond the borders of drug screening and testing. In Europe, at least, legislative pressures such as the cosmetics directive and the retrospective REACH (Registration, Evaluation and Authorisation of Chemicals) program for 30 000 chemicals, has created a dramatic increase in industrial interest in predictive human *in-vitro* tissue culture test systems for the evaluation of cosmetics, chemicals, or nutraceuticals. Hence, this book also provides useful inside information for professionals from those areas.

Finally, the book would not exist with the outstanding creative assistance of Silke Hoffmann and Philip Saunders.

Berlin, October 2006 *Uwe Marx and Volker Sandig*

Contents

Drug Testing In Vitro: Breakthroughs and Trends in Cell Culture Technology
Edited by Uwe Marx and Volker Sandig

ISBN: 978-3-527-31488-1

List of Contributors

Shahnaz Bakand
The University of New South Wales
Chemical Safety and Applied
Toxicology (CSAT) Laboratories
School of Safety Science
Sydney, NSW 2052
Australia

Paul R. Bidez III
Drexel University
School of Biomedical Engineering,
Science and Health Systems
3141 Chestnut Street
Philadelphia, PA 19104
USA

Christa Burger
Merck KGaA
Biotechnology and Protein Chemistry
Frankfurter Strasse 250
64293 Darmstadt
Germany

Christine M. Finck
Department of Pediatrics and
Pediatric Surgery
St. Christopher's Hospital for
Children
Philadelphia, PA 19134
USA

Christoph Giese
ProBioGen AG
Cell Culture R&D and
Tissue Engineering
Goethestrasse 54
13086 Berlin
Germany

Thomas Hartung
European Centre for the Validation
of Alternative Methods
21020 Ispra
Italy

Amanda Hayes
The University of New South Wales
Chemical Safety and Applied
Toxicology (CSAT) Laboratories
School of Safety Science
Sydney, NSW 2052
Australia

Ingo Jordan
ProBioGen AG
Goethestrasse 54
13086 Berlin
Germany

Drug Testing In Vitro: Breakthroughs and Trends in Cell Culture Technology
Edited by Uwe Marx and Volker Sandig

ISBN: 978-3-527-31488-1

Mirek R. Jurzak
Merck KGaA
Central Assay Department and
Screening
Frankfurter Strasse 250
64293 Darmstadt
Germany

Martina Klemm
ProteoSys AG
Carl-Zeiss-Strasse 51
55129 Mainz
Germany

Christopher D. Koharski
Department of Commerce (USPTO)
600 Dulany Street
Alexandria, VA 22314
USA

Rolf Krebs
Grosse Gallusstrasse
60311 Frankfurt/Main
Germany

Philip Lazarovici
The Hebrew University of Jerusalem
Department of Pharmacology and
Experimental Therapeutics
School of Pharmacy
Faculty of Medicine
Jerusalem 91120
Israel

Shimon Lecht
The Hebrew University of Jerusalem
Department of Pharmacology and
Experimental Therapeutics
School of Pharmacy
Faculty of Medicine
Jerusalem 91120
Israel

Peter I. Lelkes
Drexel University
School of Biomedical Engineering,
Science and Health Systems
3141 Chestnut Street
Philadelphia, PA 19104
USA

Mengyan Li
Drexel University
School of Biomedical Engineering,
Science and Health Systems
3141 Chestnut Street
Philadelphia, PA 19104
USA

Uwe Marx
ProBioGen AG
Goethestrasse 54
13086 Berlin
Germany

Anthony Meager
The National Institute for Biological
Standards and Control
Biotherapeutics
Blanche Lane
South Mimms
Hertfordshire EN6 3QG
United Kingdom

Mark J. Mondrinos
Drexel University
School of Biomedical Engineering,
Science and Health Systems
3141 Chestnut Street
Philadelphia, PA 19104
USA

Anat Perets
Drexel University
School of Biomedical Engineering,
Science and Health Systems
3141 Chestnut Street
Philadelphia, PA 19104
USA

Ralf Pörtner
Hamburg University of Technology
Institute for Bioprocess and
Biosystems Engineering
Denickestrasse 15
21071 Hamburg
Germany

Oliver Pöschke
Merck KGaA
Central Assay Department and
Screening
Frankfurter Strasse 250
64293 Darmstadt
Germany

Andrea A. Robitzki
University of Leipzig
Department of Molecular
Biological-Biochemical Processing
Technology
Deutscher Platz 5
04103 Leipzig
Germany

Andrée Rothermel
University of Leipzig
Department of Molecular
Biological-Biochemical Processing
Technology
Deutscher Platz 5
04103 Leipzig
Germany

Volker Sandig
ProBioGen AG
Goethestrasse 54
13086 Berlin
Germany

André Schrattenholz
ProteoSys AG
Carl-Zeiss-Strasse 51
55129 Mainz
Germany

Horst Spielmann
Federal Institute for Risk Assessment
Bundesinstitut für Risikobewertung
BfR
Diedersdorfer Weg 1
12277 Berlin
Germany

Glyn N. Stacey
The National Institute for Biological
Standards and Control
Cell Biology and Imaging Division
Blanche Lane
South Mimms
Hertfordshire EN6 3QG
United Kingdom

Chris Winder
The University of New South Wales
Chemical Safety and Applied
Toxicology (CSAT) Laboratories
School of Safety Science
Sydney, NSW 2052
Australia

Part I
Emerging *In-Vitro* Culture Technologies

Drug Testing In Vitro: Breakthroughs and Trends in Cell Culture Technology
Edited by Uwe Marx and Volker Sandig

ISBN: 978-3-527-31488-1

1
Intelligent Biomatrices and Engineered Tissue Constructs: *In-Vitro* Models for Drug Discovery and Toxicity Testing

Philip Lazarovici, Mengyan Li, Anat Perets, Mark J. Mondrinos, Shimon Lecht, Christopher D. Koharski, Paul R. Bidez III, Christine M. Finck, and Peter I. Lelkes

1.1
Introduction

The rapid progress in combinatorial chemistry continues to yield a myriad of potentially bioactive compounds every day. With thousands of pharmaceutically valuable drugs at hand, there is an urgent need for engineered tissue equivalents that could serve as *in-vitro* model systems during the initial stages of drug discovery, specifically during the preclinical stages of cell/tissue-based high-throughput screening (HTS). Given the well-known problematics of using two-dimensional (2D) cell cultures as pharmacological test-beds, more realistic three-dimensional (3D) tissue constructs are required. Generation of high-fidelity engineered tissue-like constructs is based on the targeted interactions of organ-specific cells and "intelligent" biomimetic scaffolds, emulating the complex natural environment, the extracellular matrix (ECM) in which these cells develop/differentiate and function.

In this chapter, we will begin by introducing the most common natural and synthetic materials and platform biotechnologies for creating scaffolds that are in use for engineering tissue constructs, and which might be useful for pharmaceutical purposes, as models for drug discovery and *in-vitro* testing. A variety of approaches for modifying the chemistry and geometry of these materials towards rendering them useful as intelligent biomatrices for 3D tissue engineering will be presented. We then will discuss, both paradigmatically and critically, the fabrication principles and potential use of engineered constructs in cardiac, hepatic, and pulmonary tissue engineering. Finally, we will explore in more detail the current and future use of engineered models of the blood–brain barrier (BBB), where the use of novel "intelligent" biomaterials and scaffolds promises to break the current impasse in engineering high-fidelity *in-vitro* models of the BBB and thus facilitate reliable and predictive HTS of central nervous system (CNS)-active compounds during early stages of drug discovery.

Drug Testing In Vitro: Breakthroughs and Trends in Cell Culture Technology
Edited by Uwe Marx and Volker Sandig

ISBN: 978-3-527-31488-1

Although the application of bioengineered 3D *in-vitro* tissue models in pharmaceutical research is in its infancy, the interdisciplinary approach and the achievements described in this chapter provide an encouraging first step towards the accelerated development of such models for drug discovery.

1.2 Intelligent Biomaterials and Scaffolds for Tissue Engineering

Currently, a broad range of synthetic polymers are being used for scaffolding, including poly(ε-caprolactone) (PCL), poly(L-lactide-*co*-ε-caprolactone) (P(LLA-CL)), polyglycolic acid (PGA), polylactic acid (PLA), or copolymer poly (lactidic-*co*-glycolic) (PLGA) [1–4], as well as natural biomaterials such as alginate, elastin, collagen, or gelatin [5–9]. Ideally, all scaffold materials should be nontoxic, biocompatible, biodegradable, and nonimmunogenic. Moreover, for use *in vivo*, scaffolds should – with only few exceptions such as cartilage and cornea – be able to induce angiogenesis to facilitate blood supply to, and waste removal from, the newly formed tissues. The "intelligence" of a biomaterial and the ensuing scaffolds can be gauged from its competence to induce and maintain organ-specific differentiation and function of the cells growing in/on them and generating tissue-like constructs.

1.2.1 Synthetic Materials

One of the main advantages of synthetic materials is the ability to precisely control their physico-chemical properties, such as molecular weight of the polymer, strength, degradation time, mechanical properties, and hydrophobicity [10]. Amongst the most widely used polymers in tissue engineering are the poly (α-hydroxy acids) of aliphatic polyesters, such as PLA, PGA, and PLGA [11]. These synthetic polymers can be produced in numerous physical forms, including meshes, sponges, and films, and molded into many shapes, depending on the type of tissue one wishes to emulate (e.g., heart, kidney, ear). The rate of biodegradation of PLGA scaffolds depends not only on the ratio of lactide and glycolide, but also on whether the polymer is mixed with polycaprolactone, collagen, or other synthetic polymers [12–14]. Cell attachment can be improved by covalently modifying the polymers, or by passively coating the scaffolds [15]. Growth factors can also be incorporated into the matrix in order to improve biocompatibility [16].

Since their discovery some 30 years ago, electrically conductive polymers – also known as "synthetic metals" – have been used in many areas of applied chemistry and physics, such as light-emitting diodes and batteries [17]. More recently, there has also been a growing interest in conductive polymers for diverse biomedical applications, specifically as conductive scaffolds for cardiac and neural tissue engineering. The rationale for using conductive polymers is based on the fact that the eukaryotic cell plasma membrane is charged and that, specifically in neurons and myocytes, a multitude of cell functions, such as attachment, proliferation, migration and differentiation could be modulated through electrical stimulation [18–22]. Common classes of organic conductive polymers include polyacetylene, polypyrrole (PPy), polythiophene, polyaniline (PANi), and poly(para-phenylene vinylene). Some of these conductive polymers (especially PPy) have found certain biomedical applications, such as for the immobilization of proteins [23, 24]. Christine Schmidt and her co-workers were the first to employ PPy for tissue engineering purposes [18, 25–28]. Interestingly, this group most recently described a novel 12-mer peptide (T59) that selective binds to conductive PPy and promotes cell attachment [29] This peptide may become useful for immobilizing a variety of bioactive molecules on PPy and other synthetic/conductive polymers, without altering their bulk properties.

Some recent studies have used PANi, another well-characterized organic conducting polymer, as an electroactive substrate for tissue engineering applications [30–33]. PANi is biocompatible *in vitro* and in long-term animal studies *in vivo* [31]. To date, most of these studies have investigated the biological properties of PANi solvent-cast into 2D films, rather than engineered into 3D (nano) fibrous scaffolds. A few years ago, Díaz et al. [34] reported that doped, conductive PANi blended with polystyrene (PS) and/or polyethylene oxide (PEO) could be electrospun into nanofibers. In extending these studies, we recently co-electrospun PANi with gelatin, for instance denatured collagen, to yield nanofibrous scaffolds which are both highly biocompatible and electroactive, and may be suitable for applications in cardiac and cardiovascular tissue engineering [35].

1.2.2 Natural Biomaterials

Natural biomaterials for scaffold fabrication include both purified ECM proteins, such as collagen, elastin, ECM derivatives, such as Matrigel™ and small intestinal submucosa (SIS™) acellular matrix, as well as materials derived from marine plants and crustaceans, such as alginate and chitosan. Natural biomaterials more closely mimic, than synthetic polymers, both function and structure of the native extracellular environment. Natural biomaterials, such as collagens, are largely conserved among different species and provide a readily available source of materials for tissue engineering. Importantly, when used as 3D matrices – either as hydrogels or as fibrous or porous scaffolds – these materials can serve as ubiquitous (but occasionally also species-or tissue-specific) templates for cell attachment, growth and differentiation.

Matrigel™, an ECM derivative isolated from the murine Engelbreth-Holm-Swarm (EHS) sarcoma, is a complex mixture of basement membrane proteins, mostly laminin and type IV collagen, which also contains a large number of essential growth factors and cytokines. Unlike artificial synthetic scaffolds, Matrigel™ provides a natural, biocompatible environment [36], which induces organotypic differentiation of cells cultured on or in this hydrogel, because of the complexity of its composition and its viscoelastic properties. Given its biological complexity, Matrigel™ provides an excellent differentiative environment for *in-vitro* tissue engineering application; its use *in vivo* in animal models is mainly restricted to syngeneic mice. Sheets made of the acellularized porcine SIS are in clinical use as well-tolerated, xenogeneic scaffolds, inducing variable degrees of tissue-specific remodeling in the organ or tissue into which it is placed [37]. SIS is mostly composed of type I collagen, though it has some type III and type IV collagen in addition to other ECM molecules, such as fibronectin, hyaluronic acid, chondroitin sulfate A and B, heparin, and heparin sulfate, and some growth factors, such as basic fibroblast growth factor (bFGF), transforming growth factor (TGF), and vascular endothelial growth factor (VEGF) [37].

Collagen and elastin are two key structural ECM components in many tissues [8, 13]. These proteins are important modulators of the physical properties of many types of engineered scaffolds, affecting cellular attachment, growth and responses to mechanical stimuli [38, 39]. Matthews et al. [40] and Boland et al. [41] were the first to generate 3D micro- and nano-fibrous scaffolds from collagen and elastin for cardiovascular tissue engineering by electrospinning (see below). Tropoelastin, the cellular precursor of elastin, is secreted from elastogenic cells as a 60-kDa monomer that is subjected to oxidation by lysyl oxidase. Subsequent protein–protein associations give rise to massive macroarrays of elastin, for example, in the inner elastic lamina of arterial blood vessels. As a consequence, elastin is a substantially insoluble protein network that displays elasticity, resilience, and biological persistence. Soluble elastin is typically available either as fragmented elastin in the form of alpha- and kappa-elastin [42], or through expression of the natural monomer, tropoelastin [43]. Recently, tropoelastin was also electrospun into scaffolds for tissue engineering purposes [9].

Alginate hydrogels (generally 1%, w/v in water) can change their physical state towards hydrogel, depending on the cross-linker and calcium chloride concentration [44]. Due to their versatile viscoelastic properties and adjustable porosity (> 80%), alginate scaffolds have been used for a number of diverse tissue engineering applications such as the liver [45] or pancreas. Alginate scaffolds could also provide a charged surface environment to facilitate the 3D culturing of cardiac cells, and can also be used for regeneration and healing of the myocardium after heart failure [46].

The complexity of the ECM composition *in situ* makes it difficult to fully emulate the "organ-specific environment" *ex vivo*, either by design and/or synthesis. However, the use of natural biomaterials, either alone or in combination with other natural or synthetic polymers, such as collagen/glycosaminoglycans or collagen/PLGA, may improve the biocompatibility of the ensuing scaffolds by reducing inflammatory responses *in vivo* and improving initial cell attachment and differentiation. Cells growing in such an instructive environment are stimulated to remodel this "provisional matrix" in a tissue-specific fashion; thus, these "naturally intelligent matrices" provide the necessary cues for organotypic differentiation and assembly of engineered tissue constructs.

1.3 Fabrication of Scaffolds for Tissue Engineering

As discussed above, an ideal scaffold for tissue engineering must provide the necessary mechanical/structural support and contain the appropriate instructive/differentiative cues, such as the capability of inducing neovascularization, to allow the tissue-engineered construct to be integrated into the host surroundings [47]. Most scaffolds presently being investigated in animal research promote some cellular ingrowth, and may hence be quite useful for *in-vitro* applications. Overall, however, most of these scaffolds pose significant limitations to host integration *in situ*, including a host-versus-graft immunological response, and do not offer a very effective basis for organ replacement. Therefore, there is a need to develop novel scaffolds and approaches.

In addition to using hydrogel-based scaffolds made of collagen or fibrin, several novel techniques have been developed for engineering 3D "solid" scaffolds with enhanced mechanical properties. The microscopic/nanoscale structure and function of biological macromolecules constituting conventional hydrogels are important for cell physiology. However, the relatively weak mechanical properties of hydrogel scaffolds pose a major drawback, especially *in vivo*. Thus, diverse "solid" scaffolds, made of water-soluble polymers (collagen, fibrin, and alginate) with improved mechanical properties have been engineered using controlled freezing and thawing procedures, followed by crosslinking. Such scaffolds have shown excellent biocompatibility and facilitated excellent cellular ingrowth both *in vitro* and *in vivo*. More recently, solid nano/microfibrous scaffolds have been generated by electrospinning or acellularization (see Section 1.3.1). These fibrous scaffolds more realistically emulate salient structural and biological features of the natural ECM, and seem to be very well suited as substrates for 3D tissue engineering purposes, both in *in vitro* and *in vivo*. One of advantages of 3D nanofibrous scaffolds is the small diameter of the fibers, which is similar to the diameters of ECM proteins *in situ*. Such small fiber diameters provide a relatively large surface-to-volume ratio, enabling the absorption of liquids and facilitating cellular attachment and cell–cell interaction. These scaffolds also exhibit unique mechanical properties which permit better cell penetration and proliferation within the scaffolds as compared to 3D hydrogels. Recent data have suggested the possibility of generating hybrid scaffolds for cardiac tissue engineering by combining hydrogel and solid scaffolds comprised of both synthetic and natural biopolymers such as PLGA, collagen, or elastin [13].

1.3.1 Electrospinning

The process of electrospinning, which has been well known for many years in the textile industry and in organic polymer science [48–50], has recently emerged as a novel tool for generating biopolymer scaffolding for tissue engineering [51]. This is a process for the production of polymer filaments using an electrostatic force

[52, 53]. In this process, a polymer solution is introduced into the electrical field generated by a high-voltage power supply. The polymer filaments are formed from the solution traveling between two electrodes bearing electrical charges of opposite polarity. Upon ejection from a metal spinneret through a small hole, the solvent in the charged solution jet rapidly evaporates, thus generating ultrathin fibers, which then are deposited onto the collector [54]. Optimization of this technique depends to a large part on the material to be electrospun, and involves adjusting crucial parameters, such as the nature of the solvent and the concentration of the solute, as well as the potential difference and the distance between the electrodes [9]. Electrospinning is a novel, increasingly important platform technology for producing nanofibrous scaffolds from a variety of polymer materials, including synthetic polymers, natural proteins, and blends of natural and synthetic materials [9, 35, 55]. Figure 1.1 illustrates typical (autofluorescent) light-microscopic images of electrospun elastin (Fig. 1.1A) and gelatin (Fig. 1.1B), as well as scanning electron microscopy (SEM) images of PLGA (1.1C), and a blend of PANi-gelatin fibers (Fig. 1.1D).

The topology of these electrospun scaffolds closely mimics that of the native ECM; it is particularly striking in the case of the wavy appearance of elastin, reminiscent of the elastic lamina in blood vessels. Depending on the spinning conditions, fibers with diameters in the range from several micrometers to less

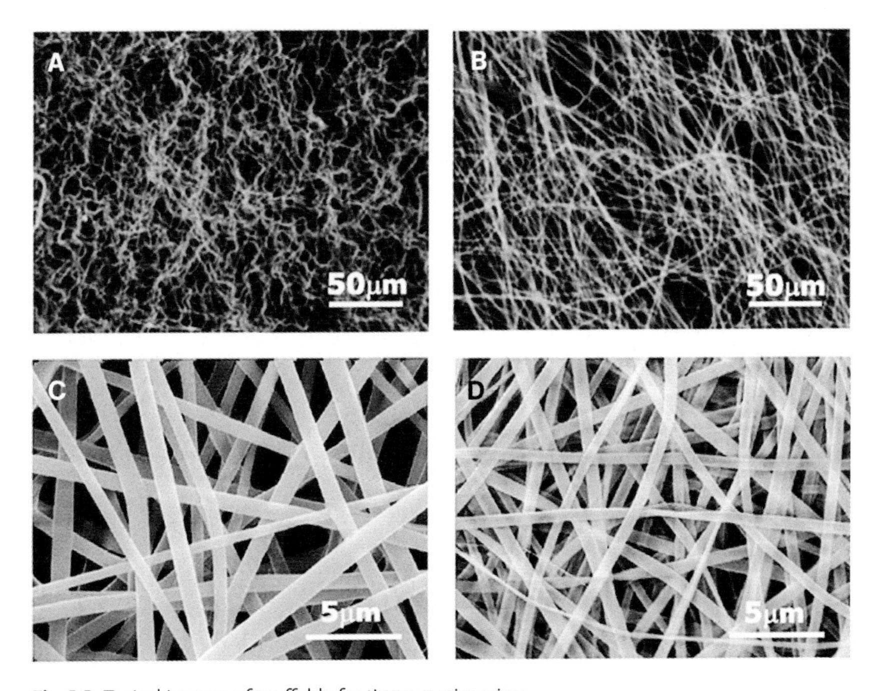

Fig. 1.1 Typical images of scaffolds for tissue engineering. Electrospun elastin (A) and gelatin (B) fibers. Scanning electron microscopy micrographs of electrospun PLGA (C) and PANi-gelatin (D) fibers.

than 100 nm are obtained. These fibrous scaffolds have a very high surface area to mass ratio, and can be electrospun into 3D scaffolds with very high porosity. These biomimetic matrices facilitate cell attachment, support cell growth, and regulate cell differentiation [56].

1.3.2 Controlled Lyophilization

Lyophilization, also called "freeze-drying", is a process by which the material is rapidly frozen and dehydrated under high vacuum [57, 58]. Controlled lyophilization techniques in which the rate of freezing and thawing can be tightly controlled, is a useful platform technology for engineering porous 3D scaffolds [59, 60]. The pore size and porosity of the ensuing scaffolds will depend primarily on the composition/concentration of the scaffold materials and the different parameters of the lyophilization protocol. Complex ECM extracts, such as Matrigel™ and other hydrogels, contain numerous differentiation cues, which are either not present in synthetic scaffolds or destroyed from natural proteins through conventional scaffold preparation techniques. Lyophilization, and to some extent also electrospinning, are rather gentle methods for creating 3D scaffolds that retain the complexity of the ingredients and the functionality of natural ECM proteins. Moreover, lyophilized ECM extracts can be used either directly as a porous scaffold or further processed by electrospinning to yield nanofibrous scaffolds, which retain bioactive proteins and growth factors necessary for tissue-specific cell proliferation and differentiation. Figure 1.2 illustrates SEM micrographs of lyophilized gelatin (Fig. 1.2A) and Matrigel™ (Fig. 1.2B) scaffolds, indicating the highly connected regular porous structure, which is particularly well suited for engineering vital organs, such as the liver.

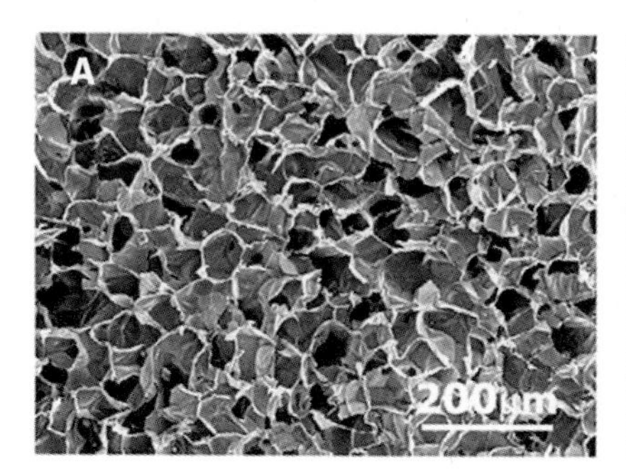

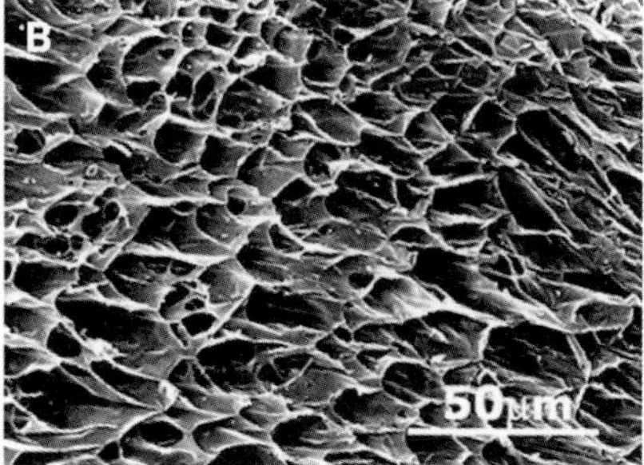

Fig. 1.2 Typical images of natural scaffolds for tissue engineering. Scanning electron microscopy micrographs of (A) lyophilized gelatin and (B) Matrigel™ scaffolds.

1.3.3 Acellularization

A subset of specialized natural scaffolds that may hold promise as organotypic matrices are acellularized tissues. The process of acellularization has been studied extensively, with acellularized tissues as scaffolds for tissue engineering currently in clinical use, for example as natural matrices for cardiac valve replacement [61, 62]. Similarly, acellularized tissue-based scaffolds made from SIS are currently being considered for the repair of damaged myocardium after cardiac heart failure [63]. All of these acellularized scaffolds are created by using a combination of hypotonic lysis buffers, detergents, and/or proteolytic enzymes to remove cells and their constituents from native tissues, while leaving behind the maximal amount of intact ECM proteins and preserving as much as possible the native organizational structure [64]. The nature and bioactivity of the residual ECM proteins retained upon acellularization depend largely on the specific methodologies used to generate these acellularized tissues. Indeed, if performed carefully, acellularized tissue can also retain vascular conduits, thus facilitating homogeneous seeding of the constructs through the vascular pedicles. For all these approaches, different methodologies are being used to disrupt cell membranes, to rupture the intracellular organelles, and finally to remove residual intracellular debris and nucleic acids. As an example, Zhong et al. [65] used a two-step approach, in which detergent (sodium dodecyl sulfate, SDS) extraction was followed by a proteolytic (trypsin) digestion yielding acellularized porcine aortic scaffolds, comprised mainly of collagen and elastin. Such acellularized blood vessels are currently being tested in preclinical studies as scaffolds for vascular grafts [66]. Using a similar approach, we have recently generated acellularized murine hearts and lungs. Figure 1.3 shows light microscopy (Fig. 1.3A) and SEM (Fig. 1.3B) images of acellularized heart tissue. The complete acellularization of heart tissue is demonstrated in Figure 1.3A. Of interest, this gentle gradual acellularization process retains the nanofibrous morphology of the cardiac ECM (Fig. 1.3B) and tissue specificity in terms of suitability for cell seeding, as assessed by the mechanical properties and the patterns of cell growth and differentiation (C. D. Koharski et al., unpublished results).

In conclusion, a number of novel platform technologies have been developed in recent years that permit fine tuning of the physico-chemical characteristics (fibrous versus porous, mechanical properties, porosity, etc.) of natural and synthetic scaffolds for tissue engineering purposes. The choice of these technologies and of the particular scaffold properties is driven by the need optimally to emulate the native ECM in a particular target tissue.

Following the introduction of some of the most frequently used natural and synthetic biomaterials and platform technologies for creating "intelligent" matrices for tissue engineering, we will now discuss four different applications (liver, heart, lung, BBB), in which the use of these scaffolds is critical for engineering high-fidelity tissue constructs, which might serve as 3D *in vitro* models for drug discovery and toxicity testing.

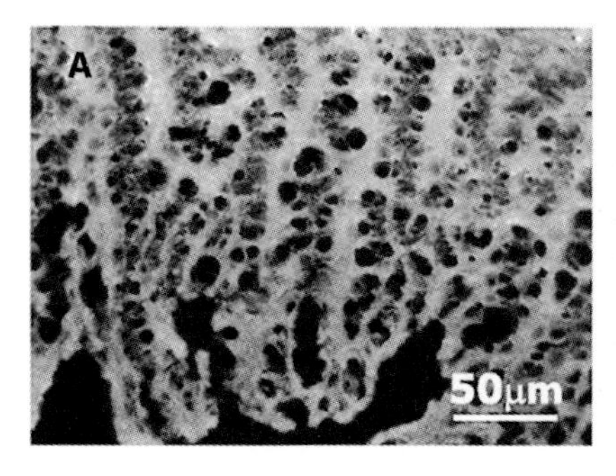

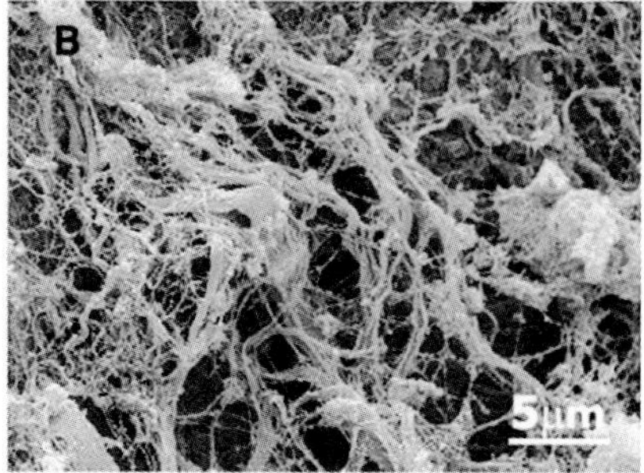

Fig. 1.3 Typical images of acellularized scaffolds for tissue engineering. (A) Nuclear staining with bisbenzimide, (B) scanning electron microscopy image of acellularized heart tissue.

1.4 Progress and Achievements in Liver Tissue Engineering

1.4.1 The Liver

The liver is a complex soft organ composed of parenchymal cells, such as hepatocytes, hepatocyte precursor cells (oval cells or Ito cells), stellate cells, Kupffer cells, epithelial cells, sinusoidal epithelial cells, biliary epithelial cells, and fibroblasts [67]. Liver cells are mitotically quiescent, although they are all capable of undergoing division, for example after partial resection of the liver. Hepatocytes are instrumental in detoxification and drug metabolism, involving a large group of enzymes coupled to different families of cytochrome P-450 proteins. Other crucial functions of the liver include the production/synthesis of albumin, blood clotting factors and α/β globulins [10]. In spite of the fact that a (healthy) liver has a large capacity for regeneration, and that liver transplants have become more feasible (i.e., through live donors), there is a continual need for engineered liver tissue for implantation in those patients with hepatic failure. Engineered liver tissue may be also useful for the design of *in-vitro* pharmacological models that could assist in the discovery and characterization of novel compounds, which target liver diseases or for predicting the liver metabolism of newly developed drugs.

Hepatocytes, the basic cellular units of the liver, represent 60% of all the cells in the liver; therefore, most attempts at engineering liver tissue focus primarily on their use. A first critical step in liver tissue engineering is the development of an appropriate scaffold (matrix) on which cultured hepatocytes can attain and retain the full repertoire of their biological functions. To achieve these goals, different 3D scaffolds were developed which mimicked the liver environment and allowed suitable growth and function of hepatocytes.

1.4.2
Scaffolds for Liver Tissue Engineering

In general, either mixed cultures of primary isolated cells [68] or freshly isolated purified primary hepatocytes are applied to the scaffold [69]. In order to generate a functional liver construct, the liver cells, once seeded onto the scaffolds, must migrate into the scaffold interior, expand, and populate the new tissue constructs. To facilitate these processes the scaffolds must provide a biocompatible surface suitable for cell recognition and adherence. Following attachment, the cells must proliferate and finally differentiate in order to perform their physiological function.

Some of the critical issues in designing a liver scaffold are listed in Table 1.1.

The most frequently used scaffolds for liver tissue engineering are made of synthetic biodegradable polymers such as PGA, PLA and PLGA, natural polymers such as collagen, alginate and chitosan, as well as several combinations of synthetic and/or natural polymers [70, 71]. Based on a more complete understanding of the importance of cell-selective adhesion motifs [71], some of the above polymers have been modified chemically to increase their biocompatibility [72]. In this context, biocompatibility is defined as the ability of a given material to perform with an appropriate host response in a specific biological application [73]. While natural proteins, such as alginate and chitosan possess relative good biocompatibility, the biocompatibility of PLGA is transient and biphasic: over a short time period, PLGA is well tolerated, but it may produce inflammatory responses in the long term. Depending on its biological provenance and type or way of preparation, collagen may also induce an inflammatory response. In the context of liver tissue engineering it is also important that the scaffolds will be *biodegradable* – that is, they will be degraded and eliminated from the body, and also be *bioerodible* – that is, their degradation products will lack toxicity or immunogenicity (Table 1.1).

Scaffold porosity (see Table 1.1) is crucial for nutrient and gas exchange, which are essential for hepatocyte survival. Pores of 100 to 500 μm diameter are optimal for hepatocyte growth, as they increase the internal surface area of the liver scaffold for optimized cell attachment and increased penetration of blood vessels [74]. Other important factors to be considered in the design of liver scaffolds are mechanical properties, such as elasticity and stability (Table 1.1).

Recently, the requirements for the design of liver scaffolds have become more sophisticated. It is increasingly recognized that in aspiring to fully mimic and recreate *in vitro* the tissue environment, scaffolds, besides providing structural support for cell growth, are an integral part of bioactive, complex systems. Such "high-fidelity" biomimetic systems for liver tissue engineering include blends of different natural and/or synthetic polymers [75], heterotypic cell populations [68], and a variety of growth factors required both for growth and differentiation of the hepatocytes and promotion of blood vessels ingrowth from the host into the scaffold to ensure long-term survival and function of the constructs [76]. This approach is exemplified by the results presented in Figure 1.4.

Table 1.1 Characterization of scaffolds commonly used for liver engineering.

Scaffold (polymer)	Physical properties (pores shape and size)	Chemical and mechanical properties	Bio-compatibility	Pharmacokinetics (bioerodible/ biodegradability)	Advantages and disadvantages	Ref.
Polyglycolic acid (PGA); poly-L-lactic acid (PLLA); polylactide-*co*-glycolide (PLGA)	High porosity; pore size can be controlled	Synthetic; hydrophobic polymers; insoluble in water; shape and porosity can be easily adjusted; suitable for chemical modifications	Synthetic, different from natural ECM FDA approved	Yes Scaffold fabrication involves toxic reagents; degraded upon hydrolytic attack of the ester bond, resulting in random degradation products	Reproducible; manufactured; easy scale-up; can be used for printing scaffolds; cell seeding is limited to scaffold periphery; change physical structure in medium or after implantation. Tendency to crumble upon degradation; induced prolonged inflammatory response	[68] [83] [221] [222] [223] [224] [225]
Alginate hydrogel	Highly inter-connecting porous structure (> 90%); pore size: 100–200 μm; pore can be modulated by freezing temperature	Hydrophilic, high water retention; convenient preparation method (The gelatin-freeze-dry method); easily isolated from brown algae; can be modified to increase cell attachment	Contain ECM-like materials	Yes	Able to hold microspheres for controlled release of growth factors (VEGF, FGF, etc.) enhancing scaffold angiogenesis	[44] [45] [71] [77] [79] [226] [227]
Collagen	High porosity (80%); can shape as fibers with diameter of ~0.5–2.0 mm; pore size range: 200–1650 μm	Isolation is complicated and expensive	Naturally derived ECM, but immunogenic	Yes	Collagen layers play an important role in preventing dedifferentiation of hepatocytes in long-term culture. Provide good cell adhesion; contracted during cell culture	[228] [229] [230] [231] [232]
Chitosan	Highly porous structure (> 90%); pore size: 50–200 μm	Hydrophilic – positively charged; high strength and flexibility; isolation by deacetylation from chitin (arthropod cuticle).	Contains ECM-like materials	Yes Slowly biodegraded by lysozyme. Chitosan and its biodegradation product *in vivo*, glucosamine, are not toxic.	Can be modified to achieve a good attachment. Preparation requires acetic acid extraction; therefore, is difficult to neutralize and remove completely the protons from the scaffold	[233] [234]

ECM = extracellular matrix.

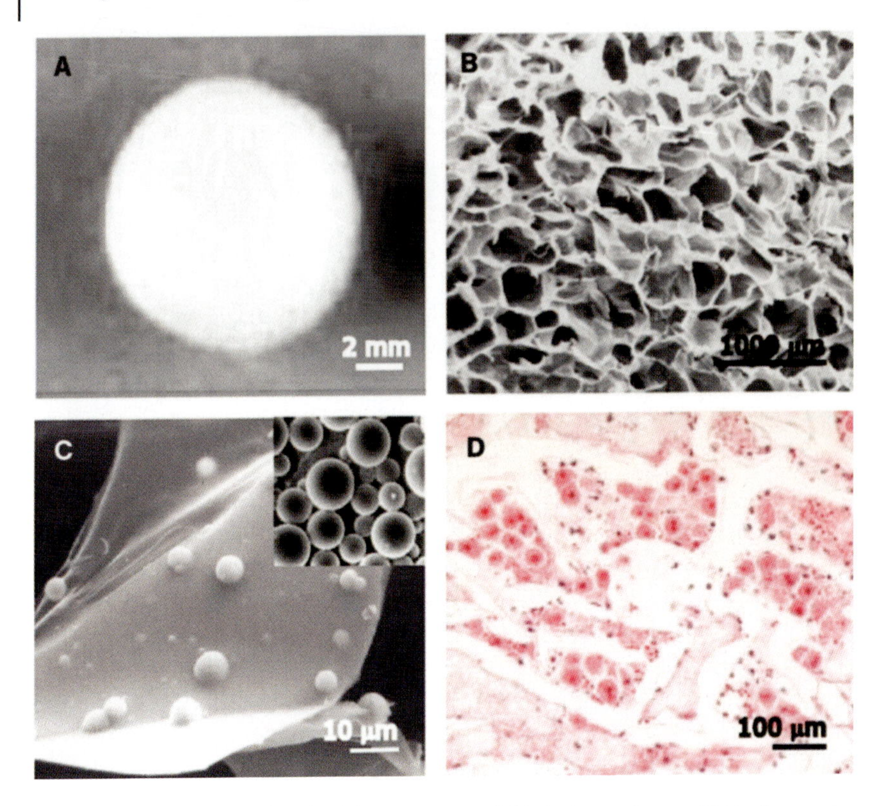

Fig. 1.4 The technology and applicability of alginate composite scaffolds for liver tissue engineering. (A) Macroscopic image of freeze-dried scaffold before population with hepatocytes. (B) Microspheres embedded into the scaffold with minimal reduction of scaffold porosity. (C) The microspheres become an integral part of the freeze-dried scaffold; the insert shows a SEM micrograph of PLGA microspheres (round spheres) before incorporation into the alginate scaffold. (D) A hematoxylin and eosin-stained section through the alginate composite scaffold center, demonstrating the survival and hepatic-like organization of the seeded hepatocytes 3 days after implantation.

Macroscopic scaffolds (Fig. 1.4A) were generated by controlled lyophilization; the porous nature of these scaffolds is shown in Figure 1.4B. PLGA microspheres containing angiogenic growth factors such as bFGF [77] or VEGF [78, 79] (Fig. 1.4C, insert) were incorporated into alginate scaffolds during the lyophilization process, without interfering with the porosity of the scaffold; in this way they became an integral part of the alginate matrix (Fig. 1.4C). These scaffolds were then implanted on one of the liver lobes of Lewis rats, enabling a short initial angiogenic process. After one week of prevascularization, primary hepatocytes [80] were seeded into the implanted VEGF-releasing scaffolds. The hepatocytes in these constructs survived for up to two weeks, and organized in a histiotypic fashion *in vivo* (Fig. 1.4D) [76]. These proof-of-concept studies attest to the feasibility of liver tissue engineering.

However, the long-term goal of extending hepatocyte survival and function in the implanted constructs requires further investigation. Furthermore, the results of these studies suggest that engineered liver constructs could also provide a suitable model for discovery and *in-vitro* testing of liver-specific drugs and therapeutic modalities.

1.4.3
Pharmaceutical Applications of Tissue-Engineered Liver Models

Throughout the drug discovery/development phase there is a continual need for rapidly evaluating the pharmacokinetic properties (absorption, distribution, metabolism and elimination; ADME) of novel drugs. At present, the effects of novel compounds on hepatic metabolism are initially screened in 2D cultures of hepatocytes, followed by testing of a reduced/limited number of promising candidates in animal experiments. Since the number of promising compounds that can be tested in animals is very small, in comparison to the thousands of compounds generated daily in medicinal chemistry laboratories, this method is not efficient for large-scale screening but, rather, is reserved for selected lead compounds. A possible solution would be to accelerate the elimination process by using high-fidelity 3D *in-vitro* tissue models.

In 2D hepatic cultures the expression of the different cytochrome P-450 (CYPs) isoforms is limited, and other relevant functions are missing [81]. Therefore, it is assumed that more complex (and hence more realistic) 3D hepatic constructs may be more relevant for high-throughput drug metabolism screening in pharmaceutical research and development programs. Although measurements of cytochrome P-450 isoform expression/activity constitute important criteria for assessing the functionality of hepatocytes grown on various scaffolds [82, 83], to date very few studies have considered the pharmaceutical applications of engineered liver constructs as pharmacological models in assessing drug metabolism. However, the few published studies do reveal several pharmaceutical trends:

- In hepatic spheroids grown on a 3D peptide scaffold, the expression of CYP isoforms, such as CYP 1A1, 1A2, 1E1, can be induced by 3-methylcholanthrene. The ability to induce these CYP isoforms indicates that the chosen scaffold elicits and/or supports a physiologically correct cell response; hence, this particular scaffold might be useful for pharmaceutical evaluations [81].

- Newer engineering technologies are available to grow hepatocytes in 3D hollow-fiber bioreactors [83]. These assemblies may be useful for pharmaceutical screening purposes, and also serve as clinically relevant extracorporeal bioartificial liver (BAL) devices which can support patients with liver failure [84].

- Recently, small and large animal models of liver disease have been developed, which employ implanted hepatocytes seeded on engineered scaffolds [85, 86]. These models might be useful in the initial steps of hepatic drug discovery.

1.4.4
Conclusions and Novel Trends in Liver Tissue Engineering

Until now, the development of liver tissue engineering has focused primarily on generating functional liver constructs for *in-vivo* use (replacement of diseased liver), or on the engineering of bioartificial hybrid BAL systems for detoxification. Novel types of scaffold that have been designed more recently for hepatic constructs, are well suited not only for implantation but also as matrices for engineering liver tissue-equivalents for basic and applied research, including *in-vitro* models for drug discovery and testing (e.g., for developing hepatic vaccines).

The liver is a complex organ that harbors many types of cell; therefore, more than one cell type is required to develop a truly liver-like engineered tissue. More effort is required to provide a better understanding of the underlying molecular physiology, specifically the differentiative and functional role of cell–cell communication in liver development, as well as the cell–biomaterial interactions if we desire to engineer functional liver tissue constructs for implantation and/or for *in-vitro* drug discovery. This goal can be facilitated by creating complex scaffolds composed of more than one material, assuming that each material will provide support for the different types of liver cells. In order to optimize cell–biomaterials interactions, such as adhesion, migration, proliferation and differentiation of the liver cells on the scaffolds, the matrix must be able to release growth factors in a temporal and spatial gradient-controlled fashion; this will then provide the necessary inductive and differentiative environment for the different types of hepatic cells in a given construct. Continual and controlled release of the growth factors may be necessary for creating durable tissue constructs *in vitro*, in which case these constructs will allow drug testing for extended periods of time, similar to *in-vivo* experiments. We believe that in the future, liver tissue engineering will include, among others, novel platform technologies, such microelectromechanical systems (MEMS) for engineering highly vascularized liver constructs [84], micro-patterned surfaces for the co-culture of hepatocytes and Kupffer cells [87], and integrated hepatocyte spheroid constructs using electrospun nanofibrous scaffolds [88]. The challenges and potential for liver tissue engineering are vast, and some of the present achievements have provided great encouragement to continue the quest.

1.5
Cardiac Tissue Engineering: Cells and Models

1.5.1
Cardiac Tissue Engineering

Cardiac tissue engineering is emerging as a promising approach to replace or support the failing heart [89]. The implantation of cells, native tissues, or *in-vitro*-reconstituted tissue constructs has been employed for more than a decade, and might provide an effective therapy for patients with end-stage heart disease.

Cardiac tissue engineering has been motivated by the need to create functional tissue equivalents for scientific studies and cardiac tissue repair. The ideal cardiac tissue constructs should faithfully replicate the functional and morphological properties of native heart muscle, and also remain viable after implantation. Mechanical, electrical, and functional integration into the organ architecture should result in improved systolic and diastolic function of the diseased myocardium. Thus, constructs should be: (1) contractile; (2) electrophysiologically stable; (3) mechanically robust and flexible; (4) vascularized after implantation; and (5) autologous [90]. As with the other cell-based models discussed in this chapter, high-fidelity *in-vitro* cardiac constructs would also be invaluable as model systems in the early stages of cardiac drug discovery and development.

For clinical applications, cardiac tissue engineering/regenerative medicine can be divided into two categories: (1) the direct implantation of isolated cells [91]; and (2) the implantation of *in-vitro*-engineered tissue constructs [92, 93]. In the case of the former category, most studies indicate that cell implantation in animal models of myocardial infarction can improve contractile function [94]. Indeed, several clinical studies are currently under way to investigate the safety and feasibility of cell implantation in patients [95]. For the latter category, cardiac tissue engineering involves the construction of cardiac tissue equivalents from donor cells, such as myocytes or stem cells, which are first seeded onto/into 3D engineered scaffolds. The constructs are then cultured and mechanically/electrically preconditioned *in vitro* prior to implantation into the diseased heart for reparative purposes.

1.5.2
Cells used in Cardiac Tissue Engineering

Crucial concerns in cardiac tissue engineering are the identification of suitable cell sources and the characterization of the composition of cell populations used in engineering high-fidelity *in-vitro* constructs. For tissue engineering purposes, the heart may be divided into two functions: mechanical, and electrical. Arguably, the main mechanical function (contractility) is mediated by cardiac myocytes, while the electrical activities of the heart are mediated by specific neurons. In the past, the main focus of cardiac tissue engineering has been on the myocytes only. These cells are mechanically and electrically active, have high metabolic rates, and are characterized by prolonged survival. However, under certain pathological conditions, they undergo hypertrophy or a variety of other diseases and defects causing a decrease in the contractile properties of the heart muscle [96]. Therefore, the important design criterion for cardiac engineered tissues is to have a template scaffold that facilitates induction and maintenance of the physiological properties mentioned above in order to ensure proper myocyte development and function towards *in-vivo* implantation.

The restoration of heart function by replacement of the diseased myocardium with functional cardiac myocytes is an attractive strategy. Indeed, cardiac myocytes and other cell types have been successfully implanted into the hearts of larger animals [97], improving contractile function after myocardial infarction [46].

However, a major limitation of cardiac tissue engineering is the inability of cardiac myocytes to proliferate [98]. Since cardiac myocytes are terminally differentiated cells, they cannot compensate for cell loss that occurs during myocardial infarction or chronic heart failure. Endothelial cells, fibroblasts, smooth muscle cells, and neuronal cells comprise some 70% of the total cell number in the working myocardium [99], and play important roles in cardiac tissue development and function [100, 101]. The exact contribution of each single cell type to tissue-formation has not yet been thoroughly analyzed, but in theory the formation of a 3D cardiac tissue-like construct needs to consider the presence of both cardiac myocytes and non-myocytes, ideally at similar ratios as in their native physiological environment [93].

Although current tissue engineering approaches mainly utilize myocytes, it appears that primary cardiac cells will never be used as a cell source for cardiac tissue engineering in patients [93]. In contrast, embryonic and adult stem cells demonstrate significant proliferation capacity and provide great hope for cardiac replacement therapy [102, 103]. The capacity of murine embryonic stem (ES) cells to differentiate into cardiac myocytes upon homing to the heart tissue has been demonstrated previously [104]. Given the current debate about ES cells, autologous, post-embryonic stem cells derived from umbilical cord blood, bone marrow, or from adult tissues would in theory be the ideal choice. Provided that some recent indications for the unexpected plasticity of adult stem cells can indeed be validated, post-embryonic stem cells would be excellent candidates for use in tissue engineering [105]. The observation that circulating or injected adult stem cells home into the injured myocardium and locally transdifferentiate into cardiac myocytes [106] is of significant interest because it opens the way to mimic those conditions *in vitro*. Recent *in-vivo* studies have also demonstrated that the local cardiac environment might drive the differentiation of implanted immature cardiac myocytes [107]. The conditions required for stem cells to efficiently differentiate into mature myocytes are not yet clear, and require elucidation.

1.5.3
Culture Models of Cardiac Tissue-Engineered Constructs

Early models of engineered cardiac tissue constructs relied on static conditions for *in-vitro* culture prior to *in-vivo* implantation. In static culture, with no fluid mixing, large diffusive gradients are formed between the outside and the inside of the constructs. Therefore, the cells in the center of the constructs lack sufficient nutrient supply, the toxic metabolites are poorly removed, and many – if not most – of the cells die [108]. Oxygen and carbon dioxide diffusion into the cardiac scaffold is the main limiting factor for gas exchange under static culture conditions [109]. A simple approach that may increase the diffusion of nutrients and gases and also enhance waste removal is to grow cells on a 3D polymer scaffolds that is placed in a dynamic environment, such as that provided by a perfused bioreactor [110].

Bioreactors have been utilized for a variety of diverse applications for cardiac tissue engineering, such as: (1) cell expansion; (2) production of 3D tissues from

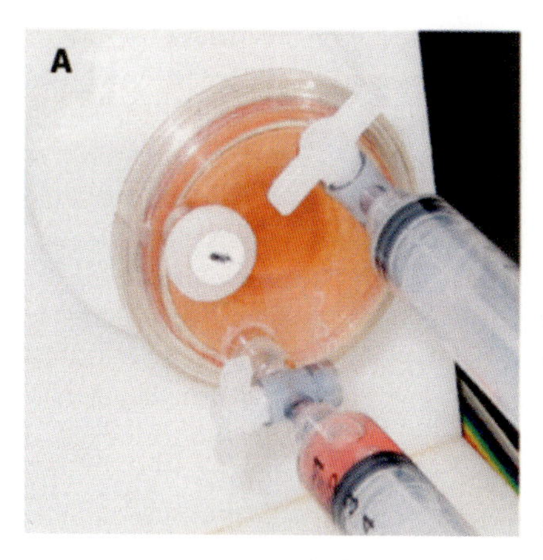

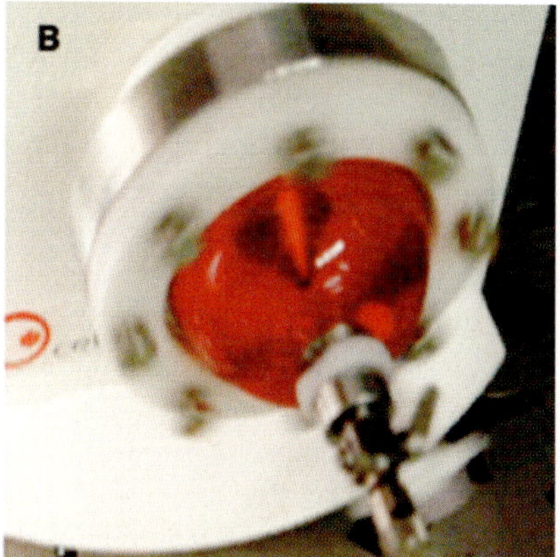

Fig. 1.5 Dynamic culture of cells or tissue constructs in rotating wall vessel bioreactors; (A) a high-aspect ratio vessel (HARV), (B) a hydrodynamic focusing bioreactor (HFB).

isolated cells *in vitro*; (3) production of 3D tissue constructs with cells on the scaffolds *in vitro*; and (4) directly as organ support devices [108, 111–114]. Amongst the dynamic bioreactors, different types of rotating wall vessel (RWV) bioreactors, such as the STLV (slow-turning lateral vessel), the HARV (high aspect ratio vessel; Fig. 1.5A) and the HFB (hydro-dynamic focusing bioreactor; Fig. 1.5B), have been used for cardiac tissue engineering.

These venues provide improved mass transport [113], leading to enhanced metabolic activity [110] and electrophysiological and molecular properties [115] of the constructs. It has been shown that the pO_2, pH and pCO_2 were better and more consistently maintained in these bioreactors than under static conditions, and facilitated aerobic respiration for constructs with higher cell densities [109, 116].

Recent studies have indicated that applied shear stress has a beneficial effect on the quality and quantity of the generated cardiac tissues [117]. Furthermore, a new generation of RWV bioreactors has been designed to impose mechanical stretch [56] on engineered heart constructs. The development of engineered cardiac tissue is modulated by mechanical signals [118]. The organization, composition, and function of the engineered cardiac tissue can be achieved by application of physiological regimens of cyclic strain [119]. Previously, Vandenburgh et al. [120] demonstrated that physical stimuli improved the proliferation and distribution of the seeded human heart cells throughout the scaffold structure, and further stimulated the formation and organization of ECM, which was responsible for improvements in the mechanical strength of the cardiac graft and population with cells. Future bioreactors for cardiac tissue engineering should combine both perfusion and mechanical stimuli, for example by allowing for adjustable pulsatile flow and varying levels of pressure [103, 110].

1.5.4
Specific Scaffolds Developed for Cardiac Tissue Engineering

For cardiac tissue engineering, the ideal scaffolds should consider mixtures of different polymers to achieve the following properties: (1) high porosity (large interconnected pores) to facilitate mass transport; (2) bioadhesiveness to enhance cell attachment; (3) structural stability to withstand both static and dynamic *in-vitro* cultivation; (4) biodegradability to ensure tissue grafting; (5) elasticity to enable the transmission of contractile forces; and (6) conductivity to facilitate electrical stimulation of the constructs. These properties apply to both synthetic polymer scaffolds and blends of natural and synthetic materials, as well as to acellularized natural cardiac tissues.

Leor et al. [46] have shown that cardiomyocyte seeding within porous alginate scaffolds yield 3D high-density cardiac constructs with a uniform cell distribution. As an alternative to seeding the cells on a preformed scaffold, Zimmermann et al. utilized Matrigel™ or Matrigel™ mixed with collagen to generate scaffolds that were uniformly populated with cells [93, 121]. A static sponge scaffold with open interconnected pores, composed of poly(dl-lactide-*co*-caprolactone), PLGA, and type I collagen demonstrated stable cardiomyocyte growth, a higher expression of cardiac markers, and better contractile properties than sponge scaffolds composed of either collagen or PLGA [13]. This typical example highlights the fact that, for cardiac tissue engineering, a mixture of polymers is preferred to an individual compound in order to generate scaffolds with several of the above-mentioned properties. The electrospinning of a mixture of biodegradable PLA- and PGA-based PLGA to generate porous, nanofibrous scaffolds was used to culture primary cardiomyocytes and generate cardiac tissue-like constructs. This indicated that the structure and function of these engineered cardiac tissue scaffolds can be modulated by the chemistry and geometry of the nano- and micro-textured surfaces [55]. Most recently, PANi – an electrical conductive polymer – was blended with a natural protein, gelatin, and co-electrospun into conductive, biocompatible nanofibers as cardiac tissue engineering scaffolds to support the attachment and proliferation of cardiac myoblasts [35]. Pedrotty et al. [20] cultured myoblasts on 3D PGA porous scaffolds and reported increased cell proliferation caused by electrical stimulation, or by a culture medium that had been conditioned by mature cardiomyocytes.

Tissue-engineered or acellularized heart valves have already been implanted in pigs [122]. In this approach, acellular scaffolds from heart valves offer unique advantages over synthetic polymers for cardiac valve engineering applications because they retain biologically active ECM molecules to support cellular ingrowth [123]. As a caveat, Rieder et al. [124] examined the immune response of acellularized porcine and human heart valves by measuring the migratory response of human monocytes, and identified species-selective immune reactions. Thus, caution is required in choosing the appropriate heart tissue for acellularization.

Another technique recently considered in cardiac tissue engineering which may accelerate and optimize engineered myocardial assembly is that of "organ printing" [125–127]. This uses a commercially available ink jet printer-like device, which overlays polymers by a "printing" methodology on a template of desired pattern/organization; in this way layers of different thickness are deposited one on top of another to generate homogeneous/heterogeneous scaffolds. Thereafter, the 3D structure is generated by printing, layer-by-layer, a rapidly solidifying thermoreversible gel containing cells in a fashion analogous to multiple layer printing on paper. Time will tell whether computer-aided, jet-based 3D tissue engineering of a living patch of human cardiac muscle is indeed a promising approach for the manufacture of large numbers of functional tissue equivalents *in vitro* and *in vivo*.

In considering pharmaceutical applications of cardiac constructs for the HTS of novel drugs to treat cardiomyopathies, arrhythmias, cardiac heart failure and myocardial infarction, there is a clear lack of suitable *in-vitro* cardiac tissue models. The future technology for cardiovascular tissue engineering for implantation and/or pharmaceutical application will combine integrative approaches of intelligent scaffolds with human stem cells. The latter will most probably be engineered to secrete a variety of growth factors required for the integration of heart tissue, and represent a combination of biomaterial, tissue engineering, cell therapy, gene therapy, and cardiology.

1.6 *In-Vitro*-Engineered Pulmonary Tissue Models: Progress and Challenges

1.6.1 Lung Tissue Engineering: The Current State of Play

The biological function of the lung is to facilitate exchange of oxygen and carbon dioxide between the cardiovascular circulation and the environment. The lung is a physiologically and anatomically complex vital organ that is hallmarked by a compartmentalized tissue architecture. It is convenient to divide the lung into two anatomically and functionally distinct units, namely the proximal and distal airways. The proximal airways (and proximal pulmonary arteries) are smooth muscle-lined conduits that control the amount of air (and blood) reaching the distal gas exchange region; they are also known as the alveoli. The functional components of the alveoli are the distal epithelium and microvasculature.

The use of tissue engineering to generate respiratory tissue models for both clinical and basic science applications has focused primarily on reconstruction of the trachea and proximal airway structures (bronchial tissue). These models consisted of collagen gels pre-seeded with tissue-derived fibroblasts on top of which an epithelial layer is seeded following a fibroblast conditioning period. The selection of collagen type I as a matrix is straightforward because the cell–cell interactions between fibroblasts and epithelial cells are critical for the formation of an organotypic basement membrane and the differentiated epithelium. Depending on the cell source used, engineered bronchial tissue models are capable of recapitulating both normal physiological as well as asthmatic features [128–130]. Models of proximal lung tissue constructs might be used for therapeutic purposes; an example would be as bronchial tissue patches to repair damaged bronchial tissue, and/or for HTS of respiratory toxicants enhancing lung permeability, or for developing drugs to treat respiratory diseases.

In the distal lung, the alveolar epithelium is composed of two distinct cell types, alveolar type I (AE1) and alveolar type II (AE2) cells. Although both cell types are found in approximately equal numbers, AE1 cells cover 95% of the alveolar surface, which makes their presence especially pertinent to the study of transport across the air–blood barrier. The essential physiological function of AE1 cells is establishment of the air–blood interface along with capillary endothelial cells (ECs). The majority of the AE2 cells lie basal to the alveolar surface, and perform numerous essential functions in the alveoli, including the production and secretion of surfactant lipids and proteins into the alveolus (i.e., the lubricant required for alveolar protection, gas exchange and antimicrobial defense). AE2 cells are critical in repair and remodeling following injury, serving as "progenitor cells" that divide, migrate and differentiate in response to insult or injury. In attempting to design an organotypic alveolar tissue model, the situation is further complicated by the contributions of the microvascular ECs and interstitial fibroblasts to the microenvironment.

In traditional 2D culture, AE2 cells lose many of their specialized features such as the ability to produce surfactant [131], which suggests that AE2 cells rapidly dedifferentiate when not provided with an appropriate cellular microenvironment. The optimization of ECM composition [132], growth factor [133] and hormone [134] composition of the medium has improved the maintenance of differentiated AE2 functions *in vitro*.

During the past year, the first reports have emerged of lung alveolar tissue models generated by tissue engineering approaches. Chen et al. [135] reported the formation of lung-alveolar-like structures following extended *in-vitro* culture of rat fetal pulmonary cells on collagen-chrondroitin-6-sulfate composite 3D scaffolds fabricated by freeze-drying and crosslinking. The use of a collagen base modified with chondroitin sulfate – an element of the ECM known to be required for epithelial morphogenesis [136] – is a paradigmatic example of engineering scaffolds with the goal of recapitulating the organotypic ECM. Recently, we have developed a tissue engineering approach for the generation of 3D alveolar tissue constructs with appropriate tissue morphology and cytodifferentiation using diverse biomatrices in combination with a growth factor-supplemented defined serum-free medium [133]. The goal was to recapitulate the native tissue architecture and cellular organization (Fig. 1.6A), and Matrigel™ and type I collagen hydrogels were used successfully to generate 3D alveolar tissue constructs with appropriate tissue architecture. Examples of this were branching epithelial morphogenesis and differentiation (Fig. 1.6B), as well as the establishment of a primitive microvascular network found in apposition to alveolar epithelial cells (Fig. 1.6C).

In more recent studies we have shown that micro and nanofibrous scaffolds electrospun from natural ECM proteins, such as elastin, in cooperation with

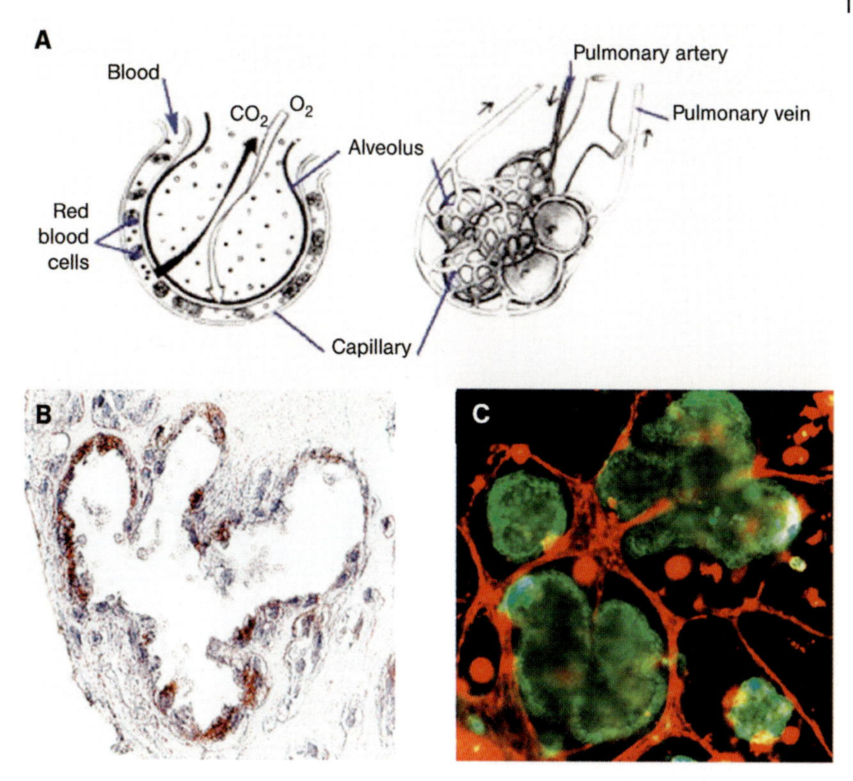

Fig. 1.6 Schematic of alveolar structure and immunophenotyping of engineered distal lung tissue constructs. (A) Diagram of alveolar structure and function (adopted from the NIH (NHLI) website). (B) Cytokeratin staining (brownish-red) of alveolar-forming structure following 7 days of 3D culture of fetal pulmonary cells in Matrigel™. (C) Cytokeratin (epithelial cells, green) and isolectin B4 (endothelial cells, red) double-labeling of fetal pulmonary cell construct generated in type I collagen gel with tissue-specific growth factors.

tissue-specific growth factors, support histiotypic co-morphogenesis of epithelial and endothelial tissue components, as required for engineering a high-fidelity alveolar tissue model [137].

Natural ECM proteins (e.g., laminin) promote AE2 morphology and differentiation *in vitro*. A plethora of *in-vitro* studies has been conducted on the effect of the ECM substrate used for culture on alveolar epithelial cells, and from these investigations much information can be applied to the development of appropriate 3D biomatrices for lung tissue engineering. However, future studies on how these matrices affect other cell types of the alveoli, such as the microvascular ECs, will be required in order to develop a truly organotypic tissue culture model. A summary of some relevant findings on the effects of various ECM constituents on alveolar epithelial cell differentiation and morphogenesis in diverse 2D and 3D *in-vitro* culture systems is provided in Table 1.2.

Table 1.2 Important effects of the extracellular matrix (ECM) on alveolar cell and tissue culture models generated *in vitro*.

Cell source	ECM/Scaffold	Major findings	Ref.
Mixed primary mouse fetal lung cells	Matrigel™, PLGA foams, PLLA fibers	Branching morphogenesis and AE2 differentiation for up to 4 weeks in Matrigel™ with growth factor-defined medium. Synthetic polymers did not support AE2 differentiation	[131] [135]
Mixed primary rat fetal lung cells	Gelatin sponge matrix	Formation of alveolar-like structures for up to 6 weeks with progressive AE2 de-differentiation	[149]
–	Fibronectin or laminin	Epithelial cells and fibroblasts expressed $\alpha_5\beta_1$, $\alpha_6\beta_1$, and the 65 kDa laminin-elastin receptor. Enhanced adhesion of epithelial cells on laminin was observed	[235]
–	Collagen-GAG (chondroitin-6-sulfate) porous scaffold	Mixed primary cultures formed histiotypic alveolar-like structures for up to 3 weeks *in vitro*. Matrices contracted over time	[133]
Fetal rat lung organ explants	N/A	Chlorate disruption of chondroitin sulfate proteoglycans in the tissue ECM resulted in arrested epithelial branching morphogenesis and reduced expression of AE2 genes	[134]
Primary fetal rat AE2 cells	Laminin, fibronectin, vitronectin, collagen, or elastin	ECM proteins supported ERK activation upon mechanical stimulation; however, laminin induced surfactant protein C expression and AE2 differentiated phenotype	[236]
Primary neonatal and adult rat AE2 cells	MDCK cell-derived ECM	Combination of MDCK basement membrane, dexamethasone and cyclic AMP maintained surfactant protein expression in neonatal, but not adult AE2 cells	[132]
SV40-T2 immortalized adult rat AE2 cell line	Fibrillar collagen matrix imbibed with Matrigel™	SV40-T2 cells formed a continuous lamina densa containing laminin, collagen IV, entactin and perlecan with Matrigel™ supplementation, but not without	[137]
Primary adult rat AE2 cells	Acellularized human alveolar tissue	AE2 cells flattened and took on an AE1-like phenotype, while AE2 cells on amnionic membrane maintained differentiation and morphology	[237]
–	Pulmonary endothelial-derived ECM	Mitogenic stimulation, well spreading, loss of lamellar bodies (AE1-like phenotype)	[220]
–	Type I collagen gels	AE2 cells proliferation and alveolar-like structures differentiation	[150]

Table 1.2 (continued)

Cell source	ECM/Scaffold	Major findings	Ref.
–	Fibronectin or type I collagen	AEC migration enhanced on fibronectin. $\alpha_V\beta_3$ antibody inhibited fibronectin migration, while α_2 inhibited migration on type I collagen	[238]
–	Fibronectin or Matrigel™	ECM modulates gap junction expression. Fibronectin increases connexin43, while Matrigel™ promotes connexin26	[239]
–	Mixture of type I collagen and Matrigel™	KGF in combination with this ECM mixture supported maximal surfactant protein A and D secretion as compared to HGF and FGF-10 media supplementation	[152]
–	Fibronectin	FGF-1 increases DNA synthesis and AE2 attachment to fibronectin	[240]
–	Collagen I, fibronectin, and laminin-5	Both collagen type I and laminin-5 were needed for prolonged maintenance of the AE2 phenotype. In the absence of laminin-5, AE2 cells tended to take on an AE1-like phenotype	[241]
Adult rat AE2/AE1 mixed cultures	Fibronectin-collagen type I-laminin-5	Culture on this ECM combination maintained mixed cultures of AE2 and AE1 cells with ratios similar to those found in human alveoli *in situ*	[130]
Guinea pig AE2 cells	Acellularized human amnionic membrane	Acellularized human amnionic membrane and fibroblast co-culture or KGF supplementation maintained AE2 morphology and differentiation	[242]
Day 29 fetal rabbit AE2 cells	Matrigel™	Basement membrane ECM maintained AE2 morphology and differentiation for 3 weeks, while cells dedifferentiated on plastic by 5 days	[138]

Abbreviations:
AE1 = alveolar type I
AE2 = alveolar type II
TGF-β = transforming growth factor beta
ECM = extracellular matrix
ERK = extracellular signal-related kinase
FGF = fibroblast growth factor
GAG = glycosaminoglycan
KGF = keratinocyte growth factor (a.k.a. fibroblast growth factor-7)
MDCK = Madin-Darby canine kidney

Review of this information is helpful in the selection and design of biomaterials and biomatrices, respectively, for engineering alveolar tissue models *in vitro*.

In the developing lung, the complex basal lamina on which the epithelium resides is composed mainly of type IV collagen, laminin and a tissue-specific mix of glycosaminoglycans [138]. The importance of the ECM composition (specifically laminin) for modulating AE2 cell structure and function has been well documented [139]. Early studies by others [140], as well as our data, have shown that AE2 cells grown two-dimensionally on Matrigel™ retain their differentiated form when compared with cells cultured on plastic surfaces. In contrast to the differentiating effects of Matrigel™, non-specific ECM produced by ECs failed to maintain an AE2 cell phenotype [141]. Results from our laboratory indicate that mixed populations of fetal pulmonary cells cultured on synthetic scaffolds in conventional serum-enriched medium do not maintain AE2 cell differentiation for extended periods [133]. Taken together, these findings indicate that the ECM composition is a key parameter in the design of an optimal 3D alveolar tissue model for pharmaceutical purposes.

1.6.2
Existing *In-Vitro* Pulmonary Cell and Tissue Culture Biological Models

The heterocellular nature and complex architecture of the alveoli presents a challenging problem for the development of physiologically relevant 3D *in-vitro* models. Traditionally, *in-vitro* models of alveolar tissue have consisted of 2D cultures of alveolar epithelial cells, either homotypic AE2 cells or heterotypic AE1/AE2 cultures [142–144]. Early experiments utilized either pulmonary epithelial cell lines or primary cultures of isolated alveolar cells. Alveolar epithelial cell lines, such as A549 have been derived from a lung adenocarcinoma [145], or generated by immortalization [146]. Although A549 cells bear morphological similarities to AE2 cells, including the presence of lamellar bodies and microvilli, they lack several key characteristics of the differentiated *in-vivo* phenotype that is retained in cultures of primary isolates from normal tissues. While monotypic cultures of A549 cells can be used as a model for some of the type II cell functions, establishing appropriate heterocellular tissue models will require the use of primary or stem cell-derived AE2 cells [147–149].

The culture of primary AE2 cells *in vitro* often results in transdifferentiation of AE2 cells into AE1-like cells, as evidenced by morphological changes that include a loss of microvilli, cell spreading resulting in the formation of thin cytoplasmic attenuations [150], and reactivity to AE1 cell-specific membrane components [151].

Following the earliest reports of *in-vitro* 3D organotypic models of lung alveolar tissue [140, 152], only a few reports of such 3D organotypic models have since emerged, presumably due to the complexity of the system. For example, Sugihara et al. [153] reported the generation of differentiated 3D alveolus-like structures *in vitro* utilizing a type I collagen gel and growth factor- and hormone-defined medium. This report was promising, but limited in that only purified AE2 cells were used in the absence of the microvascular and connective tissue components. Successful engineering of proximal airway tissues *in vitro* have demonstrated the importance of sustained cell–cell interactions between epithelial and mesenchymal tissue components [128–130]. Only recently have these principles been applied toward generating distal lung tissue constructs *in vitro* [133, 135].

1.6.3
Potential of Alveolar Tissue Models as Disease Models in Pharmaceutical Sciences

Engineered alveolar tissue models that recapitulate physiological and pharmacological characteristics of native distal lung tissue would be invaluable research tools, with broad applications as pharmacological models of normal and pathological tissues, and novel venues for investigation of pulmonary toxicology and infectivity. The refinement of traditional 2D models used to investigate drug permeability in lung tissue [142, 154] will result in a more physiologically relevant model to study drug permeation [132, 155]. Tissue engineering approaches can be used to generate complex 3D alveolar tissue models containing the epithelial, microvascular, and connective tissue components. Specifically, these approaches will yield high-fidelity models of pediatric pulmonary diseases such as bronchopulmonary dysplasia and cystic fibrosis, and adult pulmonary diseases such as emphysema and particulate lung disease. Some neonatal pulmonary diseases, occurring secondary to pulmonary hypoplasia, are hallmarked by the aberrant development of epithelial and vascular components of the lung [156]. Therefore, a more realistic 3D alveolar tissue model would be useful for investigating pathological developmental processes that are at the root of pediatric chronic lung disease.

In addition to developing disease models of (pediatric) pulmonary diseases, existing *in-vitro* 2D cell culture models of pulmonary toxicology [157] could be extended to 3D tissue models. In a recent investigation of a patient who succumbed to the fatal H5N1 avian influenza virus, the alveolar epithelial cells were found to be a major site of H5N1 replication [158]. Recent reports focus on the use of 2D respiratory epithelial cell culture models for studying pathogenesis of influenza viruses, as well as environmental toxicants [159]. We contend that extending these 2D systems to 3D alveolar tissue constructs will provide a more realistic and physiologically correct pharmacological model for developing of antiviral drugs or vaccines targeting alveolar epithelial cells.

1.6.4
The Future: Toward Engineered 3D Alveolar Tissue for Cell Therapy and Pharmacological Models

The delivery of various pharmaceuticals, such as small molecules and macromolecules via the inhalation route, is an important method of drug delivery. Two potential examples of carrier systems used for pulmonary drug delivery via inhalation are large porous microparticles and liposomes [160]. Organotypic alveolar tissue models generated using the cell sources described above, in combination with appropriate biomatrices and humoral factors, would be useful for studying the transport of drugs into alveolar epithelial cells and/or across the air–blood barrier, as well as the efficacy of novel drug carrier systems. We believe that tissue-engineered model systems of 3D alveolar tissue could expedite the formulation and testing of pharmaceuticals for diverse applications in pulmonary medicine. Future studies will involve novel technologies of cell and gene therapy for refining 3D alveolar tissue constructs as pharmacological models of lung diseases and/or for cell therapy purposes.

1.7
In-Vitro Models of the Blood–Brain Barrier (BBB)

1.7.1
The BBB, a Neurovascular Physiological Unit: The Concept

The BBB, which separates the blood from the brain, and vice versa, was first described in 1885 by Paul Ehrlich and confirmed later in 1909 by Edwin Goldman. Both investigators showed that, following intravenous injection, trypan blue (an albumin-binding dye) was dispersed throughout the whole body, except for the brain. Conversely, injection of this dye into the brain subarachnoidal space selectively stained only the brain. Reese and Karnovsky [161], Brightman and Reese [162] and others, demonstrated that movement of the markers from the bloodstream to the brain, and from the brain to the bloodstream, was stopped at the level of the endothelial junctions [163].

The anatomic structure of the BBB is composed of the cerebral microvascular endothelium which, together with astrocytes, pericytes, neurons, and the extracellular matrix, constitutes a "neurovascular physiological entity" (Fig. 1.7A) that is essential for the health and function of the CNS [164].

Astrocytic projections [165] and neurons almost completely cover the basolateral surface of the brain microvessel ECs, which form capillaries with tight junctions. The tight junction is an intricate complex of transmembrane (junctional adhesion molecule-1, occludin, claudin, etc.) and cytoplasmic proteins (zonula occludens-1,2, cingulin, AF-6, 7H6, etc.) linked to the cytoskeleton. These tight junctions represent the physical barrier which limits the penetration of most chemical compounds into the brain [166].

The BBB functions primarily as a *physical barrier* to the passage of small molecules by a paracellular route (between the cells) (Table 1.3).

Although astrocytes via their "end-feet" – which are in direct contact with the ECs – also participate in the BBB structure, they take no part in the physical barrier. However, it is this interaction of the astrocytes that induces the unique BBB endothelial phenotype in the brain. The major characteristics of this barrier are: (1) small intercellular space between adjacent ECs (< 10 Å) [162]; (2) high electrical resistance ($\geq 2000\ \Omega \times \mathrm{cm}^{-2}$) [167]; and (3) low paracellular permeability [168]. The tight junctions between ECs of the BBB restrict the paracellular diffusion

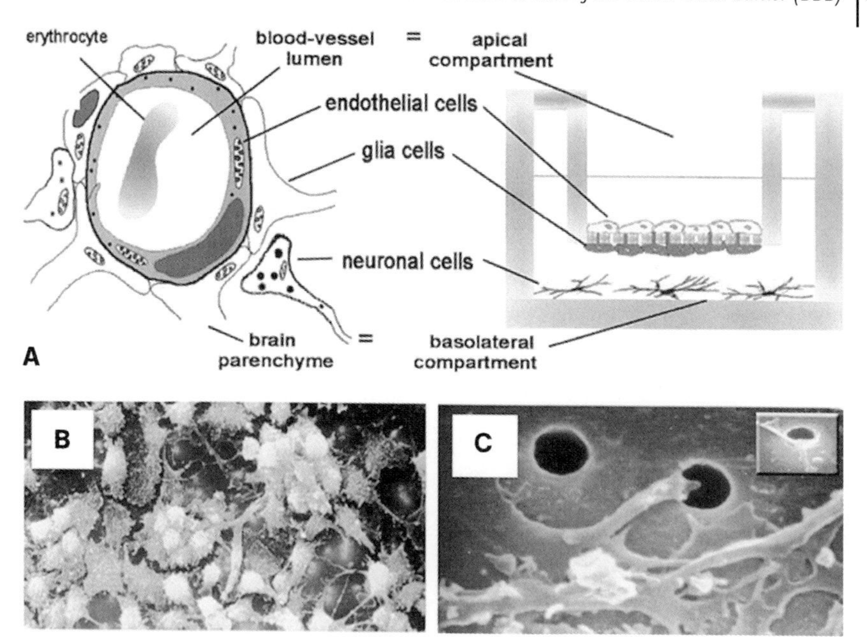

Fig. 1.7 The principle and technology of the *in-vitro* blood–brain barrier model. (A) A single neurovascular barrier unit modeled *in vitro* using transwell inserts. (B) SEM micrograph of confluent monolayer of endothelial cells on the top of the transwell insert membrane. (C) SEM micrograph of glial cell projection entering a pore in the basolateral (glia) side; the insert shows glial projections transversing the transwell insert membrane and extending to the apical (endothelial) side.

of water-soluble chemicals. Passive permeation of the BBB is generally related to the degree of the solute's lipophilicity.

The microvascular endothelium in the BBB serves as an active, *energy-dependent barrier*, with unique transport properties. In line with their high enzymatic and metabolic activities, brain ECs possess a large number of mitochondria, fuelling an elaborate system of transport proteins (influx and efflux carriers and pumps). The asymmetric distribution of the transporters on the luminal and abluminal EC membrane allows for the vectorial exchange of selected chemicals into or out of the brain. This *biochemical-transport barrier* ensures a selective permeability of nutrients, neurotransmitter precursors, and xenobiotics. As an *enzymatic-barrier*, BBB microvascular ECs express a large variety of metabolizing enzymes which serve as biotransformation and detoxification systems, metabolizing and excreting lipophilic endogenous and exogenous chemicals which may have invaded the brain environment.

The ECM of the BBB is mainly composed of fibronectin, laminin, and collagen type IV [169]. Developmentally, the matrix constitutes a pivotal biological platform for the growth of brain microvascular ECs. The cells interact with the ECM proteins through specific integrin receptors [170] which, upon activation, induce important

Table 1.3 The evolution of *in-vitro* blood–brain barrier (BBB) models.

BBB model (generation)	Endothelium (abbreviation)	Astroglia	Neurons/ Pericytes	Major finding	Ref.
1	BBMEC	–	–	Polycarbonate membranes are highly suitable for *in-vitro* BBB models	[243]
	HUVEC-304	–	–	Elevated intracellular cAMP levels – increases TEER	[244]
	MBCEC	–	–	Addition of hydrocortisone – increases TEER	[245]
2	PBCEC	C_6-glioma	–	Co-culture – increases γ-GT and Na^+/K^+ ATPase activity	[246]
	PBCEC	Rat astrocytes and C_6	–	Co-culture – increases γ-GT and ALP activities	[247]
	HCEC	–	–	Astrocytes conditioned media – increases TEER	[248]
	RBE4	C_6-glioma	–	C_6-glioma conditioned media or matrix – increases P-gp activity and levels	[249]
	b.End5	C_6-glioma	–	Co-culture – increases TEER	[170]
	BAEC	C_6-glioma	–	Co-culture – increases TEER	[250]
	RBMEC	Rat astrocytes	–	Co-culture – increases TEER and γ-GT activity	[189]
	b.End3	C_6-glioma	–	Co-culture – increases TEER and ALP activity	[251]
	PBCEC	Porcine astrocytes	–	Co-culture – increases TEER and decreases inulin permeability	[252]
	RBE4	U-373 MG	–	Co-culture – increases TEER	[253]
	BAEC	C_6-glioma	–	Addition of dexamethasone – increases TEER and decreases sucrose permeability	[254]

Table 1.3 (continued)

BBB model (generation)	Endothelium (abbreviation)	Astroglia	Neurons/ Pericytes	Major finding	Ref.
3	PBEC	–	Murine cortical neurons	Co-culture – increases γ-GT and Na^+/K^+-ATPase activities	[246]
	RBMEC	Rat astrocytes	B14	Tri-culture decreases sucrose permeability	[255]
	RBE4.B	–	Rat neurons	Neurons regulate occludin localization	[256]
	RBE4.B	–	Rat neurons	Neurons induce selective exclusion of dopamine, but allow L-tryptophan and L-DOPA permeability	[190]
	RBE4.B	Rat astrocytes	Rat neurons	Neurons and astrocytes synergistically induce localization of occludin in endothelial plasma membrane	[257]
	RBE4.B	Rat astrocytes	Rat neurons	Tri-culture decreases sucrose permeability	[198]

Abbreviations:
BBMEC = bovine brain microvascular endothelial cells
RBMEC = rat brain microvascular endothelial cells
PBCEC = porcine brain capillary endothelial cells
BAEC = bovine aortic endothelial cells
MBCEC = mouse brain capillary endothelial cells
HCEC = human capillary endothelial cells
bEnd3 and 5 = mouse brain endothelial cells
RBE4 and 4B = rat brain endothelial cells
TEER = trans-endothelial electrical resistance

genetic and phenotypic changes [171]. The ECM proteins have a crucial role in inducing and maintaining the barrier properties of ECs by contributing to their differentiation processes.

Astrocytic projections, also known as "end-feet", are instrumental for maintaining the barrier characteristics of cerebral ECs [165]. The barrier-generating effect of astrocytes can be separated into: (1) the secretion of a set of humoral growth factors that encourage barrier generation by acting on ECs [172]; and (2) an extension of "end-feet" bulbs to make direct contact with the ECs. Both these factors synergistically affect formation of the BBB-specific barrier function [173].

Passive paracellular permeability relies on simple diffusion. Only small molecules with a molecular weight of < 400–500 Da [174] and a molecular diameter < 20 Å [175] can cross the barrier through the paracellular route. In contrast to most other tissues, the mean distance between cerebral capillaries is rather short, on the average ~40 µm [176]. For small gaseous molecules (e.g., oxygen) and volatile anesthetics, which penetrate the BBB passively by the paracellular route, the high capillary density facilitates almost immediate equilibration throughout the brain parenchyma. Normally, the paracellular pathway has little therapeutic relevance for neurologically active small molecular weight drugs, such as L-dopa, due to an inability to accumulate therapeutically effective drug levels inside the cerebral milieu. By contrast, for most drugs the *transcellular* pathway (permeability *through* the cells) is the major route of entry into the brain. Transcellular transport can be divided into two general subgroups: (1) *passive* transcellular permeability, which depends mainly on the lipophilic properties of a given drug; and (2) *active* transcellular permeability, which utilizes a variety of transporters with bidirectional activity into and out of the brain [177].

The BBB is the bottleneck in the development of neurotherapeutics, as more than 98% of all potential CNS-targeted drugs do not cross the BBB [178]. The recognition that "poor permeability into the brain is one of the reasons for the failure of CNS-targeted drug candidates" has revolutionized neurological drug discovery research. Thus, given the large number of drug candidates generated by modern combinatorial chemistry-based approaches, poor brain-penetrating candidates need to be identified and eliminated at the early stages of drug discovery [179]. Due to high costs, time-consuming assays and ethical considerations, previously accepted *in-vivo* methods for measuring BBB permeability are no longer applicable for such enormous numbers of candidate drugs. Permeability evaluation through the BBB should be relatively simple, widely applicable, robust, and able to comply with HTS methods [180]. This calls for the generation of novel, *in-vitro* BBB models, which can provide the pharmacologically validated multitude of HTS screening tests required throughout the lead compound selection process.

1.7.2
In-Vitro BBB Models: Cells and Devices

A minimal approach to an *in-vitro* BBB model focuses mainly on brain microvascular ECs (see Table 1.3). For the past three decades, capillaries have been prepared from the brains of various species of embryonic or adult animals. Brain ECs have been isolated and primary ECs were cultured *in vitro*. The disadvantages of using primary cultures include:

- the extensive use of animals,
- tedious, complicated protocols for cell isolation,
- slow cell proliferation rates,
- the limitation in passaging these cells.

Cultured primary ECs quickly (within a few population doublings) lose the majority of their barrier-related characteristics, such as expression levels of transporters. Similarly, the transendothelial electrical resistance (TEER) is high immediately after isolation but declines significantly in culture [180]. Endothelial cell lines, which are obtained from tumors or immortalized upon infection with viruses or viral constructs, and can be propagated for many generations, appear to be more convenient for generating *in-vitro* BBB models. However, as inferred from electrical resistance and permeability measurements, transformed EC-based BBB models exhibit barrier properties inferior to BBB models employing primary brain capillary ECs. Moreover, due to the high variability between different cell lines, each cell line-based BBB model must be characterized extensively [180–182].

High TEER values are considered to be a physiological marker of an intact, functional BBB indicative of the organotypic differentiation state of brain ECs. Remarkably, some ECs – irrespective of their origin – have the potential to differentiate into a BBB phenotype in culture. This plasticity depends on the culture conditions [183]. Thus, a systematic search for differentiating growth factors should be performed in order to optimize the barrier properties of ECs.

Two major cell banks (www.atcc.org; www.ecacc.org) provide some of the currently available commercial EC lines for BBB models. Primary capillary ECs are mostly investigator-generated by isolation and purification from brain capillaries of a variety of species (e.g., porcine, bovine, rat, mouse, monkey, human). To date, no uniform protocols exist for the isolation and culture of these primary cultures. Although comparisons between different laboratories are difficult, it seems clear that primary ECs show more prominent barrier properties (see Table 1.3), exhibiting higher TEER values and lower paracellular permeability. However, the role of TEER as a predictor of the tightness of an EC monolayer is rather questionable and uninformative with regard to the paracellular permeability [184].

More sophisticated *in-vitro* BBB models must include, in addition to ECs, also astrocytes. Astrocytes can be obtained either from primary cultures of dissociated brain cells, or they exist as established cell lines. In co-culture with ECs, astrocytes, cell lines and primary cultures, are each capable of inducing endothelial-monolayer tightening/differentiation toward the BBB phenotype by secreting as yet unidentified growth factors, and establishing cell–cell contact-based interactions of the glial feet extensions with the ECs [185–187]. The most commonly used astrocytic cell line is C_6-glioma, a cell line isolated from a rat glioma (Table 1.3). However, a major disadvantage of C_6 glioma cells – as well as of other astrocytomas – is their ability to release VEGF (also called vascular permeability factor, VPF), which counteracts the endothelial BBB barrier functions in the *in-vitro* co-culture model [188–190].

BBB experiments *in vitro* are most frequently conducted with Transwell™ or similar filter systems. These are special tissue culture plates that enable separation of the ECs from glia by a thin, porous, polymeric membrane. This system was originally developed for modeling gastrointestinal permeability [191], and later adopted for other *in-vitro* permeability studies, including BBB models. A simple, static model of the BBB is generated by growing a confluent endothelial monolayer on top of the insert, while on the apposing side glial cells are grown (Fig. 1.7B). Several manufacturers produce transwells for pharmacological models; the most useful for BBB studies are filters made of transparent, polyethylene terephthalate (PET) with a pore size of 0.4 or 1.0 μm. These inserts allow for unimpaired exchange of small molecular nutrients and test compounds with a molecular weight < 1000 Da. At the same time, cells seeded on both sides of these inserts can interact with each other through direct cell–cell contacts via "end-feet" – that is, cell extensions penetrating through the pores (Fig. 1.7C and insert). As a first approximation of the BBB, this configuration mimics compartmentalization between the apical side (blood vessel lumen) and basolateral (brain parenchyma, for example, glia and neurons).

The ECs respond in different ways to the polymeric substrates on which they are grown [171], and consequently the type of the filter may influence endothelial behavior. This issue is extremely important when generating *in-vitro* BBB models. In many experimental situations, it is important to visually examine the interactions of the endothelial monolayer, and therefore only completely transparent filters, such as PET, should be chosen. In addition, the pore diameter of the filters has important implications for establishing cell–cell contacts in co-cultures [192]. Finally, the cost of each insert must also be taken into consideration, especially for large-scale permeability screenings.

The TEER across intact endothelial monolayers is a simple, fast, reproducible, and nondestructive measurement of monolayer tightness and drug-induced changes therein. The electrical resistance is generated because of the restriction by the tight junctions of the flow of small ions through the monolayer. There is a reciprocal relationship between electrical resistance and ion flow; for instance, the tighter the monolayer, the lower the flow, and, hence, the higher the electrical resistance across the BBB. *In vivo*, TEER values can exceed $2\ \text{k}\Omega \times \text{cm}^{-2}$, whereas in most *in-vitro* models, TEER values of $< 100\ \Omega\ \text{cm}^{-2}$ are obtained.

In addition to the TEER, the permeability of diverse compounds through cellular monolayers, and in particular *in-vitro* BBB models, can also be assessed from their permeability coefficient, P [193]. Unfortunately, for a given EC type the P-values seem to depend as much on the nature of the filters used, as on the physicochemical properties of the compounds investigated. The commonly accepted assumption has been that polymeric filters are inert and have no influence on the permeability of *in-vitro* BBB models. This assumption is inaccurate, however. The studies by Yu and Sinko [194] and Lauer et al. [195] indicate that the barrier properties of EC lines might change according to the composition of the filters in the transwell systems. Therefore, novel biocompatible supports/inserts for *in-vitro* BBB models need to be developed. Likewise, the protocol for designing and testing *in-vitro* BBB models should be strictly defined and standardized between individual laboratories, in order to facilitate stringent pharmacological studies.

1.7.3
BBB *In-Vitro* Models: From First to Third Generation; the Biological Approach

The first generation of *in-vitro* models of the BBB was represented by monocultures of ECs, isolated from diverse vessels and different animal species. By culturing ECs on top of various transwell inserts, this approach was useful to establish the growth requirements of the cells on the inserts. Apparently, confluent 2D monolayers were then used to measure TEER values and to investigate the permeability of low and high molecular-weight markers (see Table 1.3). The major conclusion from these studies was that such *in-vitro* systems, comprised of only EC monolayers, could only partially mimic the properties of the *in-vivo* BBB as the permeabilities were too high and the TEER values too low. Therefore, the pharmacological relevance of these models is rather limited.

The second generation of *in-vitro* models of the BBB (Table 1.3) is comprised of co-cultures of ECs (primary isolates or immortalized ECs lines derived from bovine, porcine, rodent or human brain capillaries) that were seeded on the top of the transwell inserts, and of rodent or human astrocytes, primary or transformed (C_6 glioma), grown on the other side of the insert [196]. While the ECs, immortalized with SV40 or polyoma virus large T-antigen or adenovirus E1A genes, continued to express a variety of endothelial markers and responded to astroglial factors, most of these models still exhibited low TEER values and relatively high paracellular permeability to small molecules [196], thus limiting their use for pharmaceutical studies. However, these studies were very important in defining the minimal size of the transwell pores (> 0.8 μm diameter) necessary to allow penetration of glial feet onto the endothelial monolayer, and also to demonstrate the presence and activity of a variety of transporters and tight junction proteins in BBB-type ECs [197]. Furthermore, these experiments confirmed that C_6 glioma cells are ill-suited as substitutes for primary glial cells [198], and also led to the conclusion that the specific astrocytic factors involved in inducing the BBB phenotype in ECs remain to be discovered [199].

The third generation of *in-vitro* BBB models (Table 1.3) begins to address the contribution of other BBB cellular components such as neurons and pericytes on the organization and function of brain endothelium, and the multidirectional interactions between these components. Despite their localization close to capillaries and tight interactions with glia, very little is known about the contribution of neurons to the properties of the BBB [200]. For example, co-cultures of RBE4 ECs with rat primary neurons indicated a reduction in the transmonolayer dopamine flux [193]. In a recent, unique study, a three-cell type *in-vitro* BBB was generated by culturing primary brain capillary ECs on top of the transwell insert, while neurons and astrocytes from the same animal were grown on the opposite side. This configuration synergistically enhanced the expression of the tight-junction protein occludin, while paradoxically showing an increase in paracellular permeability [201]. Pericytes have a fundamental role *in vivo* to stabilize all capillaries, including those in the brain. In a recent *in-vitro* BBB model, it appears however that pericytes reduce endothelial paracellular permeability [202], most probably by continuous production of TGF-β [203].

The information obtained with these three generations of *in-vitro* BBB models, while providing important data on fundamental properties of such a model, also highlights the limitations of current models. This lack of a valid, high-fidelity BBB model highlights the urgent need for innovative approaches towards engineering pharmaceutically relevant novel *in-vitro* BBB models using an integrated concept. Such a concept should include modern biomaterials and biotechnological approaches in conjunction with a heightened understanding of the specific and unique roles that each of the cellular component plays in establishing and maintaining the BBB.

1.7.4
Trends in Tissue Engineering: Realistic *In-Vitro* BBB Pharmacological Models

The ECM is crucial for the culture of brain microcapillary ECs and the establishment of tight junctions [204]. In most *in-vitro* BBB models the surface of the transwell inserts is coated with rat tail type I collagen, to support and promote EC monolayer formation [205–207]. However, other ECM molecules, such as type IV collagen, fibronectin and laminin, either alone or in combination, were also shown significantly to elevate the TEER of low-resistance porcine brain endothelial monolayers [171]. Unfortunately, the precise profile of integrin receptors in microvascular ECs of different brain regions and the composition of the ECM in these specific regions has not yet been reported. Therefore, is not surprising that in the available *in-vitro* BBB models ECM proteins coating the transwell inserts have been used according to availability, rather than based on physiological criteria. In order to generate refined, more realistic *in-vitro* BBB pharmacological models, there is an urgent need for a detailed analysis of the (region-specific?) composition of the brain capillary ECM and for translating this new knowledge into the technology of engineering BBB models. Alternatively, the emergence of new generations of synthetic biomaterials for use as 3D extracellular microenvironments, which closely mimic natural ECMs [208], may result in novel, synthetic, brain-biocompatible, ECM-like substrates that might help to tighten *in-vitro* BBB models. For example, hyaluronic acid hydrogels modified with laminin [209], procyanidolic oligomers [210] and similar compounds may be used for decreasing the permeability and increasing the TEER values of *in-vitro* BBB models. Recently, the first attempt to generate an *in-vitro* BBB chip was made by co-culturing for > 14 days bovine brain ECs and rat astrocytes on ultrathin, highly porous nanofabricated silicon nitride membranes. These membranes were about 10-fold thinner and twofold more porous than commercial membrane inserts, and also included a spun-on crosslinked collagen layer, which helped to improve astrocyte attachment [211]. However, due either to the lack of *primary* isolates of the cells or to the "wrong" type of collagen coating, these ultrathin constructs still yielded TEER values orders of magnitude lower than *in vivo*. Nevertheless, we strongly believe that such synthetic biomaterials will be indispensable as biomimetic matrices, providing appropriate instructive microenvironmental cues for tissue engineering of future *in-vitro* BBB models.

Today, the pharmaceutical industry is seeking novel, high-fidelity *in-vitro* BBB pharmacological models which may be used as a first preclinical step in screening and characterizing the BBB-permeability properties of novel candidate compounds. Ideally, as a validation step, any high-fidelity *in-vitro* BBB model will demonstrate a close correlation of the permeability of known drugs with their known brain extraction pharmacokinetic properties *in vivo*.

The first *in-vitro* BBB model in which transcellular permeability was measured by the movement of ^{14}C-labeled sucrose was established from bovine brain gray matter [212]. The bovine brain contains large amounts of microcapillary ECs which should, in theory, suffice for a large-scale, high-throughput pharmacological screening [213].

One of the best *in-vitro* BBB models, a co-culture of cloned early-passaged bovine capillary ECs and rat glia, demonstrated TEER values of 500–800 $\Omega \times cm^{-2}$ and low sucrose/mannitol permeability [214]. The highest experimental TEER value (~2000 $\Omega \times cm^{-2}$), close to the physiological values measured *in situ*, was obtained in a bovine system; however, no concomitant studies on the monolayer permeability were performed [215].

Mouse brains yield low numbers of microcapillary ECs, and consequently few murine *in-vitro* BBB models have been reported [213]. The availability of an *in-vitro* mouse BBB model would provide a unique opportunity for using transgenic and knockout brain capillary ECs (as well as glia and neurons) to investigate the contribution of different cellular proteins to TEER and permeability. Permeability measurements in rat brain-derived *in-vitro* BBB models have also been published [213]. Similarly to mice, the rat brains yield limited numbers of ECs. According to published reports, the barrier properties of rat brain-derived BBB models are much better than those of mouse origin, but are still lower than some of the near-physiological levels obtained with bovine and porcine models.

To date, the best *in-vitro* BBB models, with high barrier resistance (1800–2000 $\Omega \times cm^{-2}$) and low sucrose permeability ($P_{app} = 0.2–1.8 \times 10^{-6}$ cm s^{-1}), utilize porcine brain microcapillary ECs grown in serum-free culture conditions and treated with corticosteroids [216, 217]. The large amounts of EC that can be obtained from a porcine brain, and the close similarity between porcine and human cardiovascular physiology, suggest that, in the absence of a readily available human *in-vitro* BBB model, the porcine system might represent a reasonable *in-vitro* model for high-throughput drug screening.

In spite of the limited access to human brain tissues, a few human *in-vitro* BBB models have been established [218]. These models were less robust then porcine or bovine BBB models, and provided relatively low barrier electrical resistance and high permeability values; the highest reported TEER value was ~500 $\Omega \times cm^{-2}$ and the lowest P_{app} value for inulin 3.6×10^{-6} cm s^{-1} [213]. However, recent biotechnological developments have allowed the isolation and culture of stem cells from human brain, bone marrow or umbilical cord blood, which then can be differentiated towards endothelial and, possibly also astrocytic and neuronal phenotypes, thereby raising the expectation that these cells will provide better cellular partners for engineering a human *in-vitro* BBB model for pharmaceutical purposes.

Whilst all of the above-mentioned biological and pharmaceutical *in-vitro* BBB models are cultured under static conditions, it must be remembered that *in vivo* the BBB, like every other endothelium, is exposed to blood flow and hydrostatic pressure. These "mechanical" parameters significantly influence endothelium physiology. Any truly high-fidelity *in-vitro* pharmacological model of the BBB must therefore incorporate these hemodynamic forces, both physiological as well pathological, such as hypertension. As a first dynamic model, Janigro and co-workers recently incorporated fluid shear stress into their *in-vitro* BBB model [219, 220], which is based on the principle of hollow-fiber bioreactors. Here, an endothelial cell monolayer, grown on the inside of a porous hollow fiber, is continually exposed to fluid-shear stress and interacts with astrocytes seeded onto the outside of the fibers. Thus, this model, for the first time, approximates the tubular morphology and hemodynamic forces of a brain capillary forming the BBB *in vivo*. Importantly, the TEER values in this system, though still lower than *in situ*, were increased by fluid shear stress. However, while this approach demonstrated the need for a dynamic concept, the permeability of drugs in this model has not yet been investigated and, in its present design, the model is ill-suited to high-throughput drug screening.

1.7.5
Conclusions for BBB *In-Vitro* Models

Recent advances in combinatorial chemistry and HTS for pharmacological activity have greatly expanded the number of drug candidates for the therapy of neurological diseases. Rapid screening for BBB penetration early in the drug discovery phase can streamline the quest for promising lead compounds, and provide guidance for the rational design and synthesis of novel compounds targeting the CNS. The tight barrier of the specialized endothelial cells forming/lining the BBB will prevent most drugs from entering the brain. However, those compounds that do penetrate the BBB enter the CNS by transcellular passive diffusion. Over the past three decades, several *in-vitro* BBB biological models have been developed using transwell tissue culture inserts, which can accommodate co-cultures of brain microcapillary endothelial and glial cells, primary or cloned, from different origins. Whilst measurements of the TEER are useful as a concept, their value in these simple *in-vitro* BBB models in terms of predicting the *in-vivo* BBB permeability of specific drugs is very limited. Hence, the reliability of these first generations of *in-vitro* BBB models is questionable. At present, two *in-vitro*

BBB biological models of bovine and porcine origin, are available, which show highly restrictive paracellular permeability properties and are more closely related to the physiological parameters measured in the BBB *in situ*. However, at the time of writing, no model – either static or dynamic – can reliably and reproducibly provide adequate TEER- and P-values of CNS drugs. In our view, despite all past and current efforts, no optimal *in-vitro* BBB pharmaceutical model exists to date, which not only reflects our lack of detailed knowledge of the BBB but also indicates the need to use novel tissue engineering approaches to generate such a model *in vitro*.

These novel approaches will comprise: (1) intelligent scaffolds mimicking the structure/function and regional heterogeneity of the brain ECM; (2) novel HTS devices, which will incorporate flow and hydrostatic pressure as well as online TEER measurements; and (3) on-line concomitant sampling of the media, on both sides of the insert, to measure drug permeability.

There is an urgent need for high-quality, human cells to generate the *in-vitro* BBB model. This need can most likely be met by utilizing human brain endothelial, glial and neuronal stem/precursor cells. In our view, the optimal future model will comprise a hemodynamic system that will integrate synthetic CNS-compatible biomaterials as insert coating, together with human stem cell-derived glia and neurons and brain microvessel ECs, which can be induced to stably express the BBB phenotype (including abundant tight junctions). This future model will have to be validated for a large number of known CNS drugs, and to be compared with *in-vivo* measurements to establish the obligatory pharmaceutical *in-vitro–in-vivo* correlation (IVIVC) in order to be adopted by pharmaceutical companies and the Food and Drugs Administration (FDA). During the next decade of tissue engineering, we anticipate the emergence of fascinating, new, high-fidelity *in-vitro* BBB models, together with *in-silico* artificial membrane permeation assays (PAMPA-BBB [221]), and the generation of vast databases of CNS drugs. These advances will finally open the current bottleneck in the development of novel, more effective CNS drugs for the benefit of patients.

References

1 Shin, M., Ishii, O., Sueda, T., and Vacanti, J. P. *Biomaterials* **2004**, *25*, 3717–3723.

2 Kim, K., Yu, M., Zong, X., Chiu, J., Fang, D., Seo, Y. S., Hsiao, B. S., Chu, B., and Hadjiargyrou, M. *Biomaterials* **2003**, *24*, 4977–4985.

3 Bhattarai, S. R., Bhattarai, N., Yi, H. K., Hwang, P. H., Cha, D. I., and Kim, H. Y. *Biomaterials* **2004**, *25*, 2595–2602.

4 Mo, X. M., Xu, C. Y., Kotaki, M., and Ramakrishna, S. *Biomaterials* **2004**, *25*, 1883–1890.

5 Dar, A., Shachar, M., Leor, J., and Cohen, S. *Biotechnol. Bioeng.* **2002**, *80*, 305–312.

6 Hench, L. L. and Polak, J. M. *Science* **2002**, *295*, 1014–1017.

7 Toshima, M., Ohtani, Y., and Ohtani, O. *Arch. Histol. Cytol.* **2004**, *67*, 31–40.

8 Ntayi, C., Labrousse, A. L., Debret, R., Birembaut, P., Bellon, G., Antonicelli, F., Hornebeck, W., and Bernard, P. *J. Invest. Dermatol.* **2004**, *122*, 256–265.

9 Li, M., Mondrinos, M. J., Gandhi, M. R., Ko, F. K., Weiss, A. S., and Lelkes, P. I. *Biomaterials* **2005**, *26*, 5999–6008.

10 Vacanti, J. P. and Langer, R. *Lancet* **1999**, *354* (Suppl 1), SI32–SI34.

11 Vacanti, J. P., Langer, R., Upton, J., and Marler, J. *J. Adv Drug Deliv. Rev.* **1998**, *33*, 165–182.

12 Takagi, M., Fukui, Y., Wakitani, S., and Yoshida, T. J. *Biosci. Bioeng.* **2004**, *98*, 477–481.

13 Park, H., Radisic, M., Lim, J. O., Chang, B. H., and Vunjak-Novakovic, G. *In Vitro Cell Dev. Biol. Anim.* **2005**, *41*, 188–196.

14 Tang, Z. G., Callaghan, J. T., and Hunt, J. A. *Biomaterials* **2005**, *26*, 6618–6624.

15 Yashiki, S., Umegaki, R., Kino-Oka, M., and Taya, M. J. *Biosci. Bioeng.* **2001**, *92*, 385–388.

16 Casper, C. L., Yamaguchi, N., Kiick, K. L., and Rabolt, J. F. *Biomacromolecules* **2005**, *6*, 1998–2007.

17 MacDiarmid, A. G. *Angew Chem. Int. Ed. Engl.* **2001**, *40*, 2581–2590.

18 Schmidt, C. E., Shastri, V. R., Vacanti, J. P., and Langer, R. *Proc. Natl. Acad. Sci. USA* **1997**, *94*, 8948–8953.

19 Collier, J. H., Camp, J. P., Hudson, T. W., and Schmidt, C. E. *J. Biomed Mater. Res.* **2000**, *50*, 574–584.

20 Pedrotty, D. M., Koh, J., Davis, B. H., Taylor, D. A., Wolf, P., and Niklason, L. E. *Am. J. Physiol. Heart Circ. Physiol.* **2005**, *288*, H1620–H1626.

21 Starovoytov, A., Choi, J., and Seung, H. S. *J. Neurophysiol.* **2005**, *93*, 1090–1098.

22 Klauke, N., Smith, G. L., and Cooper, J. M. *IEEE Trans. Biomed. Eng.* **2005**, *52*, 531–538.

23 Arslan, A., Kiralp, S., Toppare, L., and Yagci, Y. *Int. J. Biol. Macromol.* **2005**, *35*, 163–167.

24 Marquette, C. A., Imbert-Laurenceau, E., Mallet, F., Chaix, C., Mandrand, B., and Blum, L. *J. Anal. Biochem.* **2005**, *340*, 14–23.

25 Kotwal, A. and Schmidt, C. E. *Biomaterials* **2001**, *22*, 1055–1064.

26 Jiang, X., Marois, Y., Traore, A., Tessier, D., Dao, L. H., Guidoin, R., and Zhang, Z. *Tissue Eng.* **2002**, *8*, 635–647.

27 Wan, Y., Wu, H., and Wen, D. *Macromol. Biosci.* **2004**, *4*, 882–890.

28 Castano, H., O'Rear, E. A., McFetridge, P. S., and Sikavitsas, V. I. *Macromol. Biosci.* **2004**, *4*, 785–794.

29 Sanghvi, A. B., Miller, K. P., Belcher, A. M., and Schmidt, C. E. *Nat. Mater.* **2005**, *4*, 496–502.

30 Kamalesh, S., Tan, P., Wang, J., Lee, T., Kang, E. T., and Wang, C. H. *J. Biomed. Mater. Res.* **2000**, *52*, 467–478.

31 Mattioli-Belmonte, M., Giavaresi, G., Biagini, G., Virgili, L., Giacomini, M., Fini, M., Giantomassi, F., Natali, D., Torricelli, P., and Giardino, R. *Int. J. Artif. Organs* **2003**, *26*, 1077–1085.
32 Guterman, E., Cheng, S., Palouian, K., Bidez, P. R., Lelkes, P. I., and Wei, Y. *Polym. Prepr. Am. Chem. Soc., Div. Polym. Chem.* **2003**, *43*, 766–767.
33 Bidez, P. R., III, Li, S., MacDiarmid, A. G., Venancio, E. C., Wei, Y., and Lelkes, P. I. *J. Biomater Sci. Polym. Ed.* **2006**, *17*, 199–212.
34 Diaz, M., Pinto, N. J., Gao, J., and MacDiarmid, A. G. Proceeding of The National Conference on Undergraduate Research (NCUR), Lexington, Kentucky, **2001**.
35 Li, M., Guo, Y., Wei, Y., MacDiarmid, A. G., and Lelkes, P. I. *Biomaterials* **2006**, *27*, 2705–2715.
36 Castillo, G. M., Cummings, J. A., Ngo, C., Yang, W., and Snow, A. D. *J. Biochem. (Tokyo)* **1996**, *120*, 433–444.
37 Ruiz, C. E., Iemura, M., Medie, S., Varga, P., Van Alsine, W. G., Mack, S., Deligio, A., Fearnot, N., Beier, U. H., Pavnik, D., Hijazi, Z. M., and Kiupel, M. *J. Thorac. Cardiovasc. Surg.* **2005**, *130*, 477–484.
38 Lu, Q., Ganesan, K., Simionescu, D. T., and Vyavahare, N. R. *Biomaterials* **2004**, *25*, 5227–5237.
39 Buijtenhuijs, P., Buttafoco, L., Poot, A. A., Daamen, W. F., van Kuppevelt, T. H., Dijkstra, P. J., de Vos, R. A., Sterk, L. M., Geelkerken, B. R., Feijen, J., and Vermes, I. *Biotechnol. Appl. Biochem.* **2004**, *39*, 141–149.
40 Matthews, J. A., Wnek, G. E., Simpson, D. G., and Bowlin, G. L. *Biomacromolecules* **2002**, *3*, 232–238.
41 Boland, E. D., Matthews, J. A., Pawlowski, K. J., Simpson, D. G., Wnek, G. E., and Bowlin, G. L. *Front. Biosci.* **2004**, *9*, 1422–1432.
42 Vrhovski, B. and Weiss, A. S. *Eur. J. Biochem.* **1998**, *258*, 1–18.
43 Martin, S. L., Vrhovski, B., and Weiss, A. S. *Gene* **1995**, *154*, 159–166.
44 Zmora, S., Glicklis, R., and Cohen, S. *Biomaterials* **2002**, *23*, 4087–4094.
45 Glicklis, R., Shapiro, L., Agbaria, R., Merchuk, J. C., and Cohen, S. *Biotechnol. Bioeng.* **2000**, *67*, 344–353.
46 Leor, J., Aboulafia-Etzion, S., Dar, A., Shapiro, L., Barbash, I. M., Battler, A., Granot, Y., and Cohen, S. *Circulation* **2000**, *102*, III56–III61.
47 Leor, J. and Cohen, S. *Ann. N. Y. Acad. Sci.* **2004**, *1015*, 312–319.
48 Baumgarten, P. K. *J. Colloid Interface Sci.* **1971**, *36*, 71–79.
49 Jin, H. J., Fridrikh, S. V., Rutledge, G. C., and Kaplan, D. L. *Biomacromolecules* **2002**, *3*, 1233–1239.
50 Huang, Z. M., Zhang, Y. Z., Kotaki, M., and Ramakrishna, S. *Composites Science and Technology* **2003**, *63*, 2223–2253.
51 Buchko, C. J., Chen, L. C., Shen, Y., and Martin, D. C. *Polymer* **1999**, *40*, 7397–7407.
52 Warner, S. B., Buer, A., Grimler, M., and Ugbolue, S. C. National Textile Center Annual Report, **1999**.

53 Li, W. J., Laurencin, C. T., Caterson, E. J., Tuan, R. S., and Ko, F. K. *J. Biomed. Mater. Res.* **2002**, *60*, 613–621.

54 Bognitzki, M., Czado, W., Frese, T., Schaper, A., Hellwig, M., Steinhart, M., Greiner, A., and Wendorff, J. H. *Adv. Mater.* **2001**, *13*, 70–73.

55 Zong, X., Bien, H., Chung, C. Y., Yin, L., Fang, D., Hsiao, B. S., Chu, B., and Entcheva, E. *Biomaterials* **2005**, *26*, 5330–5338.

56 Min, B. M., Lee, G., Kim, S. H., Nam, Y. S., Lee, T. S., and Park, W. H. *Biomaterials* **2004**, *25*, 1289–1297.

57 Nail, S. L., Jiang, S., Chongprasert, S., and Knopp, S. A. *Pharm. Biotechnol.* **2002**, *14*, 281–360.

58 Kett, V., McMahon, D., and Ward, K. *Curr. Pharm. Biotechnol.* **2005**, *6*, 239–250.

59 Tamada, Y. *Biomacromolecules* **2005**, *6*, 3100–3106.

60 O'Brien, F. J., Harley, B. A., Yannas, I. V., and Gibson, L. J. *Biomaterials* **2005**, *26*, 433–441.

61 Steinhoff, G., Stock, U., Karim, N., Mertsching, H., Timke, A., Meliss, R. R., Pethig, K., Haverich, A., and Bader, A. *Circulation* **2000**, *102*, III50–III55.

62 Kim, W. G., Lee, W. Y., Mi Kim, J., and Moon, H. J. *Int. J. Artif. Organs* **2004**, *27*, 501–508.

63 Badylak, S., Obermiller, J., Geddes, L., and Matheny, R. *Heart Surg. Forum* **2003**, *6*, E20–E26.

64 Spina, M., Ortolani, F., Messlemani, A. E., Gandaglia, A., Bujan, J., Garcia-Honduvilla, N., Vesely, I., Gerosa, G., Casarotto, D., Petrelli, L., and Marchini, M. *J. Biomed. Mater. Res. A.* **2003**, *67*, 1338–1350.

65 Zhong, H. B., Lu, S. B., Hou, S. X., and Zhao, Q. *Zhonghua Wai Ke Za Zhi* **2003**, *41*, 60–63.

66 Schaner, P. J., Martin, N. D., Tulenko, T. N., Shapiro, I. M., Tarola, N. A., Leichter, R. F., Carabasi, R. A., and Dimuzio, P. J. *J. Vasc. Surg.* **2004**, *40*, 146–153.

67 Lee, H. and Vacanti, J. P. Tissue engineering of the liver. In: Atala, A., Mooney, D. (Eds.), *Synthetic biodegradable polymer scaffolds*. Birkhauser, Boston, **1997**.

68 Kaihara, S., Kim, S., Kim, B. S., Mooney, D. J., Tanaka, K., and Vacanti, J. P. *J. Pediatr. Surg.* **2000**, *35*, 1287–1290.

69 Glicklis, R., Merchuk, J. C., and Cohen, S. *Biotechnol. Bioeng.* **2004**, *86*, 672–680.

70 Zavan, B., Brun, P., Vindigni, V., Amadori, A., Habeler, W., Pontisso, P., Montemurro, D., Abatangelo, G., and Cortivo, R. *Biomaterials* **2005**, *26*, 7038–7045.

71 Seoa, S.-J., Akaikeb, T., Choia, Y.-J., Shirakawac, M., Kangd, I.-K., and Choa, C.-S. *Biomaterials* **2005**, *26*, 3607–3615.

72 De Bartolo, L., Morelli, S., Lopez, L. C., Giorno, L., Campana, C., Salerno, S., Rende, M., Favia, P., Detomaso, L., Gristina, R., d'Agostino, R., and Drioli, E. *Biomaterials* **2005**, *26*, 4432–4441.

73 Williams, K. R. and Blayney, A. W. *Biomaterials* **1987**, *8*, 254–258.
74 Kim, S. S., Sundback, C. A., Kaihara, S., Benvenuto, M. S., Kim, B. S., Mooney, D. J., and Vacanti, J. P. *Tissue Eng.* **2000**, *6*, 39–44.
75 Fukuda, M., Peppas, N. A., and McGinity, J. W. *Int. J. Pharm.* **2006**, *310*, 90–100.
76 Kedem, A., Perets, A., Gamlieli-Bonshtein, I., Dvir-Ginzberg, M., Mizrahi, S., and Cohen, S. *Tissue Eng.* **2005**, *11*, 715–722.
77 Perets, A., Baruch, Y., Weisbuch, F., Shoshany, G., Neufeld, G., and Cohen, S. *J. Biomed. Mater. Res. A.* **2003**, *65*, 489–497.
78 Perets, A., Baruch, Y., Spira, G., and Cohen, S. *Proc. Intl. Symp. Control. Rel. Bioact. Mater.* **1998**, 25–26.
79 Perets, A., Kedem, A., and Cohen, S. (unpublished results).
80 Dvir-Ginzberg, M., Gamlieli-Bonshtein, I., Agbaria, R., and Cohen, S. *Tissue Eng.* **2003**, *9*, 757–766.
81 Semino, C. E., Merok, J. R., Crane, G. G., Panagiotakos, G., and Zhang, S. *Differentiation* **2003**, *71*, 262–270.
82 Gutsche, A. T., Lo, H., Zurlo, J., Yager, J., and Leong, K. W. *Biomaterials* **1996**, *17*, 387–393.
83 Monga, S. P., Hout, M. S., Baun, M. J., Micsenyi, A., Muller, P., Tummalapalli, L., Ranade, A. R., Luo, J. H., Strom, S. C., and Gerlach, J. C. *Am. J. Pathol.* **2005**, *167*, 1279–1292.
84 Kulig, K. M. and Vacanti, J. P. *Transpl. Immunol.* **2004**, *12*, 303–310.
85 van de Kerkhove, M. P., Hoekstra, R., van Gulik, T. M., and Chamuleau, R. A. *Biomaterials* **2004**, *25*, 1613–1625.
86 Palmes, D. and Spiegel, H. U. *Biomaterials* **2004**, *25*, 1601–1611.
87 Zinchenko, Y. S. and Coger, R. N. *J. Biomed. Mater. Res. A.* **2005**, *75*, 242–248.
88 Chua, K. N., Lim, W. S., Zhang, P., Lu, H., Wen, J., Ramakrishna, S., Leong, K. W., and Mao, H. Q. *Biomaterials* **2005**, *26*, 2537–2547.
89 Naito, H., Takewa, Y., Mizuno, T., Ohya, S., Nakayama, Y., Tatsumi, E., Kitamura, S., Takano, H., Taniguchi, S., and Taenaka, Y. *Am. Soc. Artif. Intern. Organs J.* **2004**, *50*, 344–348.
90 Zimmermann, W. H., Melnychenko, I., and Eschenhagen, T. *Biomaterials* **2004**, *25*, 1639–1647.
91 Kim, B. O., Tian, H., Prasongsukarn, K., Wu, J., Angoulvant, D., Wnendt, S., Muhs, A., Spitkovsky, D., and Li, R. K. *Circulation* **2005**, *112*, I96–I104.
92 Carrier, R. L., Papadaki, M., Rupnick, M., Schoen, F. J., Bursac, N., Langer, R., Freed, L. E., and Vunjak-Novakovic, G. *Biotechnol. Bioeng.* **1999**, *64*, 580–589.
93 Zimmermann, W. H. and Eschenhagen, T. *Heart Fail. Rev.* **2003**, *8*, 259–269.
94 Zimmermann, W. H., Didie, M., Wasmeier, G. H., Nixdorff, U., Hess, A., Melnychenko, I., Boy, O., Neuhuber, W. L., Weyand, M., and Eschenhagen, T. *Circulation* **2002**, *106*, I151–I157.

95 Hoerstrup, S. P., Sodian, R., Daebritz, S., Wang, J., Bacha, E. A., Martin, D. P., Moran, A. M., Guleserian, K. J., Sperling, J. S., Kaushal, S., Vacanti, J. P., Schoen, F. J., and Mayer, J. E., Jr. *Circulation* **2000**, *102*, III44–III49.

96 Shimizu, T., Yamato, M., Isoi, Y., Akutsu, T., Setomaru, T., Abe, K., Kikuchi, A., Umezu, M., and Okano, T. *Circ. Res.* **2002**, *90*, e40.

97 Tomita, S., Li, R. K., Weisel, R. D., Mickle, D. A., Kim, E. J., Sakai, T., and Jia, Z. Q. *Circulation* **1999**, *100*, II247–II256.

98 Pasumarthi, K. B. and Field, L. J. *Circ. Res.* **2002**, *90*, 1044–1054.

99 Nag, A. C. and Zak, R. *J. Anat.* **1979**, *129*, 541–559.

100 Shah, A. M., Grocott-Mason, R. M., Pepper, C. B., Mebazaa, A., Henderson, A. H., Lewis, M. J., and Paulus, W. J. *Prog. Cardiovasc. Dis.* **1996**, *39*, 263–284

101 Gray, M. O., Long, C. S., Kalinyak, J. E., Li, H. T., and Karliner, J. S. *Cardiovasc. Res.* **1998**, *40*, 352–363.

102 Thomson, J. A., Itskovitz-Eldor, J., Shapiro, S. S., Waknitz, M. A., Swiergiel, J. J., Marshall, V. S., and Jones, J. M. *Science* **1998**, *282*, 1145–1147.

103 Bauwens, C., Yin, T., Dang, S., Peerani, R., and Zandstra, P. W. *Biotechnol. Bioeng.* **2005**, *90*, 452–461.

104 Doetschman, T. C., Eistetter, H., Katz, M., Schmidt, W., and Kemler, R. *J. Embryol. Exp. Morphol.* **1985**, *87*, 27–45.

105 Blau, H. M., Brazelton, T. R., and Weimann, J. M. *Cell* **2001**, *105*, 829–841.

106 Condorelli, G., Borello, U., De Angelis, L., Latronico, M., Sirabella, D., Coletta, M., Galli, R., Balconi, G., Follenzi, A., Frati, G., Cusella De Angelis, M. G., Gioglio, L., Amuchastegui, S., Adorini, L., Naldini, L., Vescovi, A., Dejana, E., and Cossu, G. *Proc. Natl. Acad. Sci. USA* **2001**, *98*, 10733–10798.

107 Müller-Ehmsen, J., Whittaker, P., Kloner, R. A., Dow, J. S., Sakoda, I., Long, T. I., Laird, P. W., and Kedes, L. *J. Mol. Cell Cardiol.* **2002**, *34*, 107–116.

108 Shachar, M. and Cohen, S. *Heart Fail. Rev.* **2003**, *8*, 271–276.

109 Radisic, M., Malda, J., Epping, E., Geng, W., Langer, R., and Vunjak-Novakovic, G. *Biotechnol. Bioeng.* **2006**, *93*, 332–343.

110 Kofidis, T., Lenz, A., Boublik, J., Akhyari, P., Wachsmann, B., Mueller-Stahl, K., Hofmann, M., and Haverich, A. *Biomaterials* **2003**, *24*, 5009–5014.

111 Barron, V., Lyons, E., Stenson-Cox, C., McHugh, P. E., and Pandit, A. *Ann. Biomed. Eng.* **2003**, *31*, 1017–1030.

112 Gonen-Wadmany, M., Gepstein, L., and Seliktar, D. *Ann. N. Y. Acad. Sci.* **2004**, *1015*, 299–311.

113 Martin, Y. and Vermette, P. *Biomaterials* **2005**, *26*, 7481–7503.

114 Feng, Z., Matsumoto, T., Nomura, Y., and Nakamura, T. *IEEE Eng. Med. Biol. Mag.* **2005**, *24*, 73–79.

115 Bursac, N., Papadaki, M., White, J. A., Eisenberg, S. R., Vunjak-Novakovic, G., and Freed, L. E. *Tissue Eng.* **2003**, *9*, 1243–1253.

116 Lichtenberg, A., Dumlu, G., Walles, T., Maringka, M., Ringes-Lichtenberg, S., Ruhparwar, A., Mertsching, H., and Haverich, A. *Biomaterials* **2005**, *26*, 555–562.

117 Liu, J. S., Lu, P. C., Lo, C. W., Lai, H. C., and Hwang, N. H. *Am. Soc. Artif. Intern. Organs J.* **2005**, *51*, 336–341.

118 Boisseau, M. R. *Clin. Hemorheol. Microcirc.* **2005**, *33*, 201–207.

119 Boublik, J., Park, H., Radisic, M., Tognana, E., Chen, F., Pei, M., Vunjak-Novakovic, G., and Freed, L. E. *Tissue Eng.* **2005**, *11*, 1122–1132.

120 Vandenburgh, H. H., Solerssi, R., Shansky, J., Adams, J. W., and Henderson, S. A. *Am. J. Physiol.* **1996**, *270*, C1284–C1292.

121 Zimmermann, W. H., Schneiderbanger, K., Schubert, P., Didie, M., Munzel, F., Heubach, J. F., Kostin, S., Neuhuber, W. L., and Eschenhagen, T. *Circ. Res.* **2002**, *90*, 223–230.

122 Cui, J., Li, J., Mathison, M., Tondato, F., Mulkey, S. P., Micko, C., Chronos, N. A., and Robinson, K. A. *Cardiovasc. Revasc. Med.* **2005**, *6*, 113–120.

123 Cushing, M. C., Jaeggli, M. P., Masters, K. S., Leinwand, L. A., and Anseth, K. S. *J. Biomed. Mater. Res. A.* **2005**, *75*, 232–241.

124 Rieder, E., Seebacher, G., Kasimir, M. T., Eichmair, E., Winter, B., Dekan, B., Wolner, E., Simon, P., and Weigel, G. *Circulation* **2005**, *111*, 2792–2797.

125 Mironov, V., Boland, T., Trusk, T., Forgacs, G., and Markwald, R. R. *Trends Biotechnol.* **2003**, *21*, 157–161.

126 Sun, W., Darling, A., Starly, B., and Nam, J. *Biotechnol. Appl. Biochem.* **2004**, *39*, 29–47.

127 Wilson, W. C. Jr., and Boland, T. *Anat. Rec. A. Discov. Mol. Cell. Evol. Biol.* **2003**, *272*, 491–496.

128 Chakir, J., Page, N., Hamid, Q., Laviolette, M., Boulet, L. P., and Rouabhia, M. *J. Allergy Clin. Immunol.* **2001**, *107*, 36–40.

129 Paquette, J. S., Tremblay, P., Bernier, V., Auger, F. A., Laviolette, M., Germain, L., Boutet, M., Boulet, L. P., and Goulet, F. *In Vitro Cell Dev. Biol. Anim.* **2003**, *39*, 213–220.

130 Le Visage, C., Dunham, B., Flint, P., and Leong, K. W. *Tissue Eng.* **2004**, *10*, 1426–1435.

131 Dobbs, L. G. *Am. J. Physiol.* **1990**, *258*, L134–L147.

132 Olsen, C. O., Isakson, B. E., Seedorf, G. J., Lubman, R. L., and Boitano, S. *Exp. Lung Res.* **2005**, *31*, 461–482.

133 Mondrinos, M. J., Koutzaki, S., Jiwanmall, E., Li, M., Dechadarevian, J. P., Lelkes, P. I., and Finck, C. M. *Tissue Eng.* **2006**, *12*, 717–728.

134 Bates, S. R., Gonzales, L. W., Tao, J. Q., Rueckert, P., Ballard, P. L., and Fisher, A. B. *Am. J. Physiol. Lung Cell Mol. Physiol.* **2002**, *282*, L267–L276.

135 Chen, P., Marsilio, E., Goldstein, R. H., Yannas, I. V., and Spector, M. *Tissue Eng.* **2005**, *11*, 1436–1448.

136 Shannon, J. M., McCormick-Shannon, K., Burhans, M. S., Shangguan, X., Srivastava, K., and Hyatt, B. A. *Am. J. Physiol. Lung Cell Mol. Physiol.* **2003**, *285*, L1323–L1336.

137 Mondrinos, M. J., Li, M., Koutzaki, S., Finck, C. M., and Lelkes, P. I. Biomedical Engineering Society Annual Fall Meeting, Electronic Proceedings, **2005** (www.bmes.org).

138 Wasowicz, M., Biczysko, W., Marszalek, A., Yokoyama, S., and Nakayama, I. *Folia Histochem. Cytobiol.* **1998**, *36*, 3–13.

139 Furuyama, A. and Mochitate, K. *J. Cell Sci.* **2000**, *113*, 859–868.

140 Blau, H., Guzowski, D. E., Siddiqi, Z. A., Scarpelli, E. M., and Bienkowski, R. S. *J. Cell Physiol.* **1988**, *136*, 203–214.

141 Adamson, I. Y. and Young, L. *Am. J. Physiol.* **1996**, *270*, L1017–1022.

142 Mathias, N. R., Kim, K. J., Robison, T. W., and Lee, V. H. *Pharm. Res.* **1995**, *12*, 1499–1505.

143 Lehr, C. M. *Cell Culture Models of Biological Barriers.* Taylor and Francis, London, **2002**.

144 Steimer, A., Haltner, E., and Lehr, C. M. *J. Aerosol. Med.* **2005**, *18*, 137–182.

145 Giard, D. J., Aaronson, S. A., Todaro, G. J., Arnstein, P., Kersey, J. H., Dosik, H., and Parks, W. P. *J. Natl. Cancer Inst.* **1973**, *51*, 1417–1423.

146 Pasternack, M., Floerchinger, C. S., and Hunninghake, G. W. *Exp. Lung Res.* **1996**, *22*, 525–539.

147 Ali, N. N., Edgar, A. J., Samadikuchaksaraei, A., Timson, C. M., Romanska, H. M., Polak, J. M., and Bishop, A. E. *Tissue Eng.* **2002**, *8*, 541–550.

148 Rippon, H. J., Ali, N. N., Polak, J. M., and Bishop, A. E. *Cloning Stem Cells* **2004**, *6*, 49–56.

149 Coraux, C., Nawrocki-Raby, B., Hinnrasky, J., Kileztky, C., Gaillard, D., Dani, C., and Puchelle, E. *Am. J. Respir. Cell. Mol. Biol.* **2005**, *32*, 87–92.

150 Cheek, J. M., Evans, M. J., and Crandall, E. D. *Exp. Cell Res.* **1989**, *184*, 375–387.

151 Danto, S. I., Zabski, S. M., and Crandall, E. D. *Am. J. Respir. Cell. Mol. Biol.* **1992**, *6*, 296–306.

152 Douglas, W. H., Moorman, G. W., and Teel, R. W. *In Vitro* **1976**, *12*, 373–381.

153 Sugihara, H., Toda, S., Miyabara, S., Fujiyama, C., and Yonemitsu, N. *Am. J. Pathol.* **1993**, *142*, 783–792.

154 Forbes, B. and Ehrhardt, C. *Eur. J. Pharm. Biopharm.* **2005**, *60*, 193–205.

155 Isakson, B. E., Seedorf, G. J., Lubman, R. L., and Boitano, S. *In Vitro Cell Dev. Biol. Anim.* **2002**, *38*, 443–449.

156 Bland, R. D. *Biol. Neonate* **2005**, *88*, 181–191.

157 Castell, J. V., Donato, M. T., and Gomez-Lechon, M. *J. Exp. Toxicol. Pathol.* **2005**, *57* (Suppl 1), 189–204.

158 Uiprasertkul, M., Puthavathana, P., Sangsiriwut, K., Pooruk, P., Srisook, K., Peiris, M., Nicholls, J. M., Chokephaibulkit, K., Vanprapar, N., and Auewarakul, P. *Emerg. Infect. Dis.* **2005**, *11*, 1036–1041.

159 Jaspers, I., Ciencewicki, J. M., Zhang, W., Brighton, L. E., Carson, J. L., Beck, M. A., and Madden, M. C. *Toxicol. Sci.* **2005**, *85*, 990–1002.

160 Cryan, S. A. *Am. Assoc. Pharm. Sci. J.* **2005**, *7*, E20–E41.

161 Reese, T. S. and Karnovsky, M. J. *J. Cell Biol.* **1967**, *34*, 207–217.
162 Brightman, M. W. and Reese, T. S. *J. Cell Biol.* **1969**, *40*, 648–677.
163 Janzer, R. C. *J. Inherit. Metab. Dis.* **1993**, *16*, 639–647.
164 Hawkins, B. T. and Davis, T. P. *Pharmacol. Rev.* **2005**, *57*, 173–185.
165 Prat, A., Biernacki, K., Wosik, K., and Antel, J. P. *Glia* **2001**, *36*, 145–155.
166 Vorbrodt, A. W. and Dobrogowska, D. H. *Brain Res. Brain Res. Rev.* **2003**, *42*, 221–242.
167 Crone, C. and Olesen, S. P. *Brain Res.* **1982**, *241*, 49–55.
168 Bundgaard, M. *Brain Res.* **1982**, *241*, 57–65.
169 Tilling, T., Engelbertz, C., Decker, S., Korte, D., Huwel, S., and Galla, H. *J. Cell Tissue Res.* **2002**, *310*, 19–29.
170 Iivanainen, E., Kahari, V. M., Heino, J., and Elenius, K. *Microsc. Res. Tech.* **2003**, *60*, 13–22.
171 Tilling, T., Korte, D., Hoheisel, D., and Galla, H. J. *J. Neurochem.* **1998**, *71*, 1151–1157.
172 Ramsohoye, P. V. and Fritz, I. B. *Neurochem. Res.* **1998**, *23*, 1545–1551.
173 Tan, K. H., Dobbie, M. S., Felix, R. A., Barrand, M. A., and Hurst, R. D. *NeuroReport* **2001**, *12*, 1329–1334.
174 Pardridge, W. M. *J. Neurochem.* **1998**, *70*, 1781–1792.
175 Ooms, F., Weber, P., Carrupt, P. A., and Testa, B. *Biochim. Biophys. Acta* **2002**, *1587*, 118–125.
176 Duvernoy, H., Delon, S., and Vannson, J. L. *Brain Res. Bull.* **1983**, *11*, 419–480.
177 de Boer, A. G., van der Sandt, I. C., and Gaillard, P. J. *Annu. Rev. Pharmacol. Toxicol.* **2003**, *43*, 629–656.
178 Pardridge, W. M. *NeuroRx* **2005**, *2*, 3–14.
179 Clark, D. E. *Drug Discov. Today* **2003**, *8*, 927–933.
180 Gumbleton, M. and Audus, K. L. *J. Pharm. Sci.* **2001**, *90*, 1681–1698.
181 Hayashi, Y., Nomura, M., Yamagishi, S., Harada, S., Yamashita, J., and Yamamoto, H. *Glia* **1997**, *19*, 13–26.
182 Kuchler-Bopp, S., Delaunoy, J. P., Artault, J. C., Zaepfel, M., and Dietrich, J. B. *NeuroReport* **1999**, *10*, 1347–1353.
183 Bauer, H. C., Bauer, H., Lametschwandtner, A., Amberger, A., Ruiz, P., and Steiner, M. *Brain Res. Dev. Brain Res.* **1993**, *75*, 269–278.
184 Lu, S., Gough, A. W., Bobrowski, W. F., and Stewart, B. H. *J. Pharm. Sci.* **1996**, *85*, 270–273.
185 Gaillard, P. J., Voorwinden, L. H., Nielsen, J. L., Ivanov, A., Atsumi, R., Engman, H., Ringbom, C., de Boer, A. G., and Breimer, D. D. *Eur. J. Pharm. Sci.* **2001**, *12*, 215–222.
186 Hurst, R. D. *NeuroReport* **2000**, *11*, L1–L2.
187 Megard, I., Garrigues, A., Orlowski, S., Jorajuria, S., Clayette, P., Ezan, E., and Mabondzo, A. *Brain Res.* **2002**, *927*, 153–167.
188 Bacic, M., Edwards, N. A., and Merrill, M. J. *Growth Factors* **1995**, *12*, 11–15.
189 Rascher, G., Fischmann, A., Kroger, S., Duffner, F., Grote, E. H., and Wolburg, H. *Acta Neuropathol. (Berl.)* **2002**, *104*, 85–91.

190 Khodarev, N. N., Yu, J., Labay, E., Darga, T., Brown, C. K., Mauceri, H. J., Yassari, R., Gupta, N., and Weichselbaum, R. R. *J. Cell Sci.* **2003**, *116*, 1013–1022.

191 Misfeldt, D. S., Hamamoto, S. T., and Pitelka, D. R. *Proc. Natl. Acad. Sci. USA* **1976**, *73*, 1212–1216.

192 Demeuse, P., Kerkhofs, A., Struys-Ponsar, C., Knoops, B., Remacle, C., and van den Bosch de Aguilar, P. *J. Neurosci. Methods* **2002**, *121*, 21–31.

193 Cestelli, A., Catania, C., D'Agostino, S., Di Liegro, I., Licata, L., Schiera, G., Pitarresi, G. L., Savettieri, G., De Caro, V., Giandalia, G., and Giannola, L. I. *J. Control. Release* **2001**, *76*, 139–147.

194 Yu, H. and Sinko, P. J. *J. Pharm. Sci.* **1997**, *86*, 1448–1457.

195 Lauer, R., Bauer, R., Linz, B., Pittner, F., Peschek, G. A., Ecker, G., Friedl, P., and Noe, C. R. *Farmaco* **2004**, *59*, 133–137.

196 Roux, F. and Couraud, P. O. *Cell. Mol. Neurobiol.* **2005**, *25*, 41–58.

197 Hosoya, K. I., Takashima, T., Tetsuka, K., Nagura, T., Ohtsuki, S., Takanaga, H., Ueda, M., Yanai, N., Obinata, M., and Terasaki, T. *J. Drug Target.* **2000**, *8*, 357–370.

198 Boveri, M., Berezowski, V., Price, A., Slupek, S., Lenfant, A. M., Benaud, C., Hartung, T., Cecchelli, R., Prieto, P., and Dehouck, M. P. *Glia* **2005**, *51*, 187–198.

199 Haseloff, R. F., Blasig, I. E., Bauer, H. C., and Bauer, H. *Cell. Mol. Neurobiol.* **2005**, *25*, 25–39.

200 Bauer, H. C. and Bauer, H. *Cell. Mol. Neurobiol.* **2000**, *20*, 13–28.

201 Schiera, G., Sala, S., Gallo, A., Raffa, M. P., Pitarresi, G. L., Savettieri, G., and Di Liegro, I. *J. Cell Mol. Med.* **2005**, *9*, 373–379.

202 Hayashi, K., Nakao, S., Nakaoke, R., Nakagawa, S., Kitagawa, N., and Niwa, M. *Regul. Pept.* **2004**, *123*, 77–83.

203 Dohgu, S., Takata, F., Yamauchi, A., Nakagawa, S., Egawa. T., Naito, M., Tsuruo, T., Sawada, Y., Niwa, M., and Kataoka, Y. *Brain Res.* **2005**, *1038*, 208–215.

204 Arthur, F. E., Shivers, R. R., and Bowman, P. D. *Brain Res.* **1987**, *433*, 155–159.

205 Raub, T. J., Kuentzel, S. L., and Sawada, G. A. *Exp. Cell Res.* **1992**, *199*, 330–340.

206 Dehouck, M. P., Meresse, S., Delorme, P., Fruchart, J. C., and Cecchelli, R. *J. Neurochem.* **1990**, *54*, 1798–1801.

207 Dehouck, M. P., Jolliet-Riant, P., Bree, F., Fruchart, J. C., Cecchelli, R., and Tillement, J. P. *J. Neurochem.* **1992**, *58*, 1790–1797.

208 Lutolf, M. P. and Hubbell, J. A. *Nat. Biotechnol.* **2005**, *23*, 47–55.

209 Hou, S., Xu, Q., Tian, W., Cui, F., Cai, Q., Ma, J., and Lee, I. S. *J. Neurosci. Methods* **2005**, *148*, 60–70.

210 Robert, A. M., Tixier, J. M., Robert, L., Legeais, J. M., and Renard, G. *Pathol. Biol. (Paris)* **2001**, *49*, 298–304.

211 Ma, S. H., Lepak, L. A., Hussain, R. J., Shain, W., and Shuler, M. L. *Lab. Chip* **2005**, *5*, 74–85.

212 Bowman, P. D., Ennis, S. R., Rarey, K. E., Betz, A. L., and Goldstein, G. W. *Ann. Neurol.* **1983**, *14*, 396–402.

213 Deli, M. A., Abraham, C. S., Kataoka, Y., and Niwa, M. *Cell. Mol. Neurobiol.* **2005**, *25*, 59–127.

214 Cecchelli, R., Dehouck, B., Descamps, L., Fenart, L., Buee-Scherrer, V. V., Duhem, C., Lundquist, S., Rentfel, M., Torpier, G., and Dehouck, M. P. *Adv. Drug Deliv. Rev.* **1999**, *36*, 165–178.

215 Zenker, D., Begley, D., Bratzke, H., Rubsamen-Waigmann, H., and von Briesen, H. *J. Physiol.* **2003**, *551*, 1023–1032.

216 Nitz, T., Eisenblatter, T., Psathaki, K., and Galla, H. J. *Brain Res.* **2003**, *981*, 30–40.

217 Franke, H., Galla, H. J., and Beuckmann, C. T. *Brain Res.* **1999**, *818*, 65–71.

218 Reichel, A., Abbott, N. J., and Begley, D. J. *J. Drug Target.* **2002**, *10*, 277–283.

219 Krizanac-Bengez, L., Kapural, M., Parkinson, F., Cucullo, L., Hossain, M., Mayberg, M. R., and Janigro, D. *Brain Res.* **2003**, *977*, 239–246.

220 Parkinson, F. E., Friesen, J., Krizanac-Bengez, L., and Janigro, D. *Brain Res.* **2003**, *980*, 233–241.

221 Di, L., Kerns, E. H., Fan, K., McConnell, O. J., and Carter, G. T. *Eur. J. Med. Chem.* **2003**, *38*, 223–232.

222 Gao, J., Niklason, L., Langer, R. *J. Biomed. Mater. Res.* **1998**, *42*, 417–424.

223 Griffith, L. G. *Ann. N. Y. Acad. Sci.* **2002**, *961*, 83–95.

224 Jiang, J., Kojima, N., Guo, L., Naruse, K., Makuuchi, M., Miyajima, A., Yan, W., Sakai, Y. *Tissue Eng.* **2004**, *10*, 1577–1586.

225 Kim, S. S., Utsunomiya, H., Koski, J. A., Wu, B. M., Cima, M. J., Sohn, J., Mukai, K., Griffith, L. G., Vacanti, J. P. *Ann. Surg.* **1998**, *228*, 8–13.

226 Park, T. G. *J. Biomed. Mater. Res.* **2002**, *59*, 127–135.

227 Shapiro, L. and Cohen, S. *Biomaterials* **1997**, *18*, 583–590.

228 Wayne, J. S., McDowell, C. L., Shields, K. J., Tuan, R. S. *Tissue Eng.* **2005**, *11*, 953–963.

229 Koebe, H. G., Pahernik, S., Eyer, P., and Schildberg, F. W. *Xenobiotica* **1994**, *24*, 95–107.

230 Sugimoto, S., Harada, K., Shiotani, T., Ikeda, S., Katsura, N., Ikai, I., Mizuguchi, T., Hirata, K., Yamaoka, Y., and Mitaka, T. *Tissue Eng.* **2005**, *11*, 626–633.

231 Terada, S., Sato, M., Sevy, A., and Vacanti, J. P. *Yonsei Med. J.* **2000**, *41*, 685–691.

232 Weiss, T. S., Jahn, B., Cetto, M., Jauch, K. W., and Thasler, W. E. *Cell Prolif.* **2002**, *35*, 257–267.

233 Woodfield, T. B., Blitterswijk, C. A., Wijn, J. D., Sims, T. J., Hollander, A. P., and Riesle, J. *Tissue Eng.* **2005**, *11*, 1297–1311.

234 Chandy, T. and Sharma, C. P. *Biomater. Artif. Cells Artif. Organs* **1990**, *18*, 1–24.

235 Ma, J., Wang, H., He, B., and Chen, J. *Biomaterials* **2001**, *22*, 331–336.

236 Caniggia, I., Liu, J., Han, R., Wang, J., Tanswell, A. K., Laurie, G., and Post, M. *Am. J. Physiol.* **1996**, *270*, L459–L468.

237 Sanchez-Esteban, J., Wang, Y., Filardo, E. J., Rubin, L. P., and Ingber, D. E. *Am. J. Physiol. Lung Cell Mol. Physiol.* **2006**, *290*, L343–L350.

238 Lwebuga-Mukasa, J. S., Ingbar, D. H., and Madri, J. A. *Exp. Cell Res.* **1986**, *162*, 423–435.

239 Kim, H. J., Henke, C. A., Savik, S. K., and Ingbar, D. H. *Am. J. Physiol.* **1997**, *273*, L134–L141.

240 Guo, Y., Martinez-Williams, C., Yellowley, C. E., Donahue, H. J., and Rannels, D. E. *Am. J. Physiol. Lung Cell. Mol. Physiol.* **2001**, *280*, L191–L202.

241 Mason, R. J., Lewis, M. C., Edeen, K. E., McCormick-Shannon, K., Nielsen, L. D., and Shannon, J. M. *Am. J. Physiol. Lung Cell Mol. Physiol.* **2002**, *282*, L249–L258.

242 Pagan, I., Khosla, J., Li, C. M., and Sannes, P. L. Exp. Lung Res. **2002**, *28*, 69–84.

243 Sakamoto, T., Hirano, K., Morishima, Y., Masuyama, K., Ishii, Y., Nomura, A., Uchida, Y., Ohtsuka, M., and Sekizawa, K. *In Vitro Cell Dev. Biol. Anim.* **2001**, *37*, 471–479.

244 Shah, M. V., Audus, K. L., and Borchardt, R. T. *Pharm. Res.* **1989**, *6*, 624–627.

245 Hurst, R. D. and Clark, J. B. *Biochem. Soc. Trans.* **1998**, *26*, S353.

246 Weidenfeller, C., Schrot, S., Zozulya, A., and Galla, H. J. *Brain Res.* **2005**, *1053*, 162–174.

247 Tontsch, U. and Bauer, H. C. *Brain Res.* **1991**, *539*, 247–253.

248 Rauh, J., Meyer, J., Beuckmann, C., and Galla, H. J. *Prog. Brain Res.* **1992**, *91*, 117–121.

249 Muruganandam, A., Herx, L. M., Monette, R., Durkin, J. P., and Stanimirovic, D. B. *FASEB J.* **1997**, *11*, 1187–1197.

250 El Hafny, B., Chappey, O., Piciotti, M., Debray, M., Boval, B., and Roux, F. *Neurosci. Lett.* **1997**, *236*, 107–111.

251 Cucullo, L., McAllister, M. S., Kight, K., Krizanac-Bengez, L., Marroni, M., Mayberg, M. R., Stanness, K. A., and Janigro, D. *Brain Res.* **2002**, *951*, 243–254.

252 Omidi, Y., Campbell, L., Barar, J., Connell, D., Akhtar, S., and Gumbleton, M. *Brain Res.* **2003**, *990*, 95–112.

253 Jeliazkova-Mecheva, V. V., Bobilya, D. J. *Brain Res. Brain Res. Protoc.* **2003**, *12*, 91–98.

254 Toimela, T., Maenpaa, H., Mannerstrom, M., and Tahti, H. *Toxicol. Appl. Pharmacol.* **2004**, *195*, 73–82.

255 Cucullo, L., Hallene, K., Dini, G., Dal Toso, R., and Janigro, D. *Brain Res.* **2004**, *997*, 147–151.

256 Stanness, K. A., Neumaier, J. F., Sexton, T. J., Grant, G. A., Emmi, A., Maris, D. O., and Janigro, D. *NeuroReport* **1999**, *10*, 3725–3731.

257 Savettieri, G., Di Liegro, I., Catania, C., Licata, L., Pitarresi, G. L., D'Agostino, S., Schiera, G., De Caro, V., Giandalia, G., Giannola, L. I., and Cestelli, A. *NeuroReport* **2000**, *11*, 1081–1084.

258 Schiera, G., Bono, E., Raffa, M. P., Gallo, A., Pitarresi, G. L., Di Liegro, I., and Savettieri, G. *J. Cell Mol. Med.* **2003**, *7*, 165–170.

2
An Overview on Bioreactor Design, Prototyping and Process Control for Reproducible Three-Dimensional Tissue Culture

Ralf Pörtner and Christoph Giese

Bioreactor systems play an important role in tissue engineering, as they enable reproducible and controlled changes to be made in specific environmental factors. They can also provide the technical means to perform controlled studies aimed at understanding specific biological, chemical, or physical effects. Furthermore, bioreactors allow for a safe and reproducible production of tissue constructs. For later clinical applications, the bioreactor system should be an advantageous method in terms of low contamination risk, ease of handling, and scalability. With respect to drug screening, the main challenge is the efficient, reproducible handling of a large quantity of tissue constructs in parallel (high-throughput screening). To date, the goals and expectations of bioreactor development have been fulfilled only to some extent, as bioreactor design in tissue engineering is very complex and still at an early stage of development, especially for use in drug screening. In this chapter, important aspects of bioreactor design are summarized, and an overview of existing concepts is provided. An artificial immunosystem will be used as an example to demonstrate how an increased fundamental understanding of biological, biochemical, and engineering aspects can significantly improve the properties of 3D tissue constructs.

2.1
Introduction

Tissue engineering, which means the generation of artificial three-dimensional (3D) tissues, is intended as a powerful tool for regenerative medicine and for drug screening [1–5]. The goal of tissue engineering can be defined as the development of cell-based substitutes to restore, maintain, or improve tissue function. These substitutes should have organ-specific properties with respect to biochemical activity, microstructure, mechanical integrity and biostability [2]. Cell-based therapy concepts include: (1) the direct transplantation of isolated cells; (2) the implantation of a bioactive scaffold for the stimulation of cell growth within the original tissue; and (3) the implantation of a 3D biohybrid structure of a scaffold and cultured

Drug Testing In Vitro: Breakthroughs and Trends in Cell Culture Technology
Edited by Uwe Marx and Volker Sandig

ISBN: 978-3-527-31488-1

cells or tissue. Furthermore, nonimplantable tissue structures can be applied as external support devices (e.g., an extracorporal liver support when a compatible donor organ is not readily available [6, 7]), or engineered tissues can be used as *in-vitro* physiological models for studying disease pathogenesis and developing new molecular therapeutics (e.g., *in-vitro* assays for drug screening [7–9]).

For drug screening based on cell models, 2D cellular assays are mostly applied, using often well-described cell line models [10, 11]. These assays can be performed quite efficiently in high-throughput screening (HTS) systems, but to date their value for predicting the clinical response of new agents, especially with respect to cancer therapy, is limited. This lack of predictability of 2D cellular assays is attributed to the fact that such systems do not mimic the response of the 3D microenvironment present in a tissue, or tumor, *in vivo* [10, 12–14]; therefore, 3D-cellular assays are required. Compared to 3D-tissue cultures intended for medical applications, here the size of the tissue construct is not a major problem, as smaller tissue constructs are also appropriate. The main challenge is the efficient, reproducible handling of a large quantity of tissue constructs in parallel.

The increasing market in highly specific biological pharmaceuticals such as antibodies, cytokines, growth factors or cells and tissue products highlighted the need for specific human relevant test systems on immunofunctions. *In-vitro* assays and transgenic animal models can be used in this situation. The development and use of immunotests *in vitro*, as well as transgenic animal models, must consider the remarkable specificity of the immune systems of different mammalian species. The investigation of effects on human patients at the research level of drug screening, in addition to tests on potency in samples of process development and production and an adapted preclinical risk assessment, requires human *in-vitro* immunotests to be conducted.

Established *in-vitro* tests based on acute lymphocyte reactions for induced cytokine release or cell proliferation of freshly prepared blood samples, such as mixed lymphocyte reaction, are inadequate for investigations into complex or long-term effects such as induced hypersensitivity and allergy.

The effects of immunogenicity or immunotoxicity must have been monitored in long-term culture under physiological and histological equivalency of secondary lymphatic organs using primary cells, tissue preparations, or immunocompetent cell lines. Recently, the use of transgenic animals mimicking isolated functions of the human immunosystem has been inadequate. In the future, immune cell-based *in-vitro* test systems may have to compete with transgenic animal models having a reconstituted human immune system.

The *in-vitro* generation of 3D tissue constructs requires not only a biological model (e.g., an adequate source of proliferative cells with appropriate biological functions; a protocol for proliferating cells while maintaining the tissue-specific phenotype) but also the further development of new culture strategies, including bioreactor concepts [7, 15, 16]. Bioreactors established for the cultivation of microbes or mammalian cells under monitored and controlled environmental and operational conditions (e.g., pH, temperature, oxygen tension, nutrient supply) are mostly inapplicable to 3D tissue constructs. Furthermore, each type of tissue (e.g., skin, bone, blood vessels, cartilage) will likely require an individualized bioreactor design [15]. Therefore, tissue-specific bioreactors should be designed on the basis of a comprehensive understanding of biological and engineering aspects. Additionally, typical engineering aspects such as reliability, reproducibility, scalability and safety should be addressed [7, 16]. In the following sections, the key technical challenges are identified and an overview of existing culture systems and bioreactors used for tissue engineering is provided. These topics have been addressed to some extent by several authors [6, 7, 15–24], and also reviewed [25]; therefore, they will be discussed only briefly at this point. Particular focus will be given to the interaction between biological and engineering aspects and the special demands of using these reactor systems for drug screening. Using an artificial immunosystem as an example, it will be shown how an increased fundamental understanding of biological, biochemical, and engineering aspects can significantly improve the properties of 3D tissue constructs.

2.2 Important Aspects for Bioreactor Design

With regard to tissue engineering, bioreactors are used for cell proliferation on a small scale (e.g., for individual patients) and on a large scale (e.g., for allogeneic therapy concepts), to generate 3D tissue constructs from isolated and proliferated cells *in vitro* and for direct organ-support devices [23]. These bioreactors should enable the control of environmental conditions such as oxygen tension, pH, temperature, and shear stress, as well as allowing aseptic operation (e.g., feeding and sampling). Furthermore, a bioreactor system should allow for automated processing steps. This is essential not only for controlled, reproducible, statistically relevant basic studies, but also for the future routine manufacture of tissues for clinical application or drug screening [7, 26]. In addition to these global requirements, specific key criteria for 3D tissue constructs based on cells and scaffolds must also be met, including the proliferation of cells, the seeding of cells onto macroporous scaffolds, nutrient (particularly oxygen) supply within the resulting tissue, and mechanical stimulation of the developing tissues [7].

The proliferation of cells represents the first step in establishing a tissue culture. Usually, cells harvested from a biopsy must be expanded by several orders of magnitude. Cell proliferation is quite often accompanied by the dedifferentiation of cells [27, 28], with small culture dishes (e.g., Petri dishes, 12-well plates, and T-flasks) being mainly used for cell expansion. As these devices allow an increase in cell number by a factor of only about 10, several subcultivation steps are required. These are considered to be a major cause of dedifferentiation of cells. Recent studies have shown that microcarrier cultures performed in well-mixed bioreactor systems can significantly improve cell expansion [29–31].

A further critical aspect of a differentiated tissue is the extracellular matrix (ECM), as described by Alison Abbott (cited from [95]): "In mammalian tissues, cells connect not only to each other, but also to a support structure called the ECM.

This contains proteins, such as collagen, elastin and laminin, that give tissues their mechanical properties and help to organize communication between cells embedded within the matrix. Receptors on the surface of the cells, in particular a family of proteins called the integrins, anchor their bearers to the ECM, and also determine how the cells interpret biochemical cues from their immediate surroundings. Given this complex mechanical and biochemical interplay, it is perhaps no surprise that researchers will miss biological subtleties if the cells they are studying grow only in flat layers. But providing an appropriate environment in which to culture cells in three-dimensions is no easy matter (...). Some researchers use simple gels consisting of collagen, whereas others make their own gels by extracting ECM material from relevant tissues. Another popular option is the commercially available Matrigel, which consists of structural proteins such as laminin and collagen, plus growth factors and enzymes, all taken from mouse tumours." [94, 95]. Further 3D (mostly macroporous) scaffolds used for tissue engineering have been discussed previously. The cell seeding of scaffolds is an important step in establishing a 3D culture in a macroporous scaffold, as not only seeding at high cell densities but also a homogeneous distribution of cells within the scaffold is essential [32–34]. Several techniques for cell seeding have been discussed by Martin et al. [7].

A sufficient supply of nutrients, together with the removal of toxic or inhibitory substances, is crucial for long-term culture to control a constant and defined environment. In 2D "flat culture", the formation of a suitable microenvironment is disturbed by convection of the culture supernatant and the periodically exchanged media. Cell migration and interaction is limited on the 2D-culture surface, and is more or less defined by the initial seeding of cells or resuspension and may be suitable for single-layer epithelioid tissues based on a high proliferative capacity. Perfusable and cell migrational voluminous matrices – the so-called 3D-matrices – support the formation of local microgradients, cell migration, cell–cell-activation, leading to coordinated proliferation aggregation, the initialization of tissue forming, and tissue polarization.

All kinds of flows needed for long-term supply, such as transfusion, perfusion, circulation and convection, disturb the formation of guiding microgradients. For an ideal microenvironment, it is vital to ensure the correct balance of minimal, yet sufficient, perfusion and a maximum of self-conditioning. Sufficient perfusion ensures an optimum of nutrient and metabolite concentrations. Inhomogeneity in oxygen tension, and local accumulations of cytokines and chemokines, triggers chemotaxis and cell activation and differentiation [101, 102].

Furthermore, the size of most engineered tissues is limited as they do not have their own blood system and the cells are supplied only by diffusion [8, 25, 35]. Oxygen supply is particularly critical, as only cell layers of 100–200 μm thickness can be supplied by diffusion [36]. However, as tissue constructs should have larger dimensions, mass-transfer limitations represent one of the greatest engineering challenges [25].

Various studies have shown that mechanical stimulation (e.g., mechanical compression, hydrodynamic pressure and fluid flow, which are important modulators of cell physiology) can have a positive impact on tissue formation [37], particularly in the context of musculoskeletal tissue engineering, cartilage formation, and cardiovascular tissues [19, 38–45]. As yet, however, little is known about the specific mechanical forces or the ranges of application, such as magnitude, frequency, continuous or intermittent, and duty cycle [7, 17]. Further studies of these factors must be coupled with quantitative and computational analyses of physical forces experienced by cells and changes in mass transport induced by the method used.

2.3 Culture Systems and Bioreactors Used in Tissue Engineering

An overview of the culture systems and bioreactors used for the engineering of 3D tissue constructs, including cell maintenance, proliferation, and tissue formation, is illustrated schematically in Figure 2.1a–c.

Culture systems developed for the monolayer culture of adherent cells (T-flasks, Petri dishes, multiwell plates) are normally used for cell maintenance and proliferation. These systems allow for sterile handling procedures and are easy to use, disposable, and inexpensive [46]. By contrast, they require individual handling, for example in stages of medium exchange and cell seeding, and their usefulness is limited when large quantities of cells are required [15], though this can be overcome to some extent by using sophisticated robotics [7]. In addition, environmental parameters such as pH, pO_2, and temperature cannot be controlled. A further drawback is the limited increase in cell number (approximately 10- to 20-fold during cultivation); consequently, the generation of a large number of cells requires several enzymatic subcultivation steps, accompanied by an increased passage number and cell dedifferentiation. In recent studies small well-mixed bioreactors (e.g., shake flasks, stirred vessels and "super spinner") have been suggested for cell proliferation in which the cells are grown on microcarriers [29–31]. These systems have been used for the cultivation of encapsulated cells [27, 28] or neural stem cells in single-cell suspension culture [47].

3D tissue cultures can be performed in fixed-bed and fluidized-bed bioreactors, with the cells being immobilized in macroporous carriers or in networks of fibers arranged in a column so that they are either packed (fixed-bed) or floating (fluidized-bed). The column is permanently perfused with a conditioned medium contained in a reservoir, mostly using a circulation loop. These types of reactor are very efficient for the long-term cultivation of mammalian cells to produce biopharmaceuticals, such as monoclonal antibodies, recombinant drugs including tissue plasminogen activator (tPA) and erythropoietin (EPO), or recombinant retroviruses for gene therapy [48–50]. For tissue engineering, the reactors have been investigated for several applications, including the cultivation of "liver" cells as an extracorporeal liver device [6, 18, 51], the proliferation of stem cells [52–54], the cultivation of cardiovascular cells [19] and cartilage cells [17], or as an *in-vitro* human placental model [55].

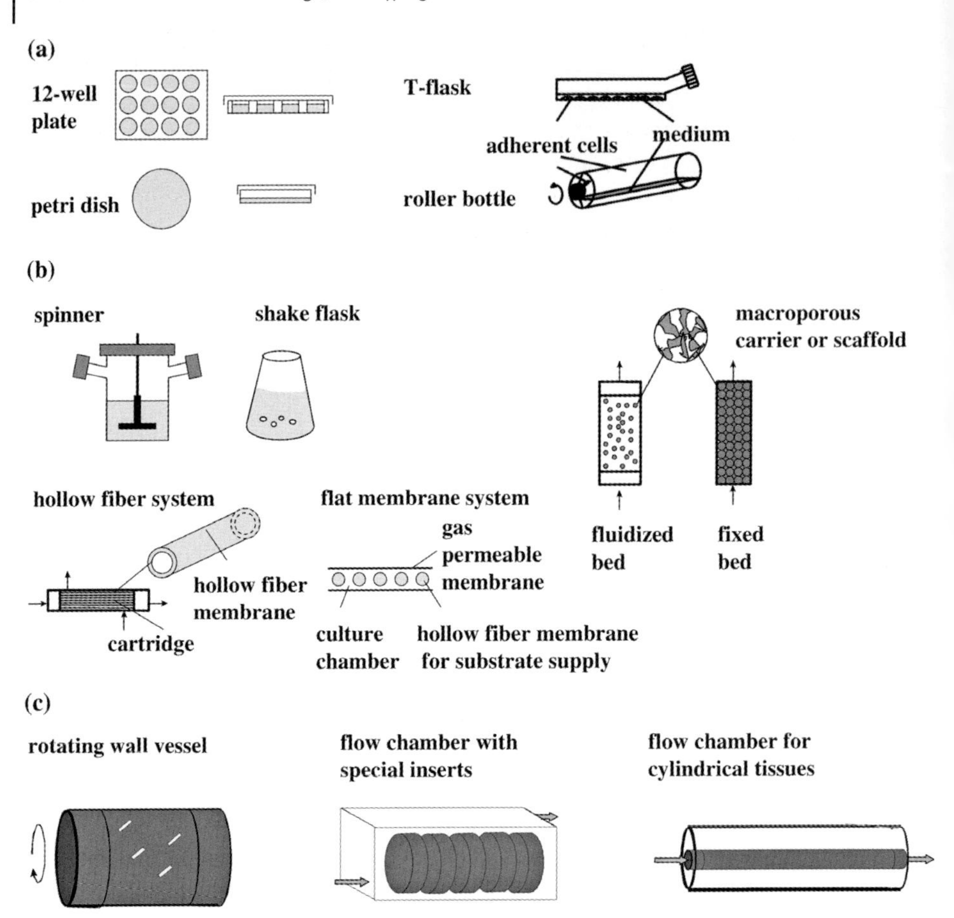

Fig. 2.1 An overview of cell culture systems used in tissue engineering (adapted and modified from [25]). (a) Systems used for routine cultivation within an incubator, where the cells grow mainly in monolayer (e.g., 12-well-plates, Petri dishes, T-flasks or roller bottles). (b) Culture systems developed mainly for cultivation of mammalian cells, which were adapted for cultivation of tissue cells in three-dimensional structures (e.g., spinner and shake flasks, membrane-based systems such as hollow-fiber reactors or fluidized- and fixed-bed reactors). (c) Culture systems designed especially for tissue engineering mimicking the special demands of a three-dimensional tissue.

In membrane bioreactors, including hollow-fiber reactors [56], the miniPerm system [57] or the tecnomouse [58], cells are cultivated at tissue-like densities in a compartment which contains one or several types of membrane for nutrient and oxygen supply and removal of toxic metabolites. Hollow-fiber systems are widely used in the production of biopharmaceuticals, including monoclonal antibodies. Several examples of modified membrane bioreactors exist for the 3D culture of tissue cells, including hepatocytes [6, 59–64], skin cells [65] or other human cells [58, 66].

Most of the culture systems and bioreactors discussed so far were first developed for the cultivation of mammalian cells, and subsequently adapted to the engineering of 3D tissue constructs. However, apart from some exceptions, they cannot easily be used in the generation of implantable tissue constructs, as each type of tissue intended for implantation (e.g., skin, heart valve, blood vessel, cartilage) requires a different geometric structure and a specific bioreactor design. One of the most prominent culture systems is the rotating-wall vessel [32], in which a construct remains in a state of free-fall through the medium with a low shear stress and a high mass transfer rate. This system has a wide range of practical applications [7, 15, 67]. A multipurpose culture system was introduced by Minuth et al. [68] for perfusion cultures under organotypic conditions. In this situation, several tissue carriers can be placed inside a perfusion container and, depending on the type of tissue-specific cell required, different supports can be selected. A perfused flow-chamber bioreactor with a new concept for aeration has been recently introduced [44, 69] in which tissue-specific inserts for various types of tissue (e.g., cartilage, skin, bone) can be applied.

In addition to these examples of multipurpose bioreactors, numerous tissue-specific culture systems have been suggested and reviewed [7, 8, 16–19, 22, 23, 35, 59, 70–72]. Unfortunately, the majority of these have been custom-made, with only very few having been commercialized.

2.4
The Operation of Bioreactors

Bioreactors allow for different process strategies including batch, fed-batch, or continuous cultivation (Figs. 2.2 and 2.3).

Continuous perfusion, in particular, enables cultivation to be carried out under constant and controlled environmental conditions [44, 51, 68, 69, 73, 74]. Martin et al. [7] summarized some of the effects of direct perfusion on tissue-specific properties such as growth, differentiation and mineralized matrix deposition by bone cells, the proliferation of human oral keratinocytes, rates of albumin synthesis by hepatocytes, the expression of cardiac-specific markers by cardiomyocytes, and glucosaminoglycan (GAG) synthesis and matrix formation by chondrocytes (Fig. 2.4) [69].

On the other hand, a bioreactor system becomes more complex when additional features such as feeding pumps, vessels for fresh and spent medium, and control strategies are required, particularly in the case of mechanical stimulation. With regard to the formation of an implantable tissue, the bioreactor system must be integrated into the entire cultivation scheme, including biopsy, proliferation (cell expansion, usually in T-flasks), cell seeding of the bioreactor, tissue formation, and delivery to the site of application (e.g., the hospital). This is particularly important with regard to the manufacture of engineered tissue constructs for clinical applications, when good manufacturing practice (GMP) requirements must also be met [15, 16].

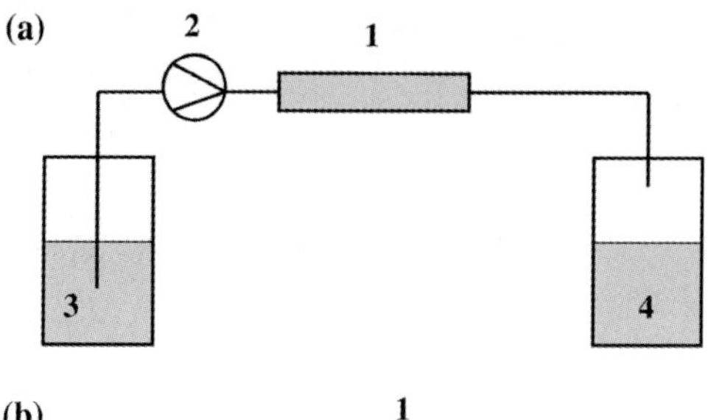

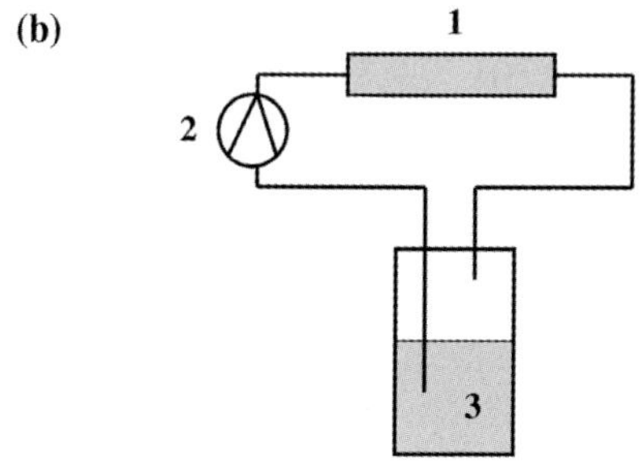

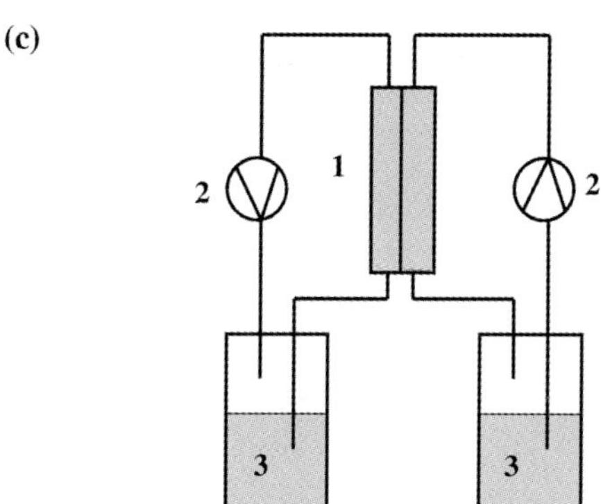

Fig. 2.2 The operation of tissue culture systems.
(a) Fresh medium is pumped continuously through the culture unit in a plug-flow. The required flow rate is determined by the substrate demands (especially oxygen) of the cells (1 = culture unit; 2 = circulation pump; 3 = feed medium; 4 = spent medium).
(b) Medium is pumped continuously from a medium vessel through the cultivation unit and back. The medium in the medium vessel can be changed at intervals (1 = culture unit; 2 = circulation pump; 3 = medium vessel).
(c) The culture unit consists of two culture chambers separated by a semipermeable membrane. Each culture chamber is supplied with fresh medium from an individual medium vessel. This technique is intended for the co-cultivation of different types of cells (1 = culture unit; 2 = circulation pumps; 3 = medium vessels).

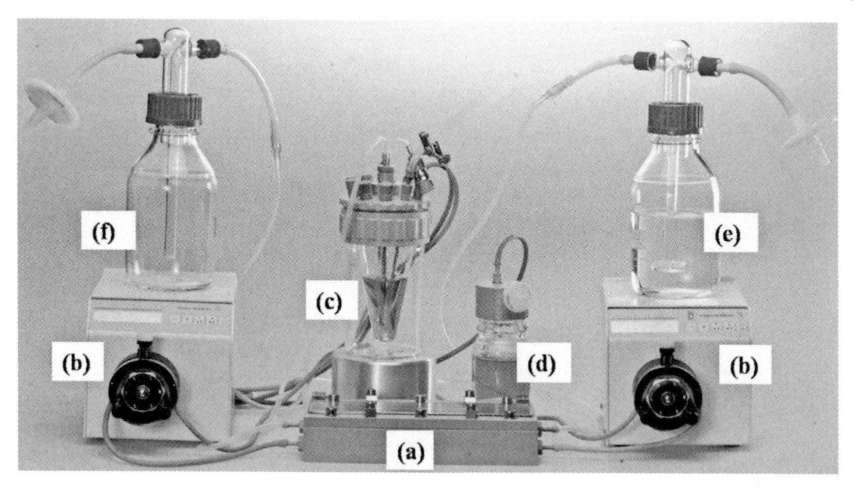

Fig. 2.3 The set-up of a flow-chamber bioreactor system to cultivate cartilage-carrier-constructs. The system consists of: (a) a flow chamber with inserts for tissue constructs; (b) peristaltic pumps for medium circulation; (c) a medium vessel; (d) a medium exchange bottle; (e) a humidifier; and (f) a flask to trap the exhaust gas. This set-up refers to a culture unit indicated in Fig. 2.1c (flow chamber with special inserts) operated according to the operation mode shown in Fig. 2.2b.

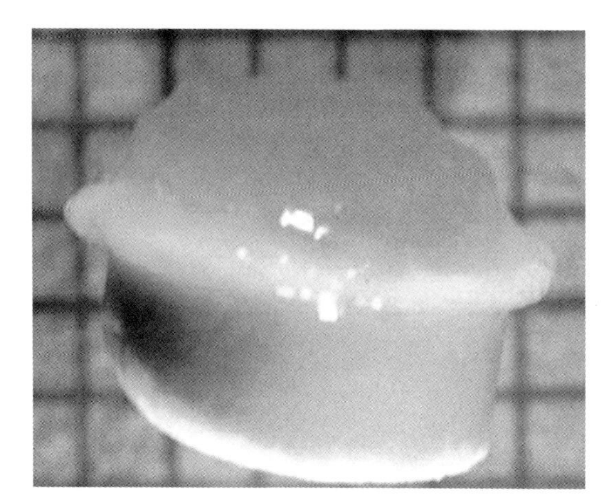

Fig. 2.4 Photograph (side view) of a cartilage-carrier-construct after cultivation in a flow chamber bioreactor on scale paper (scale 1 mm) (for comparison, see [69]).

2.5
3D Systems Used for Drug Testing

The importance of handling a large quantity of tissue constructs in parallel has already been addressed in Section 2.1. Most systems discussed so far do not fulfill this requirement; rather, they are mostly applicable for a specific tissue construct of a desired size, to provide a controlled environment, and to enable a continuous supply of nutrients and oxygen as well as the removal of metabolites. Some systems were successfully applied for *in-vitro* drug studies [45, 55, 67, 74, 75], but these proved mostly to be unsuitable for handling a magnitude of culture units in parallel, as the need for additional devices such as pumps, tubes, medium reservoirs and aeration units ballooned.

As an alternative, tissue culture units especially intended for HTS have been developed, and included the multicellular spheroid model [11, 77, 78], microelectrode arrays ("Biology on a Chip" [79–82]), the cultivation of organ slices [83], or multilayered post-confluent cell culture [84]. The specific features of these techniques are discussed at length in the following sections. Most of these techniques are well suited to the cultivation of small tissue constructs, and it could be shown that tissue-specific functions were maintained satisfactorily. The lack of environmental control is seen as the main drawback of these systems; for example, the exchange of medium occurs mostly in batch mode, which leads to a constantly changing environment. A continuous exchange of medium under controlled conditions, such as in bioreactors for tissue constructs, is seldom achieved. Hence, the future challenge will be to combine the advantages of both approaches – to have a "conventional" bioreactor design for tissue constructs on the one hand, and HTS-culture techniques for 3D-constructs on the other hand.

2.6
Modeling of Bioreactor Systems for Tissue Engineering

The appropriate molecular and macroscopic architecture of 3D tissue constructs is essential when producing a phenotypically appropriate tissue [8]. The exact local conditions experienced by the cells must be understood, yet in many cases the culture systems and bioreactors used for 3D tissue culture have not been optimized in this respect. Several parameters, such as perfusion rate, flow conditions, shear stress, and compression magnitude, have been varied, quite often by using a trial-and-error approach. Furthermore, different conditions must be examined accurately with regard to their effect. For example, hydrostatic pressure applied during cartilage culture can lead to an improved mass transfer of small and large molecules into the cartilage matrix, but can also induce a mechanical stimulation of embedded cells.

As an example, mass transfer effects within a bioreactor designed for the cultivation of artificial blood vessels will be discussed in order to provide a deeper understanding of the physiological situation of the cells, especially with regards to oxygen supply within the reactor system. Oxygen can be supplied to the cells within the vessel matrix only by diffusion, as the vessel wall is not vascularized. As a direct measurement of oxygen concentration within the vessel matrix is not possible, the only way to obtain a better understanding of the oxygen concentration profile within the vessel matrix is to develop a mathematical model considering the mass transfer limitations on both sides of the vessel, as well as within the vessel matrix. The model assumptions are described in Figure 2.5a, while Figure 2.5b shows the radial oxygen profile within the vessel wall for different cell numbers. This indicates that, over the radius of the vessel, the oxygen concentration decreases rapidly, depending on the number of immobilized cells.

For cell densities of approximately 4×10^7 mL^{-1}, severe oxygen limitation must be expected, whereas for cell densities of 10^8 mL^{-1} (a tissue-like cell density) the penetration depth for oxygen is less than 100 μm. As the thickness of vessels used experimentally at present is approximately 1 mm, an appropriate cell density with sufficient oxygen supply is about 10^7 mL^{-1} of the vessel matrix. Further detailed simulations showed that the flow rate had no significant effect on the oxygen profile within the vessel wall, as the main mass transfer resistance is not in the boundary layer medium/vessel wall, but rather is within the vessel wall. Other parameters (e.g., vessel thickness, vessel inner radius and/or length) were also of minor importance. It can be concluded from these results that the number of cells within the vessel matrix and/or the thickness of the vessel matrix, which is not perfused, are strongly limited. These findings agree well with those of investigations into the cell penetration depth in macroporous carriers (using a NMR technique), and were in the range of 100–200 μm [36, 85].

To prove further how much oxygen is supplied from the outer medium (which is not flowing), a corresponding model was formulated. The vessel wall was described by a plate geometry, with the further assumption of one-dimensional diffusion without reaction. From Figure 2.5c it can be concluded that the oxygen concentration in the outer vicinity of the vessel decreases very rapidly, so that the oxygen contribution from this side may be neglected.

This small example underlines the importance of theoretical considerations regarding mass transfer effects in 3D tissue cultures. Therefore, experimental studies should always be supported by simulation methods such as computational fluid dynamics (CFD), or the finite-element approach. Several examples underline the potential of an integrated study of mechanical and biomechanical factors that control the functional development of tissue-engineered constructs [86–92], and this approach will undoubtedly significantly improve bioreactor design in the near future.

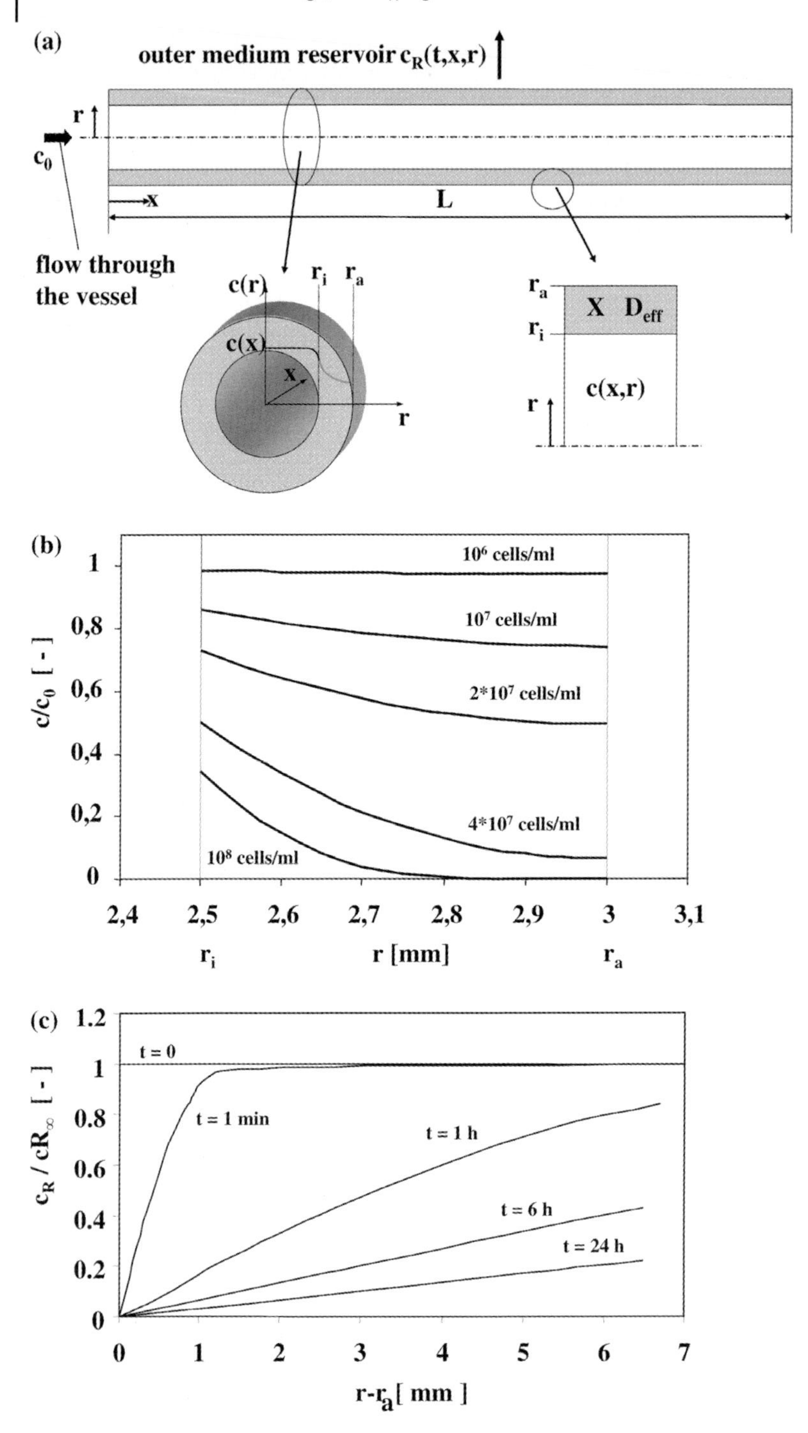

Fig. 2.5 (legend see p. 65)

2.7 The Artificial Immune System

In vivo, immunogenicity and effects of immunotoxicity are localized in primary immune organs such as bone marrow or thymus, or secondary immune organs such as the lymph nodes and spleen, and spread into the peripheral tissues of the body, such as the skin and gut [106]. Tissue engineering approaches can be used to remodel cascades of cellular interactions *in vitro*. Recently, *ex-vivo*-generated and cultivated lymphatic cells have been used for immune cell therapies (e.g., adoptive immune responses, immune tolerance), but solid lymphoid tissue might also be a new substrate for regenerative medicine, in future. The implantation of artificial tissue-engineered lymphatic constructs in mice was recently described [112].

Due to the limitations of assays using suspended lymphocytes or native tissue biopsy preparations, it is essential that immunofunctions are monitored in long-term cultures, using *in-vitro*-designed complex multicellular systems of different cell types in tissue-like structures, termed "organoids". Organoids can be defined as artificial organ structures that are formed either by step-by-step reconstruction or by induced self-organization and self-assembly, and which demonstrate integrated organ functions in *in-vivo* equivalency. These organoids are generated in 3D culture systems, assisted by matrices for the initial seeding of cells and sustained support of cell interaction. Organoids overcome the limitations of "flat biology" *in vitro* [93, 98]; embedding of the matrices supports 3D cell–cell interactions and the formation of larger cell aggregates [99].

Fig. 2.5 Theoretical considerations of the oxygen supply within perfused artificial blood vessels.
(a) The concept of the model. The vessel with an outer radius r_a and an inner radius r_i is perfused by a flow rate with an inlet oxygen concentration, c_0. The outer lumen is not perfused, and the oxygen concentration $c_R(t,x,r)$ in the outer medium depends on time and position. The oxygen concentration $c(x,r)$ in the inner medium flow and in the wall of the vessel depends on radius r and the length parameter x. Oxygen uptake within the wall may change with time due to the growth or death of cells; however, this process can be regarded as slow. It is further assumed that the cells are distributed homogeneously within the cell wall. A diffusion coefficient D_{eff} is introduced to describe diffusion within the wall.
(b) The oxygen profile within the vessel. The complex model describing the oxygen profile within the inner lumen of the vessel and the vessel wall contains the oxygen uptake kinetic of the cells, the diffusion of oxygen in the vessel wall, and the mass transfer resistance from the flow in the center of the lumen to the inner radius.
(c) The oxygen profile around the vessel. In order to prove how much oxygen is supplied from the outer medium which is not flowing, a corresponding model was formulated with the following assumptions. The vessel wall was described by a plate geometry; with the further assumption of one-dimensional diffusion without reaction, the following differential equations and boundary conditions, with D as the diffusion coefficient for oxygen in the medium, are valid. Parameters: flow rate = 6 mL min^{-1}; diffusion coefficient oxygen in membrane D_{eff} = 8.9 mm^2 h^{-1}; diffusion coefficient oxygen in medium D = 11.16 mm^2 h^{-1}.

Remodeling the cellular interactions in dynamic tissues such as lymph nodes, spleen or endothelial–blood complexes requires controlled perfusion and mixing of the suspended mobile cells with matrix-bound immobile cells. Cell mixing enhances the probability of statistically distributed, rare and highly specific but much-needed initial cell–cell contacts and interactions, and for this perfusable matrices of sufficient porosity and migrational support are needed. The cell suspension can be applied continuously, periodically, or as a single event; moreover, the cell suspension can be reused in circulation for better stochastics.

The artificial lymph node (ALN) technology mimics the lymph node physiology by supporting the highly dynamic cellular self-organization, and the intensive interaction of mobile and stationary immune-competent cells and soluble antigens. The human lymph node acts like an interface between blood flow and the lymphatic fluid, transporting different types of cell populations. The lymphatic tissue ensures effective interaction of antigen-presenting cells (APC), for example, dendritic cells (DCs) and lymphocytes [106]. Antigen-loaded DCs of all body compartments enter the lymph nodes via lymphatic fluids, and adhere to the cellular network and inner surface of macroporous artificial ECM-equivalents. Naïve or resting T and B lymphocytes are transported by arterial blood flow and penetrate the endothelial barrier by cytokine- and chemokine-directed extravasation. The T lymphocytes then migrate through the lymphatic tissue to achieve close contact

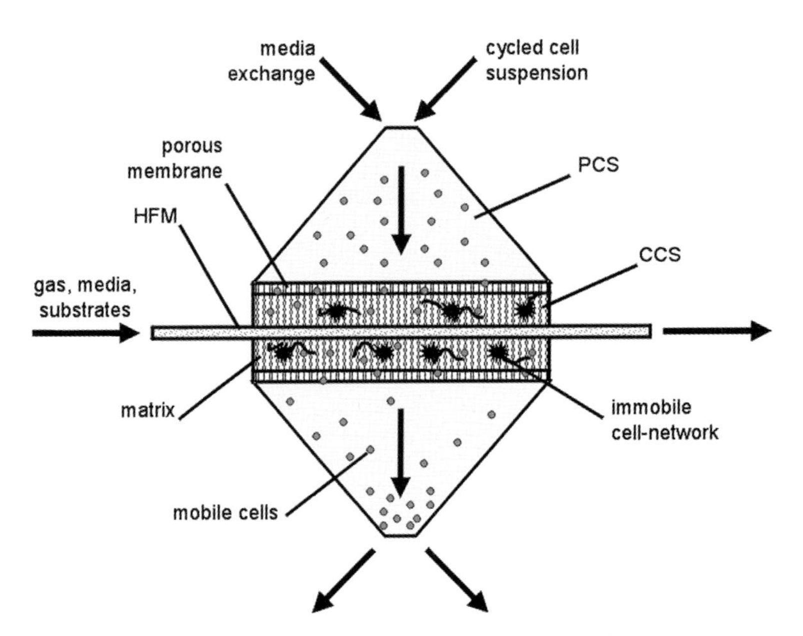

Fig. 2.6 Concept of mobile and immobile cell phases interacting in the artificial lymph node bioreactor. The suspended lymphocytes pass the central culture space (CCS) and come into close contact with the immobile dendritic cell network in the embedding matrix of the CCS. HFM = hollow-fiber membrane.

with the immobilized APC for induction. The B lymphocytes form distinct B-cell areas, where they await antigen and co-stimulation by activated T cells.

As a technical equivalent, the ALN-bioreactor integrates mobile and immobile cell phases for effective interaction (Fig. 2.6). A central culture space (CCS) of 500 μL (scalable up to 4 mL) is supplied by a planar set of microporous hollow fibers for oxygenation and pH control. The CCS is matrix-filled and supported by planar macroporous membranes that allow continuous transfusion with media and cells from the peripheral culture space (PCS). Defined perfusion rates ensure adhesion and migration, and the media and cell suspension are continuously transfused in cyclic fashion to ensure long-term cultivation.

The CCS and PCS are designed in terms of geometry and transfusion velocity for a long and effective residence time of cells in the matrix, but a short residence time in the supporting fluids.

The application of suspended T lymphocytes and B lymphocytes is later restricted to a defined period of culture time. The transfusion of cells and media exchange and dilution is controlled independently (Fig. 2.7).

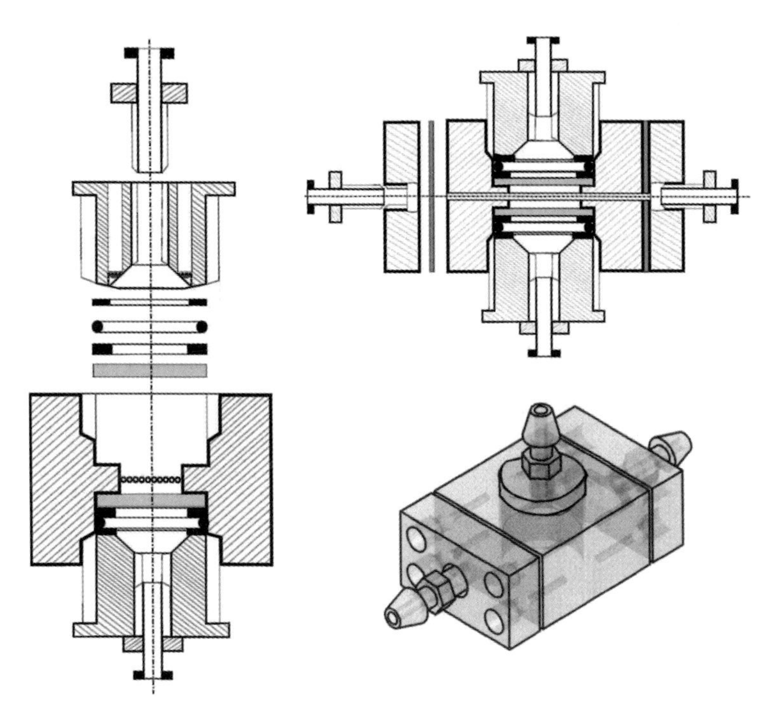

Fig. 2.7 Design of the artificial lymph node bioreactor in cross-sections and 3D shape. The central culture space (CCS) is supported by microporous hollow-fiber membranes for media and gas supply. For stabilization, the matrix-filled CCS is separated from the outer culture space by a macroporous membrane. The geometry of culture spaces and supporting fluidics is optimized for a maximum residential time in the CCS and efficient, but gentle, cell transportation in the peripheral culture space.

2.7.1
Matrices

Supporting matrices for the CCS need to be perfusable, cell-adhesive, and cell-migrational. In contrast to biohybrid implant technology (e.g., cartilage), biodegradability is irrelevant, but cell proliferation and the growth of organoids is necessary. Matrix sheets of nonwoven fibers (polyamides), hygrogels (peptides, agaroses, alginates, fibrin) and sponges (collagens) are particularly suitable for this purpose. Matrix performance is defined by the composition and structural properties on nanometer and micrometer scales.

2.7.2
Microenvironment

Cellular interaction and activation is controlled by the local microenvironment. A good balance of moderate perfusion and diffusion limitation ensures the formation of local microgradients in the CCS. As with many hormones, cytokines and growth factors have both autocrine and paracrine effects [105], but in contrast to hormones most of them function on a short track in the cellular neighborhood, or by direct cell–cell contact only. An overlay of different metabolite and factor microgradients forms distinct niches for cell differentiation, activation or energy. The differentiation and activation of hematopoietic stem cells and mature lymphocytes is biased by distinct levels of factors, such as glucose, lactate, dissolved oxygen, and pH [96, 97, 100–102].

APC immobilized in the ECM form a cellular network and secrete chemokines for the attraction of highly migrational T lymphocytes [109–111]. The communication of APC and T cell, and the formation of immunological synapses, is driven by a set of cytokines [103, 104]. Activated T lymphocytes are key players in antigen-specific co-stimulation of B cells to ensure sustained B-cell activation and antibody rearrangement and secretion.

The T-cell activation can switch into a humoral, antibody-driven or a cellular immune response, and is strongly influenced by interleukin (IL)-4, IL-10, tumor necrosis factor (TNF)-α and interferon (IFN)-γ. In addition to the given local microenvironment, immune responses can also be guided by additional supplementation.

2.7.3
Monitoring

The online monitoring of tissue morphogenesis and organoids in perfusion bioreactor culture is limited to a small set of parameters. They need to be described as "black boxes" during the culture period, and the readout is defined by histological endpoints. Process control is based on online-sensing of oxygen and pH, and samples of culture supernatant are periodically harvested (at 12 h and 24 h) to this purpose. New approaches of miniaturized fluorescence-based online sensing for

pO_2, pCO_2 and pH can be implemented [113, 114]. In perfusion culture, the pH and defined levels of dissolved oxygen, metabolites and factors in the PCS can be adjusted by media and gas exchange.

The monitoring of cytokines provides insight into cellular immunological processes. Successful antigen presentation leads to increased TNF-α concentrations, while the induced T-cell reaction is described by IFN-γ release. Shifts in T-lymphocyte populations into humoral or cellular inductors are better described by a set of cytokines (the cytokine panel). A cellular immune response (TH1-response) is detected by increased levels of IL-2, IL-12 and IFN-γ or TNF-α. In contrast, IL-4, IL-5, IL-6, IL-10 and IL-13 describe a humoral response (TH2-response) [107, 108].

Tissue formation of the generated organoids can be characterized after harvest and histological preparation. Histochemical and immune-histological staining of the matrix sheets allows investigations to be made of the tissue formation and cellular composition. To prove the development of germinal centers (GC) [106] as key structures for antibody maturation *in vivo*, hematoxylin and eosin (H&E) staining or May–Gruenwald–Giemsa-staining can be used. The immune-specific detection of cell surface markers can identify lymphocytes (CD3, CD4, CD19, CD20), and also verify the segregation of lymphocytes into T- and B-cell areas in the matrix. Ki-67 or BrdU-incorporation is indicative of cell proliferation, while key to the scheme is the histological confirmation of secreted or membrane-bound IgM/IgG and their spatial distribution.

Living cells can be prepared following the disintegration of organoids, and also for further cultivation, cell fusion and fluorescence-activated cell sorting (FACS). DNA and RNA preparations can be obtained by using microdissection technology, while new approaches in live cell-imaging technology, multiphoton microscopy and microtomography each allow the monitoring and characterization of cell migration, homing proliferation [115, 116], and tissue formation *in situ* [117].

2.8 Conclusions

An overall comparison of different culture methods clearly shows the advantages of 3D tissue constructs for drug screening. Bioreactors can provide a better process control by taking into account the different demands of cells during cultivation. Furthermore, they can provide the technical means of performing controlled studies aimed at understanding specific biological, chemical, or physical effects, or for drug screening. Moreover, bioreactors enable a safe and reproducible production of tissue constructs, and can also be used to study effects such as shear flow and/or hydrostatic pressure on the generation of tissues.

With regard to future clinical applications, the bioreactor system should be an advantageous method in terms of low contamination risk, ease of handling, and scalability. With regard to drug screening, it is important that bioreactor systems are designed which allow for the cultivation of a magnitude of small tissue samples or constructs in parallel (HTS), under controlled conditions. These devices should consist of disposable, inexpensive cultivation units implemented with hardware to control temperature and gas and medium supplies. Moreover, such devices should be capable of operating automatically and continuously, with minimal manipulation steps.

To date, the goals and expectations of bioreactor development have been fulfilled only to a limited extent, notably because bioreactor design in tissue engineering is not only very complex but is also at an early stage of development. In the future, an intimate collaboration between engineers and biologists should lead to an increased fundamental understanding of the complex issues that will impact on tissue formation in bioreactors. These advances should help to ensure that, in time, tissue engineering fulfils the expectations for revolutionizing drug screening tools and medical care.

Acknowledgments

The authors thank Dr.-Ing. Stephanie Nagel-Heyer and Katja Schmidt for their scientific input and help in the mathematical simulations.

References

1 Langer, R. Tissue engineering. *Mol. Ther.* **2000**, *1*, 12–15.
2 Langer, R. and Vacanti, J. P. Tissue engineering. *Science* **1993**, *260*, 920–926.
3 Prochazkova, J. Contribution of '*in vitro*' assays to preclinical and pre-marketing testing in immunotoxicology. *Cent. Eur. J. Public Health* **1993**, *1*(2), 101–105.
4 Petersen, J. P., Rücker, A., von Stechow, D., Adamietz, P., Pörtner, R., Rueger, J. M., and Meenen, N. M. Present and future therapies of articular cartilage defects. *Eur. J. Trauma* **2003**, *1*, 1–10.
5 Bradlaw, J. A. Evaluation of drug and chemical toxicity with cell culture systems. *Fund. Appl. Toxicol.* **1986**, *6*(4), 598–606.
6 Chamuleau, R. A. Artificial liver support in the third millennium. *Artif. Cells Blood Substit. Immobil. Biotechnol.* **2003**, *31*, 117–126.
7 Martin, I., Wendt, D., and Heberer, M. The role of bioreactors in tissue engineering. *Trends Biotechnol.* **2004**, *22*, 80–86.
8 Griffith, L. G. and Naughton, G. Tissue engineering – current challenges and expanding opportunities. *Science* **2002**, *295*, 1009–1014.
9 Bhadriraju, K. and Chen, C. S. Engineering cellular microenvironments to improve cell-based drug testing. *Drug Discov. Today* **2002**, *7*(11), 612–620.
10 Balcarcel, R. R. and Clark, L. M. Metabolic screening of mammalian cell cultures using well-plates. *Biotechnol. Prog.* **2003**, *19*(1), 98–108.

11 Kunz-Schughart, L. A., Freyer, J. P., Hofstaedter, F., and Ebner, R. The use of 3D cultures for high-throughput screening: the multicellular spheroid model. *J. Biomol. Screen.* **2004**, *9*(4), 273–285.

12 DeClerck, Y. A. and Neustein, H. B. The contribution of tissue culture to the study of solid tumors of childhood. *Perspect. Pediatr. Pathol.* **1987**, *9*, 214–243.

13 Balimane, P. V. and Chong, S. Cell culture-based models for intestinal permeability: a critique. *Drug Discov. Today* **2005**, *10*(5), 335–343.

14 Kaspers, G. J., Zwaan, C. M., Pieters, R., and Veerman, A. J. Cellular drug resistance in childhood acute myeloid leukemia. A mini-review with emphasis on cell culture assays. *Adv. Exp. Med. Biol.* **1999**, *457*, 415–421.

15 Ratcliffe, A. and Niklason, L. E. Bioreactors and bioprocessing for tissue engineering. *Ann. N. Y. Acad. Sci.* **2002**, *961*, 210–215.

16 Naughton, G. K. From lab bench to market: critical issues in tissue engineering. *Ann. N. Y. Acad. Sci.* **2002**, *961*, 372–385.

17 Darling, E. M. and Athanasiou, K. A. Articular cartilage bioprocesses and bioreactors. *Tissue Eng.* **2003**, *9*, 9–26.

18 Allen, J. W. and Bhatia, S. N. Improving the next generation of bioartificial liver devices. *Semin. Cell Dev. Biol.* **2002**, *13*, 447–454.

19 Barron, V., Lyons, E., Stenson-Cox, C., McHugh, P. E., and Pandit, A. Bioreactors for cardiovascular cell and tissue growth: a review. *Ann. Biomed. Eng.* **2003**, *31*, 1017–1030.

20 Godbey, W. T. and Atala, A. *In vitro* systems for tissue engineering. *Ann. N. Y. Acad. Sci.* **2002**, *961*, 10–26.

21 Morsi, Y. S., Birchall, I. E., and Rosenfeldt, F. L. Artificial aortic valves: an overview. *Int. J. Artif. Organs* **2004**, *27*, 445–451.

22 Park, K. D., Kwon, I. K., and Kim, Y. H. Tissue engineering of urinary organs. *Yonsei Med. J.* **2000**, *41*, 780–788.

23 Shachar, M. and Cohen, S. Cardiac tissue engineering, *ex-vivo*: design principles in biomaterials and bioreactors. *Heart Fail. Rev.* **2003**, *8*, 271–276.

24 Vunjak-Novakovic, G. The fundamentals of tissue engineering: scaffolds and bioreactors. *Novartis Found. Symp.* **2003**, *249*, 34–46.

25 Pörtner, R., Nagel-Heyer, St., Goepfert, Ch., Adamietz, P., and Meenen, N. M. Bioreactor design for tissue engineering. *J. Bioeng. Biosci.* **2005**, *100*(3), 235–245.

26 Nagel-Heyer, S., Goepfert, Ch., Morlock, M. M., and Pörtner, R. Relationship between gross morphological and biochemical data of tissue engineered cartilage-carrier-constructs. *Biotechnol. Lett.* **2005**, *27*, 187–192.

27 Malda, J., van Blitterswijk, C. A., van Geffen, M., Martens, D. E., Tramper, J., and Riesle, J. Low oxygen tension stimulates redifferentiation of dedifferentiated adult human nasal chondrocytes. *Osteoarthritis Cartilage* **2004**, *12*, 306–313.

28 Domm, C., Schünke, M., Christesen, K., and Kurz, B. Redifferentiation of dedifferentiated bovine articular chondrocytes in alginate culture under low oxygen tension. *Osteoarthritis Cartilage* **2002**, *10*, 13–22.

29 Nagel-Heyer, S., Leist, Ch., Lünse, S., Goepfert, C., and Pörtner, R. From biopsy to cartilage-carrier constructs by using microcarrier cultures as sub-process. In: *Proceedings of 19th ESACT meeting*. Harrogate, UK, **2005**, p. 139.

30 Malda, J., van den Brink, P., Meeuwse, P., Grojec, M., Martens, D. E., Tramper, J., Riesle, J., and van Blitterswijk, C. A. Effect of oxygen tension on adult articular chondrocytes in microcarrier bioreactor culture. *Tissue Eng.* **2004**, *10*, 987–994.

31 Bardouille, C., Lehmann, J., Heimann, P., and Jockusch, H. Growth and differentiation of permanent and secondary mouse myogenic cell lines on microcarriers. *Appl. Microbiol. Biotechnol.* **2001**, *55*, 556–562.

32 Freed, L. E., Langer, R., Martin, I., Pellis, N. R., and Vunjak-Novakovic, G. Tissue engineering of cartilage in space. *Proc. Natl. Acad. Sci. USA* **1997**, *94*, 13885–13890.

33 Holy, C. E., Shoichet, M. S., and Davies, J. E. Engineering three-dimensional bone tissue in vitro using biodegradable scaffolds: investigating initial cell-seeding density and culture period. *J. Biomed. Mater. Res.* **2000**, *51*, 376–382.

34 Carrier, R. L., Papadaki, M., Rupnick, M., Schoen, F. J., Bursac, N., Langer, R., Freed, L. E., and Vunjak-Novakovic, G. Cardiac tissue engineering: cell seeding, cultivation parameters, and tissue construct characterisation. *Biotechnol. Bioeng.* **1999**, *64*, 580–589.

35 Kannan, R. Y., Salacinski, H. J., Sales, K., Butler, P., and Seifalian, A. M. The roles of tissue engineering and vascularisation in the development of micro-vascular networks: a review. *Biomaterials* **2005**, *26*, 1857–1875.

36 Fassnacht, D. and Pörtner, R. Experimental and theoretical considerations on oxygen supply for animal cell growth in fixed bed reactors. *J. Biotechnol.* **1999**, *72*, 169–184.

37 Butler, D. L., Goldstein, S. A., and Guilak, F. Functional tissue engineering: the role of biomechanics. *J. Biomech. Eng.* **2000**, *122*, 570–575.

38 Carver, S. E. and Heath, C. A. Increasing extracellular matrix production in regenerating cartilage with intermittent physiological pressure. *Biotechnol. Bioeng.* **1999**, *62*, 166–174.

39 Carver, S. E. and Heath, C. A. Influence of intermittent pressure, fluid flow, and mixing in the regenerative properties of articular chondrocytes. *Biotechnol. Bioeng.* **1999**, *65*, 274–281.

40 Hall, A. C., Urban, J. P. G., and Gehl, K. A. The effects of hydrostatic pressure on matrix synthesis in articular cartilage. *J. Orthop. Res.* **1991**, *9*, 1–10.

41 Jagodzinski, M., Cebotari, S., Tudorache, I., Zeichen, J., Hankemeier, S., Krettek, C., van Griensven, M., and Mertsching, H. Tissue engineering of long bones with a vascular matrix in a bioreactor. *Orthopade* **2004**, *33*, 1394–1400.

42 Seidel, J. O., Pei, M., Gray, M. L., Langer, R., Freed, L. E., and Vunjak-Novakovic, G. Long-term culture of tissue engineered cartilage in a perfused chamber with mechanical stimulation. *Biorheology* **2004**, *41*, 445–458.

43 Yu, X., Botchwey, E. A., Levine, E. M., Pollack, S. R., and Laurencin, C. T. Bioreactor-based bone tissue engineering: the influence of dynamic flow on osteoblast phenotypic expression and matrix mineralization. *Proc. Natl. Acad. Sci. USA* **2004**, *101*, 11203–11208.

44 Nagel-Heyer, St. *Engineering aspects for generation of three-dimensional cartilage-carrier-constructs.* Books on Demand GmbH, Norderstedt, Germany, **2004**.

45 Garvin, J., Qi, J., Maloney, M., and Banes, A. J. Novel system for engineering bioartificial tendons and application of mechanical load. *Tissue Eng.* **2003**, *9*(5), 967–979.

46 Morgan, J. R. and Yarmush, M. L. *Tissue engineering methods and protocols.* Humana Press, Totowa, NJ, **1999**.

47 Sen, A., Kallos, M. S., and Behie, L. A. New tissue dissociation protocol for scaled-up production of neural stem cells in suspension bioreactors. *Tissue Eng.* **2004**, *10*, 904–913.

48 Fassnacht, D., Rössing, S., Singh, R., Al-Rubeai, M., and Pörtner, R. Influence of BCL-2 Expression on antibody productivity in high cell density hybridoma culture systems. *Cytotechnology* **1999**, *30*, 95–105.

49 Fussenegger, M., Fassnacht., D., Schwartz, R., Zanghi, J. A., Graf, M., Bailey, J. E., and Pörtner, R. Regulated overexpression of the survival factor bcl-2 in CHO cells increases viable cell density in batch culture and decreases DNA release in extended fixed-bed cultivation. *Cytotechnology* **2000**, *32*, 45–61.

50 Nehring, D., Gonzales, R., Czermak, P., and Pörtner, R. Mathematical model of a membrane filtration process using ceramic membranes to increase retroviral pseudotype vector titer. *J. Membr. Sci.* **2004**, *237*, 25–38.

51 Fassnacht, D., Rössing S., Stange, J., and Pörtner. R. Long-term cultivation of immortalised mouse hepatocytes in a high cell density fixed bed reactor. *Biotechnol. Tech.* **1998**, *12*, 25–30.

52 Noll, T., Jelinek, N., Schmid, S., Biselli, M., and Wandrey, C. Cultivation of hematopoietic stem and progenitor cells: biochemical engineering aspects. *Adv. Biochem. Eng. Biotechnol.* **2002**, *74*, 111–128.

53 Cabrita, G. J., Ferreira, B. S., da Silva, C. L., Goncalves, R., Almeida-Porada, G., and Cabral, J. M. Hematopoietic stem cells: from the bone to the bioreactor. *Trends Biotechnol.* **2003**, *21*, 233–240.

54 Schubert, H., Garrn, I., Berthold, A., Knauf, W. U., Reufi, B., Fietz, T., and Gross, U. M. Culture of haematopoietic cells in a 3D bioreactor made of Al_2O_3 or apatite foam. *J. Mater. Sci. Mater. Med.* **2004**, *15*, 331–334.

55 Ma, T., Yang, S. T., and Kniss, D. A. Development of an in vitro human placenta model by the cultivation of human trophoblasts in a fiber-based bioreactor system. *Tissue Eng.* **1999**, *5*(2), 91–102.

56 Davis, J. M. and Hanak, J. A. Hollow-fiber cell culture. *Methods Mol. Biol.* **1997**, *75*, 77–89.

57 Weichert, H., Falkenberg, F. W., Krane, M., Behn, I., Hommel, U., and Nagels, H. O. Cultivation of animal cells in a new modular minifermenter. In: Beuvery, E. C., Griffiths, J. B., and Zeijlemaker, W. P. (Eds.), *Animal Cell Technology: Developments towards the 21st century*. Kluwer Academic Publishers, The Netherlands, **1995**, pp. 907–913.

58 Nagel, A., Effenberger, E., Koch, S., Lübbe, L., and Marx, U. Human cancer and primary cell culture in the new hybrid bioreactor system tecnomouse. In: Spier, R. E., Griffiths, J. B., and Berthold W. (Eds.), *Animal Cell Technology: Products of today, prospects for tomorrow*. Butterworth-Heinemann, Oxford, **1994**, pp. 296–298.

59 De Bartolo, L. and Bader, A. Review of a flat membrane bioreactor as a bioartificial liver. *Ann. Transplant.* **2001**, *6*, 40–46.

60 Jasmund, I. and Bader, A. Bioreactor developments for tissue engineering applications by the example of the bioartificial liver. *Adv. Biochem. Eng. Biotechnol.* **2002**, *74*, 99–109.

61 Kulig, K. M. and Vacanti, J. P. Hepatic tissue engineering. *Transpl. Immunol.* **2004**, *12*, 303–310.

62 Gerlach, J. C. Development of a hybrid liver support system: a review. *Int. J. Artif. Organs* **1996**, *19*, 645–654.

63 Sielaff, T. D., Hu, M. Y., Amiot, B., Rollins, M. D., Rao, S., McGuire, B., Bloomer, J. R., Hu, W. S., and Cerra, F. B. Gel-entrapment bioartificial liver therapy in galactosamine hepatitis. *J. Surg. Res.* **1995**, *59*, 179–184.

64 Ostrovidov, S., Jiang, J., Sakai, Y., and Fujii, T. Membrane-based PDMS microbioreactor for perfused 3D primary rat hepatocyte cultures. *Biomed. Microdevices* **2004**, 6(4), 279–287.

65 Prenosil, J. E. and Villeneuve, P. E. Automated production of cultured epidermal autografts and sub-confluent epidermal autografts in a computer controlled bioreactor. *Biotechnol. Bioeng.* **1998**, *59*, 679–683.

66 Marx, U., Matthes, H., Nagel, A., and Baehr, R. V. Application of a hollow fiber membrane cell culture system in medicine. *Am. Biotechnol. Lab.* **1993**, *11*, 26.

67 Zhau, H. E., Goodwin, T. J., Chang, S. M., Baker, T. L., and Chung, L. W. Establishment of a three-dimensional human prostate organoid coculture under microgravity-simulated conditions: evaluation of androgen-induced growth and PSA expression. *In Vitro Cell Dev. Biol. Anim.* **1997**, *33*(5), 375–380.

68 Minuth, W. W., Stöckl, G., Kloth, S., and Dermietzel, R. Construction of an apparatus for perfusion cell cultures which enables *in vitro* experiments under organotypic conditions. *Eur. J. Cell. Biol.* **1992**, *57*, 132–137.

69 Nagel-Heyer, S., Goepfert, Ch., Adamietz, P., Meenen, N. M., Petersen, J.-P., and Pörtner, R. Flow-chamber bioreactor culture for generation of three-dimensional cartilage-carrier-constructs. *Bioproc. Biosyst. Eng.* **2005**, *27*, 273–280.

70 Ratcliffe, A. Tissue engineering of vascular grafts. Matrix Biol. **2000**, *19*, 353–357.

71 Risbud, M. V. and Sittinger, M. Tissue engineering: advances in *in vitro* cartilage generation. *Trends Biotechnol.* **2002**, *20*, 351–356.

72 Minuth, W. W., Strehl, R., and Schumacher, K. *Tissue engineering – from cell biology to artificial organs.* Wiley-VCH Verlag, Weinheim, **2005**.

73 Sittinger, M., Schultz, O., Keyser, G., Minuth, W. W., and Burmester, G. R. Artificial tissues in perfusion culture. *Int. J. Artif. Organs* **1997**, *20*, 57–62.

74 Nehring, D., Adamietz, P., Meenen, N. M., and Pörtner, R. Perfusion cultures and modelling of oxygen uptake with three-dimensional chondrocyte pellets. *Biotechnol. Tech.* **1999**, *13*, 701–706.

75 Koebe, H. G., Deglmann, C. J., Metzger, R., Hoerrlein, S., and Schildberg, F. W. *In vitro* toxicology in hepatocyte bioreactors-extracellular acidification rate (EAR) in a target cell line indicates hepato-activated transformation of substrates. *Toxicology* **2000**, *154*(1–3), 31–44.

76 Zeilinger, K., Sauer, I. M., Pless, G., Strobel, C., Rudzitis, J., Wang, A., Nussler, A. K., Grebe, A., Mao, L., Auth, S. H., Unger, J., Neuhaus, P., and Gerlach, J. C. Three-dimensional co-culture of primary human liver cells in bioreactors for *in vitro* drug studies: effects of the initial cell quality on the long term maintenance of hepatocyte-specific functions. *Altern. Lab. Anim.* **2002**, *30*(5), 525–538.

77 Thielecke, H., Mack, A., and Robitzki, A. A multicellular spheroid-based sensor for anti-cancer therapeutics. *Biosens. Bioelectron.* **2001**, *16*(4–5), 261–269.

78 Kelm, J. M., Ehler, E., Nielsen, L. K., Schlatter, S., Perriard, J. C., and Fussenegger, M. Design of artificial myocardial microtissues. *Tissue Eng.* **2004**, *10*(1–2), 201–214.

79 Park, T. H. and Shuler, M. L. Integration of cell culture and microfabrication technology. *Biotechnol. Prog.* **2003**, *19*(2), 243–253.

80 Stett, A., Egert, U., Guenther, E., Hofmann, F., Meyer, T., Nisch, W., and Haemmerle, H. Biological application of microelectrode arrays in drug discovery and basic research. *Anal. Bioanal. Chem.* **2003**, *377*(3), 486–495.

81 Li, N., Tourovskaia, A., and Folch, A. Biology on a chip: microfabrication for studying the behaviour of cultured cells. *Crit. Rev. Biomed. Eng.* **2003**, *31*(5–6), 423–488.

82 Hoffman, R. M. The three-dimensional question: can clinically relevant tumor drug resistance be measured *in vitro*? *Cancer Metastasis Rev.* **1994**, *13*(2), 169–173.

83 Vickers, A. E. and Fisher, R. L. Organ slices for the evaluation of human drug toxicity. *Chem. Biol. Interact.* **2004**, *150*(1), 87–96.

84 Padron, J. M., van der Wilt, C. L., Smid, K., Smitskamp-Wilms, E., Backus, H. H., Pizao, P. E., Giaccone, G., and Peters, G. J. The multilayered postconfluent cell culture as a model for drug screening. *Crit. Rev. Oncol. Hematol.* **2000**, *36*(2–3), 141–157.

85 Thelwall, P. E., Anthony, M. L., Fassnacht, D., Pörtner, R., and Brindle, K. M. Analysis of cell growth in a fixed bed bioreactor using magnetic resonance spectroscopy and imaging. In: Merten, O.-W., Perrin, P., and Griffiths, B. (Eds.), *New Developments and New Applications in Animal Cell Technology*. Kluwer Academic Publishers, The Netherlands, **1998**, pp. 627–633.

86 Lima, E. G., Mauck, R. L., Shelley, H. H., Park, S., Ng, K. W., Ateshian, G. A., and Hung, C. T. Functional tissue engineering of chondral and osteochondral constructs. *Biorheology* **2004**, *41*, 577–590.

87 Raimondi, M. T., Boschetti, F., Falcone, L., Fiore, G. B., Remuzzi, A., Marinoni, E., Marazzi, M., and Pietrabiss, R. Mechanobiology of engineered cartilage cultured under a quantified fluid-dynamic environment. *Biomech. Model. Mechanobiol.* **2002**, *1*, 69–82.

88 Connelly, J. T., Vanderploeg, E. J., and Levenston, M. E. The influence of cyclic tension amplitude on chondrocyte matrix synthesis: experimental and finite element analyses. *Biorheology* **2004**, *41*, 377–387.

89 Sengers, B. G., Oomens, C. W. J., and Baaijens, F. P. T. An integrated finite-element approach to mechanics, transport and biosynthesis in tissue engineering. *J. Biomech. Eng.* **2004**, *126*, 83–91.

90 Williams, K. A., Saini, S., and Wick, T. M. Computational fluid dynamics modelling of steady-state momentum and mass transport in a bioreactor for cartilage tissue engineering. *Biotechnol. Prog.* **2002**, *18*, 951–963.

91 Mauck, R. L., Hung, C. T., and Ateshian, G. A. Modelling of neutral solute transport in a dynamically loaded porous permeable gel: Implications for articular cartilage biosynthesis and tissue engineering. *J. Biomech. Eng.* **2003**, *125*, 602–614.

92 Begley, C. M. and Kleis, S. J. The fluid dynamic and shear environment in the NASA/JSC rotating-wall perfused-vessel bioreactor. *Biotechnol. Bioeng.* **2000**, *70*, 32–40.

93 Anonymous. Good bye, flat biology? *Nature* **2003**, *424*, 861.

94 Griffith, L. and Swartz, M. A. Capturing complex 3D tissue physiology *in vitro*. *Nat. Rev. Mol. Cell Biol.* **2006**, *7*(3), 211–224.

95 Abbott, A. Biology's new dimension. *Nature* **2003**, *424*, 870–872.

96 Wilson, A. and Trumpp, A. Bone-marrow haematopoietic stem-cell niches. *Nat. Rev. Immunol.* **2006**, *6*(2), 93–106.

97 Mohtashami, M., and Zuniga-Pflücker, J. C. Cutting edge: three-dimensional architecture of the thymus is required to maintain delta-like expression necessary for inducing T-cell development. *J. Immunol.* **2006**, *176*, 730–734.

98 Bell, E. Why 3D is better than 2D. *Nat. Rev. Immunol.* **2006**, *6*(2), 87.

99 Lutolf, M. P. and Hubbell, J. A. Synthetic biomaterials as instructive extracellular microenvironments for morphogenesis in tissue engineering. *Nat. Biotechnol.* **2005**, *23*(1), 47–55.

100 Carswell, K. S. and Papoutsakis E. T. Extracellular pH affects the proliferation of cultured human T cells and their expression of the interleukin-2 receptor. *J. Immunother.* **2000**, *23*(6), 669–674.

101 Carswell, K. S., Weiss, J. W., and Papoutsakis E. T. Low oxygen tension enhances the stimulation and proliferation of human T-lymphocytes in the presence of IL-2. *Cytotherapy* **2000**, *2*(1), 25–37.

102 Koller, M. R., Bender, J. G., Miller, W. M., and Papoutsakis, E. T. Reduced oxygen tension increases hematopoiesis in long-term culture of human stem and progenitor cells from cord blood and bone marrow. *Exp. Hematol.* **1992**, *20*(2), 264–270.

103 Dustin, M. L., Allen, P. M., and Shaw, A. S. Environmental control of immunological synapse formation and duration. *Trends Immunol.* **2001**, *22*(4), 192–194.

104 Li, Q. J., Dinner, A. R., Qi, S., Irvine, D. J., Huppa, J. B., Davis, M. M., and Chakraborty, A. K. CD4 enhances T-cell sensitivity to antigen by coordinating Lck accumulation at the immunological synapse. *Nat. Immunol.* **2004**, *5*, 791–799.

105 Thomson, A. *The Cytokine Handbook*. 3rd edn., Academic Press, London, **1998**.

106 Janeway, C. A., Travers, P., Walport, M., and Shlomchik, M. *Immunobiology: The Immune system in Health and Disease*. 6th edn. Garland Science Publishing, New York, **2005**.

107 Reddy, M., Eirikis, E., Davis, C., Davis, H. M., and Prabhakar, U. Comparative analysis of lymphocyte activation marker expression and cytokine secretion profile in stimulated human peripheral blood mononuclear cell cultures: an *in vitro* model to monitor cellular immune function. *J. Immunol. Methods* **2004**, *293*, 127–142.

108 Rodriguez-Caballero, A., Garcia-Montero, A. C., Bueno, C., Almeida, J., Varro, R., Chen, R., Pandiella, A., and Orfao, A. A new simple whole blood flow cytometry-based method for simultaneous identification of activated cells and quantitative evaluation of cytokines released during activation. *Lab. Invest.* **2004**, *84*, 1387–1398.

109 Moser, B. and Loetscher, P. Lymphocyte traffic control by chemokines. *Nat. Immunol.* **2001**, *2*(2), 123–128.

110 Bachmann, M. F., Kopf M., and Marsland B. J. Chemokines: more than just road signs. *Nat. Rev. Immunol.* **2006**, *6*(2), 159–164.

111 von Andrian, U. H. Introduction: chemokines – regulation of immune cell trafficking and lymphoid organ architecture. *Semin. Immunol.* **2003**, *15*, 239–241.

112 Suematsu, S. and Watanabe, T. Generation of a synthetic lymphoid tissue-like organoid in mice. *Nat. Biotechnol.* **2004**, *22*(12), 1539–1545.

113 Gao, F. G., Jeevarajan, A. S., and Anderson, M. M. Long-term continuous monitoring of dissolved oxygen in cell culture medium for perfused bioreactors using optical oxygen sensors. *Biotechnol. Bioeng.* **2004**, *86*(4), 425–433.

114 Kellner, K., Liebsch, G., Klimant, I., Wolfbeis, O. S., Blunk, T., Schulz, M. B., and Göpferich, A. Determination of oxygen gradients in engineered tissue using a fluorescent sensor. *Biotechnol. Bioeng.* **2002**, *80*(1), 73–83.

115 von Andrian U. H. and Mempel T. R. Homing and cellular traffic in lymph nodes. *Nat. Rev. Immunol.* **2003**, *3*, 867–878.

116 Helmchen, F. and Denk, W. Deep tissue two-photon microscopy. *Nat. Methods* **2005**, *2*(12), 932–940.

117 Müller, B., Riedel, M., and Thurner, P. J. Three-dimensional characterization of cell clusters using synchroton-radiation-based micro-computed tomography. *Microsc. Microanal.* **2006**, *12*, 97–105.

3
An Overview on Bioelectronic and Biosensoric Microstructures Supporting High-Content Screening in Cell Cultures

Andrea A. Robitzki and Andrée Rothermel

3.1
The Potential of Drug Development and Demand on High-Content Screening Systems

The successful development of novel pharmaceutical products depends heavily on the employment and consolidation of powerful, already existing technologies, such as gene technology, biotechnology, combinatorial chemistry, drug delivery, regenerative therapies, and biosensor or biosystem technology. The Human Genome Project, as well as the Proteome Project, will accelerate the development and biotechnological fabrication of novel drugs and pharmaceutical compounds, which in turn will also increase the demand for feasible and reliable functional screening tools for living cells and tissues. The strongest economic growth of the pharmaceutical market is expected in China and South-East Asia by, for example, the licensing of Chinese or Asian traditional medicine and medical products. With regard to development, therefore, the modification of existing drugs, and the search for novel drugs, will result in an expansion of high-throughput screening (HTS) systems for molecule characterization and of high-content screening (HCS) modules for functional biomonitoring worldwide. These micro- and sensor system technologies seize the opportunity to provide automated, ultra-fast, highly sensitive, reproducible online monitoring systems for analytical and functional screening of drugs and medical products.

3.1.1
Post-Genomics or Proteomics: An Analysis of Manifold Systems and Functional Monitoring of Drugs

Genomics, transcriptomics, proteomics, metabolomics, glyconomics and toponomics are keywords that describe strategies for the investigation of the complex interacting regulatory networks in biological cells, tissues, and organs. The next generation of screening devices should be able to detect and analyze the dynamic network of proteins and signal transduction pathways within cells and

Drug Testing In Vitro: Breakthroughs and Trends in Cell Culture Technology
Edited by Uwe Marx and Volker Sandig

ISBN: 978-3-527-31488-1

tissues. The biohybrid technology at the interface of molecular cell biology, tissue engineering, sensor system, microsystem technology, nanobiotechnology, and nanoelectronics is able to provide HCS devices for the detection and monitoring of proteins in, for example, single cells, tissue models or stem cells with the same genetic programme but a varying toponome (arrangement of proteins within cells) and proteome, which in turn can define the cellular differentiation and maturation state. Ultra-sensitive high-content analytical screening of the proteome pattern on the level of single cells or tissues can be realized by a new generation of large-scale protein profiling-based modules. The *W*hole *C*ell *P*rotein *F*ingerprinting system (WCPF; MelTec GmbH, Magdeburg, Germany) enables scientists to analyze and to decode the composition of the proteome on a single cell level. Along the lines of molecular mass scanning, proteins could be detected in specific subcellular areas. Each cell area corresponds to a protein fingerprint, which is finally represented in a data warehouse. Single cell analysis or proteome fingerprinting describing parameters of pathological versus normal cells could be a benefit for identifying novel targets for drugs or physiological active compounds. An automated multifunctional biohybrid screening system allows the functional real-time monitoring of the proteome and toponome, resulting in a characteristic cellular reading frame. These systems should realize the measurement, selection and comparison of proteome patterns in multidimensional data areas, predicting molecular targets, and experimental assessment. Using such biohybrid modules (cell-based biosensors or so-called "living chips"), targets and drug leads which influence, for example, the invasive process of immune cells and/or tumor cells can be identified. A coupling of these modules with other HTS and HCS systems as cell and tissue workbenches for drug discovery and identification should be possible. On the other hand, these novel screening systems represent an efficient platform for the validation of targets of genomics and proteomics studies. The importance of this biohybrid cell-based technology depends on the ability to identify targets that are specific for diseases and drug leads, and to connect various modified modules with a target factory.

3.1.2 Pharmaceutical Research and High-Technology Platforms in the Biohybrid Technology Field

In modern pharmaceutical research, knowledge of disease-relevant changes (genomics) at the molecular level is of fundamental importance. This leads to the characterization of endogenous factors or molecular targets, which can serve as a basis for novel and trend-setting drugs. One central discipline as a key in-house competence is cell- and tissue-based impedimetric HTS in the second generation of proteomics up to pharmacogenetics/-genomics. In contrast to conventional intracellular and extracellular recordings, bioimpedance measurements can also be carried out with electrophysiologically inactive cells. The steady increase in testing capacity (approximately 200 000 hits per day) demands new modules which could be integrated into robot technology and computerized analysis, as well as in automated evaluation methods. Cell-based biosensors can be used to test more than thousands of substances simultaneously, and to determine their influence on single cells, signaling pathways, and finally the intercellular interaction within a tissue. Bioelectronic cell-based assays are noninvasive and much more efficient than light-emitting, cell-based systems. The use of genomics, bioinformatics, combinatorial chemistry, and HTS methods in modern drug discovery leads to a permanently increasing number of compounds with pharmacological activity, with one of the most important reasons for failure in the development cycle often being appropriate pharmacokinetics and toxicity. Thus, the pharmaceutical industry is currently expressing great interest in screening the identified compounds early in the drug discovery cycle for pharmacokinetic parameters such as absorption, distribution, metabolism, penetration, and excretion. In this way, through secondary screening, the drop-out rate in the later – and more costly – stage of the development cycle can be reduced. Clearly, there is also a demand for measuring the physico-chemical, pharmacokinetic and pharmacotoxicity parameters under high-content or high-throughput conditions.

3.1.3
Synergy of Microchip Technology and Living Cells

During the past few years, microsystems technology or microelectronics – and especially chip fabrication techniques – have been utilized to design and produce new types of microchips that can be employed in medicine to control heart pacemakers, drug delivery microimplant systems, or as replacements for neural tissues (e.g., retinal implants and brain pacemakers). As rational screening tools for drug discovery and development, as well as for clinical diagnosis, microarrays are applied to the investigation of nucleic acids, proteins, bacteria, viruses, cells, and tissues. In particular, cell- and tissue-based microchips (biochips) are attracting increasing attention as they can provide information not only about the simple binding and activity of promising drug candidates, but also about their effect on the physiology of biological systems. In order to obtain these functional data, it is necessary to bring cells or tissues into close contact with substrate (silicon, glass or polymers) -embedded microelectrodes. Here, drug-induced alterations in the extracellular potential, which are usually generated by local ion fluxes over the cell membranes, are measured electronically against a stable reference electrode, whereby the microelectric readout can be carried out simultaneously online and in real-time on multiple microlelectrodes. Although these recordings using multielectrode arrays (MEAs) or multi-microelectrode arrays (MMEAs) represent a powerful technique to measure electrophysiological alterations of excitable single cells or even tissues [1–9], cells without any electric activity cannot be analyzed. An alternative method for the measurement of intrinsic and extrinsic cellular properties of both electrophysiologically active and inactive cells is provided by *bioimpedance spectroscopy* [10–14]. In principle, a similar multielectrode array configuration (as described above for electrically active cells) can be used for impedance spectroscopy, by which the impedance of intracellular and extracellular compartments is measured simultaneously at different frequencies on multiple electrodes. A further chip-based *in-vitro* screening system for analyzing the physiological state of cells is that of *multiparametric chips* [15, 16]. This type of biochip or biosensor is characterized by the implementation of different types of electrodes in one device. Cellular parameters such as metabolism, mitochondrial activity, and cytoskeletal integrity can be measured, for example, with ion-sensitive field effect transistors (ISFET, for measuring pH changes), interdigital electrode structures (IDES, for monitoring membrane properties), and amperometric electrodes (for measurements of glucose and oxygen consumption).

3.2 Microfabrication Techniques to Generate Miniaturized Chip Components

Based on conventional microfabrication technologies (which include photolithography technology), it is now possible to generate a large variety of complex structured biochips (e.g., planar MEAs and multi-microcapillary arrays) comprising microchannels (for hydrodynamic cell and tissue positioning), interconnects, and up to thousands of microelectrodes and/or field-effect transistors (FETs) for the detection of morphological and physiological alterations induced by drug and compound applications.

At present, a large number of tested and biocompatible materials are available. In general, glass, silicon, and polymers are utilized as substrates for the fabrication of biosensor chips. If used for biological applications, glass is the material of choice since, in contrast to silicon, it is transparent and thus the quality and growth of cells can be monitored directly on the surface of the chip by light or, if required, by fluorescence microscopy. In this way, it is possible to discriminate covered from uncovered electrodes, and also to identify genetically manipulated cells if they are marked by a reporter gene (e.g., green fluorescent protein, GFP) prior to measurement. Electrodes and interconnects – mostly noble metals such as iridium, gold, or platinum – are coated onto the surface of glass or silicon substrates, which are subsequently isolated by a thin film (passivation) of, for example, silicon nitrite, silicon oxide or organic polymers (Fig. 3.1). Subsequently, the passivation layer above the electrode is removed to provide direct contact with cells or tissues. In order to facilitate and to direct the positioning of cells on electrodes, the surfaces can be further modified by soft-lithography of repellent (e.g., silane) or attractant coatings (e.g., laminin, fibronectin) and/or by microcavities with implemented microelectrodes and/or suction holes (Fig. 3.2).

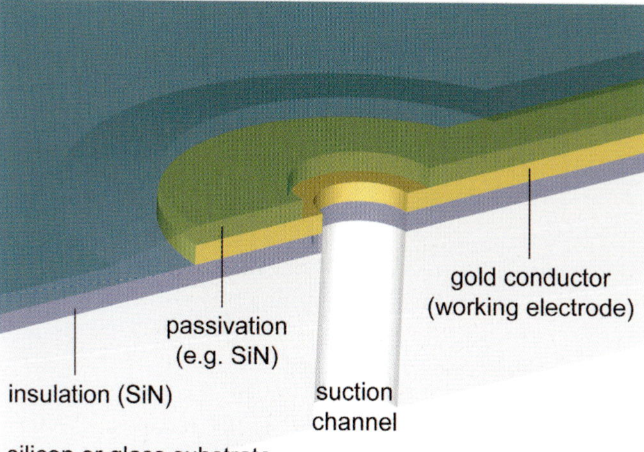

Fig. 3.1 Schematic drawing of a microstructure containing thin-film electrodes using gold or platinum in a geometry in the micrometer (μm) range. The electrode diameter is 10 μm, the suction hole diameter 6 μm; the thickness of the SiN insulation is 0.2 μm, the gold conductor 0.3 μm, and the SiN passivation 0.3 μm.
(Copyright © BBZ/Andrea Robitzki, Maik Schmidt).

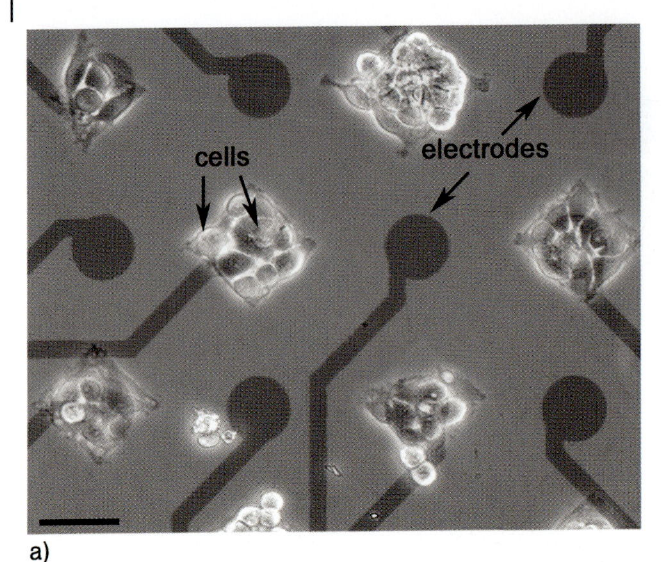

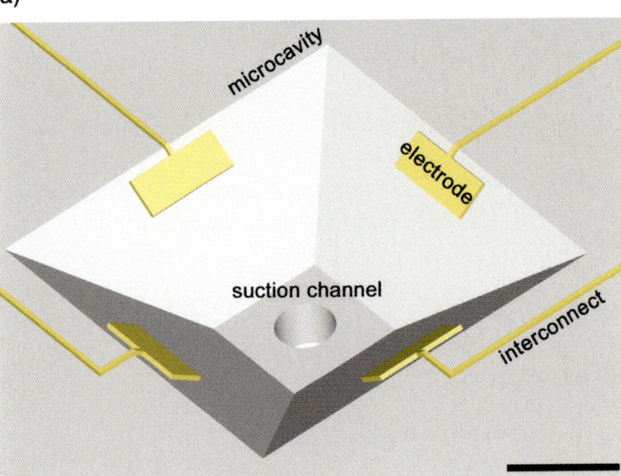

Fig. 3.2 (a) Positioning of cells in a square-like arrangement on micro-electrodes can be achieved by surface coating with repellent and attractive substances.
(b) Implementation of microcavities and suction channels on a planar chip allows the hydrodynamic positioning of suspended cells by applying a negative pressure. The tighter the contact of cells and electrodes, the higher the signal-to-noise ratio and sensitivity.
Scale bars: left, 90 μm; right 10 μm.
(Copyright © Center for Biotechnology and Biomedicine, Leipzig).

3.3
Microelectrode-Based Techniques for Analyzing Cellular Parameters: Possible Use of Real-Time and HTS of Drugs Without Labeling

3.3.1
Impedance Spectroscopy: Screening the Cellular Parameters of Electrophysiologically Inactive Cells

Impedance spectroscopy, which is also known as cellular dielectric spectroscopy (CDS) or electric impedance spectroscopy (EIS), can be used to measure frequency-dependent changes in the passive electrical properties of single cells or complex tissues by applying defined alternate currents. The bioimpedance of single cells or complex tissues, when combined with a working and counter electrode, is determined by different cellular parameters such as the electric resistance and capacitance of the (sub)cellular membranes, the resistance of the extracellular medium and cytoplasm (intracellular), and the contact between cells and cell-electrode system (Fig. 3.3).

For the measurement of impedance, an alternate voltage current is applied to a biological sample, whereby the current flows from an active working electrode through and beneath the cell or tissue to a counter electrode. Under these conditions the cell itself can act as a resistor and capacitor affecting the recorded impedance. Depending on the dielectric properties of subcellular structures, it is possible to classify three different frequency-dependent main dispersions (α-, β-, and γ-dispersion). It is assumed that the low-frequency dielectric behavior (up to 1 kHz) of current flow through ion channels mainly contributes to the α-dispersion

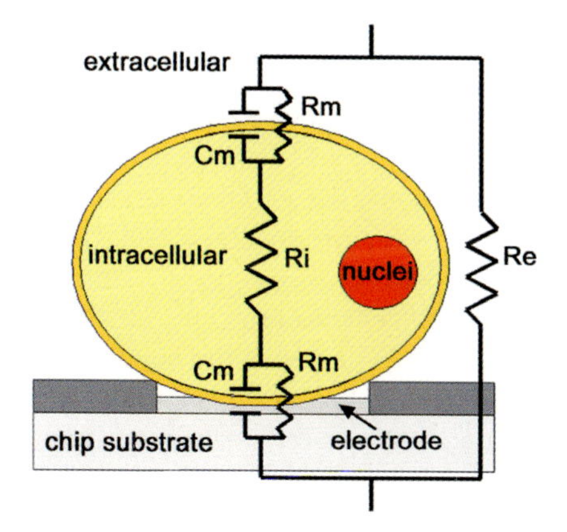

Fig. 3.3 Basic circuit model of a single cell positioned on an electrode. If an alternate current is applied to living cells or tissues, the bioimpedance is influenced by several physical parameters such as the resistances of the intracellular medium (Ri), extra-cellular medium (Re), membrane (Rm), and the membrane capacitance (Cm) [60].

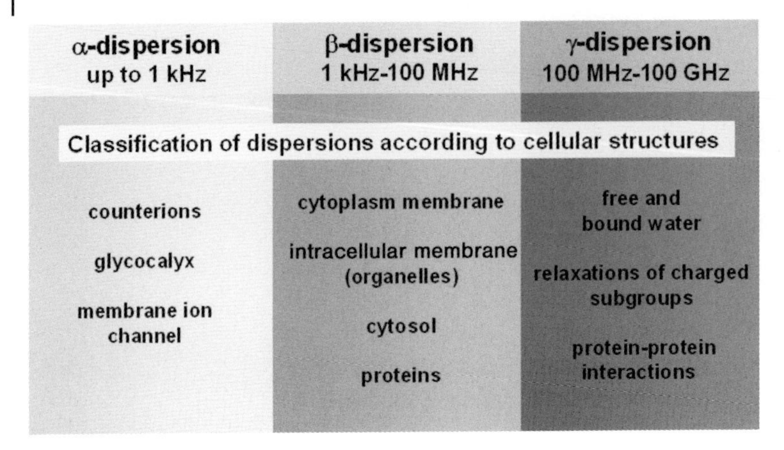

Fig. 3.4 Frequency-dependent classification of the three main dispersions according to cellular structures.

(Fig. 3.4), whereas the β-dispersion (1 kHz to 100 MHz) is associated with the dielectricity of the cytoplasm–membrane interface, intracellular membrane systems, and cytoplasm. The third dispersion range is termed γ-dispersion (100 MHz to 100 GHz), and characterizes the dissociation/association relaxation of small charged groups, protein–protein interactions and bound, as well as free, water [17–19].

The increasing popularity of impedance recording is reflected by the large number of reports published during the past few decades. This is, in all probability, due to the availability of commercial hardware and software which allows feasible and reliable impedance measurements to be made. Additionally, the still-expanding computer capacity is now sufficient for adequate data acquisition, especially in terms of HTS. Moreover, the recording of cellular parameters under noninvasive and real-time conditions makes this technique attractive for both basic research and pharmaceutical screening.

One commonly used impedance recording method is the so-called electric cell-substrate impedance sensing (ECIS), introduced by Giaever and Keese [20]. These authors were among the first to report that impedance recording is a suitable method for monitoring morphological changes of cells. In principle, for ECIS, cells must be grown on a small gold electrode implemented at the bottom of a culture dish. If an alternate current voltage is applied between a small working electrode and a large counter electrode, the impedance of a cell can be observed at a given time and on one single frequency. Over the years, ECIS has been further optimized for automated, noninvasive, real-time, and high-throughput analysis. For example, Applied Biophysics Inc. offers a complete ECIS system that provides an 8-well or 96-well format, and promises a wide range of biological screening applications such as analysis of cell attachment, signal transduction, cell–substrate interaction, barrier function, chemotaxis, toxicology, proliferation, and apoptosis (for ECIS literature, see under www.biophysics.com).

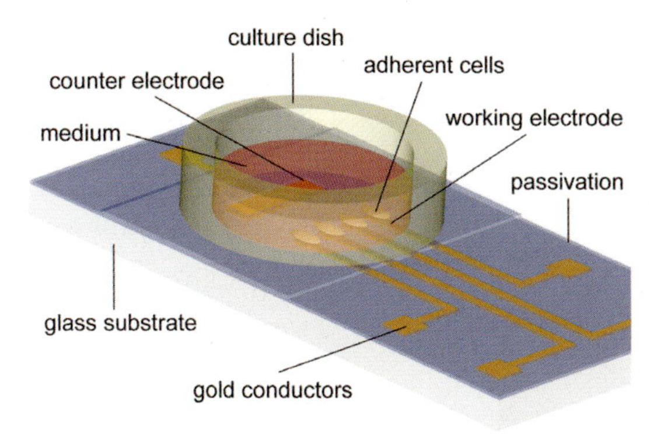

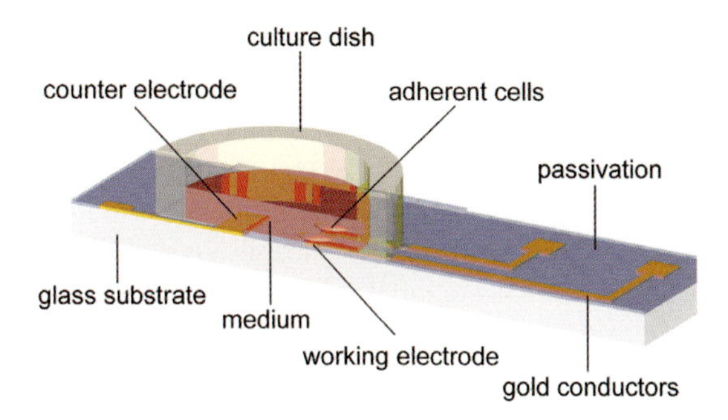

Fig. 3.5 Principle of a biosensor for electric cell-substrate impedance sensing (ECIS). (Copyright © BBZ/Andrea Robitzki, Maik Schmidt).

To illustrate the potential of ECIS, the details of a more recent study are provided below. Here, ECIS has been used to monitor the programmed cell death (apoptosis) of endothelial cells under real-time conditions [21]. In that approach, cells were cultured to confluence at thin gold-film electrodes (Fig. 3.5), whereby the impedance of the cell-electrode system was measured over an extended frequency range of 1 to 10^6 Hz.

The comparison of electric impedance of cell-covered and uncovered electrodes has shown that the contribution of cell bodies to the total impedance lies in a frequency range of 10 Hz to 100 kHz, with a maximal amplitude at 1 kHz. For that reason, the impedance of endothelial cells before and after induction of apoptosis by cycloheximide (CHX) was measured at a sampling frequency of 1 kHz. As cell bodies undergo dramatic structural changes during apoptosis, it was not surprising that the impedance of endothelial cells dramatically decreased within 12 h after CHX application. The authors of this study also pointed out that, in comparison to classical immunolabeling techniques, ECIS is apparently more sensitive to detecting apoptotic changes in real-time. If this technique is adapted to an automated work station, it probably represents a well-suited HTS system, especially in terms of drug-induced apoptosis or other adverse reactions.

Ciambrone and co-workers [11] utilized a 96-well microplate with interdigitated electrodes at the bottom of each well. Impedance measurements in a frequency range of 1 kHz to 100 MHz were carried out using adherent cell lines to detect subtype-specific activation of G-protein-coupled receptors (G_s, G_q, G_i) and protein tyrosine kinase receptors under noninvasive and labeling-free, high-throughput conditions. Sample application and buffer exchange was performed with a 96-head delivering system. As reported by the authors, the time for impedance measurements without consideration of liquid handling amounted to only 20 s for a complete 96-well plate.

Another chip-based approach for electronic impedance measurement has been developed and used by Robitzki and co-workers [10, 14, 22, 23]. These authors developed an impedance recording platform that enables the impedimetric measurement of 3-D spherical *in-vitro* tissues (e.g., tumor spheroids, neurospheres or cardiomyocyte spheres), as well as single cells. The fabricated cell-based sensors consist of either multiple microcapillaries or microcavities, each with two or four implemented electrodes where cells or 3-D *in-vitro* tissues can be hydrodynamically positioned (Fig. 3.6), without requiring time-consuming procedures for precultivating cells on planar electrodes. This innovative system allows HTS and HCS of biological samples in which tested cells are released and new cells or tissues can be repeatedly positioned and monitored. A further advantage is that each single cell or tissue sample can be addressed in such arrays for a specific monitoring of cellular or physiological and molecular changes. Although single cells produce adequate impedimetric signals, the major advantage of using 3-D histotypic sphere-like *in-vitro* tissues is that they fit more precisely into the 3-D geometry of *in-vivo* tissues, such as intracellular and extracellular properties, representing a consolidated response of thousands of cells, and allowing a more reliable impedance analysis of cellular changes. This type of tissue-based sensor is an outstanding biohybrid system for functional drug discovery, and allows the screening of one substance per minute at, for example, 30 individual sites simultaneously.

Based on the time needed for sample application, liquid handling, and the time range of the expected drug effect (e.g., 2 min plus 1 min for data reading out), it is possible to perform an automated functional impedance screening of 600 substances per day. In this way, the use of this technique clearly accelerates the complex and cost-intensive process of identifying potential leads in secondary screening. However, the chip platform is also applicable as a screening tool to test the efficiency of cytostatic drugs on cancer cell lines, or directly on biopsies obtained from patients. In the latter case, biopsies are applied immediately or after generation of tumor spheroids on the chip, and subsequently the effects of cytostatic drugs – for example, on the proliferation or apoptosis of tumor cells – can be measured in real-time via impedance spectroscopy.

In this context it has been shown by Robitzki and colleagues that nontreated tumor spheroids derived from a mamma carcinoma cell line similar to other types of tissues (e.g., heart muscle and neural tissues) showed a cellular resistance in a range of 1 kHz to 100 kHz (Fig. 3.7a). If tumor spheroids were genetically

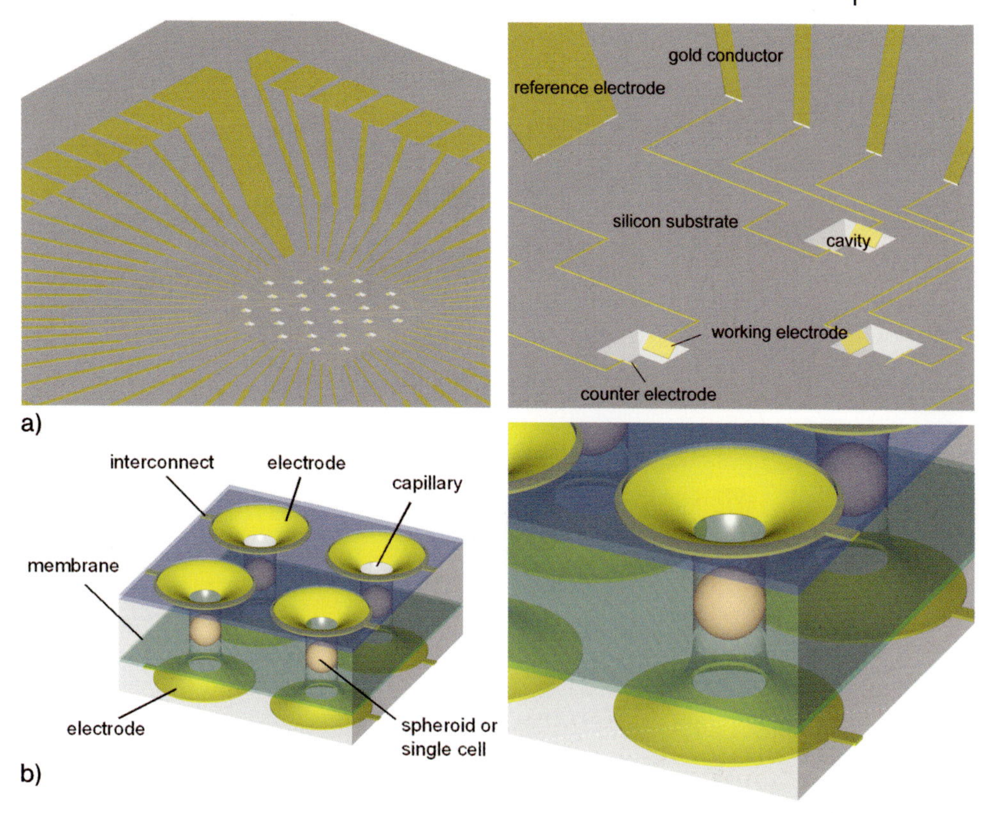

Fig. 3.6 Cell- or spheroid-based multielectrode impedance sensors. Planar glass or silicon chips can be fabricated with small cavities (a) or capillaries (b) for an improved cell or tissue positioning and electric impedance spectroscopy via two or multiple electrode arrangements. In the case of capillary multiarrays, positioning of the cells or spheroids is facilitated by a medium permeable membrane (green). Cavities can be produced with different diameters ranging from 10 to 1000 μm, whereas capillaries may have diameters of 100 μm, 200 μm, 300 μm, and 400 μm. (Copyright © BBZ/Andrea Robitzki, Maik Schmidt).

(transient gene silencing) manipulated to decrease proliferation and to induce apoptosis, the changed cell and tissue properties would correlate with a fivefold decrease (Fig. 3.7b) of the relative extracellular resistance as measured by impedimetric spectroscopy [14, 22, 23].

A further important application area of impedance spectroscopy is the testing of blood–brain barrier (BBB) function and permeability with respect to the application of novel drugs or toxicological chemicals. Here, the barrier function of endothelial or epithelial cell layers can be quantified by determining the transcellular electric resistance (TER). To measure the TER, cells are either placed directly on the surface of planar gold electrodes (similar to ECIS) [24–26] or, by an alternative approach of preculturing cells on multi-well filter inserts which then can be

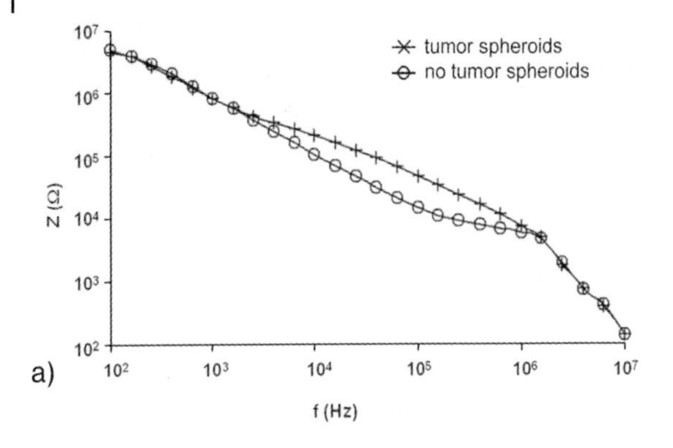

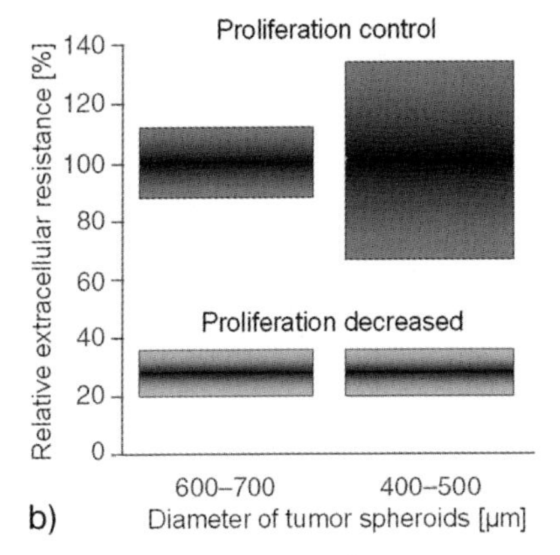

Fig. 3.7 Cellular properties contributing to the impedance of biological cells or tissues are detectable in a frequency range of 1 KHz to 1 MHz (a), as revealed by the impedance measurement of covered and noncovered recording sites. When proliferation is decreased in tumor spheroids by genetic manipulation, a fivefold decrease can be measured in the relative extracellular resistance (b).

transferred into the measuring chamber for frequency-dependent impedance recording [27]. At the bottom of this chamber a thin, gold film is evaporated as a measuring electrode, whilst a ring-shaped platinum counter electrode is dipped into the culture medium above. In seeking an automated approach, Wegener et al. [28] developed a multi-well measuring chamber that could be loaded with cell-covered filter inserts where the TER is measured on up to 24 different sites simultaneously [28]. Improved and modulated software may provide a system that enables the screening in a 96-array format

Another approach to investigating membrane properties in terms of drug transport is to use membrane transistors with lipid vesicles or lipid rafts on silicon chips. Lipid rafts or bilayers on silicon substrates represent the future of bioelectronic devices for measuring and monitoring the drug transport if the junction is sufficiently insulated (Fig. 3.8).

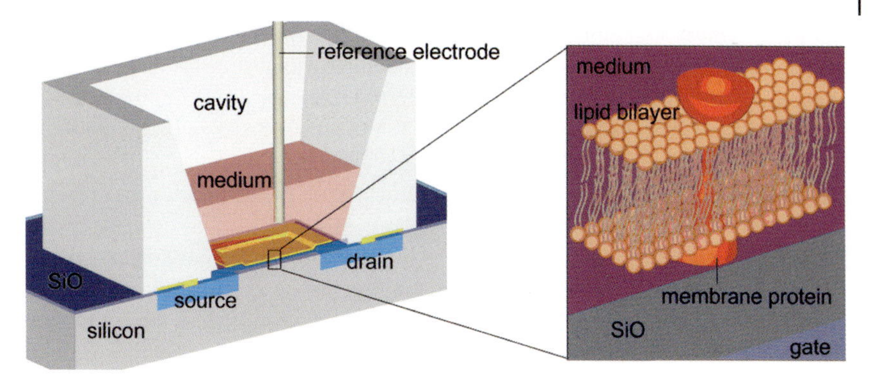

Fig. 3.8 Schematic illustration of an open gate field transistor coated with a giant lipid vesicle for testing membrane properties before and after application of drugs.
(Copyright © BBZ/Andrea Robitzki, Maik Schmidt).

An open gate of a field transistor is touched with a preformed giant lipid vesicle; the membrane is bound by a polyelectrolyte interaction [29]. At a contact area of 0.1 mm^2, the sheet resistance along the junction is 100 GΩ and the membrane resistance was above 100 GΩ. This contact area of the membrane and solid phase reflects a planar electrical core-coat conductor. The insulation of the junction can be determined by the resistance of the membrane and by the resistance of the cleft between membrane and substrate. Such a membrane-based semiconductor can be generated by: (1) depositing monomolecular lipid films or (2) by spreading lipid vesicles [30–34]. Such compound lipid–silicon microstructures are suitable for coupling semiconductor and electroactive proteins. Using such hybrid devices, biomolecules can be addressed by micro- and nanostructures in a semiconductor. In case the close contact of a bilayer and substrate interferes with the function of a biomolecule – for example, its lateral motion and conformational changes – a local blister on the lipid film can be created by a local coating of the chips via immobilization of proteins that enhances the distance between the chip and the lipid bilayer.

3.3.2 Intracellular Recording of Electroactive Cells: Chip-Based, Automated Patch-Clamp Recording

Fundamental cellular processes, such as neuronal signaling and the contraction of heart and skeletal muscle cells, are regulated by ion channels. A number of diseases which are caused by pathology of ion channels – so-called "channelopathies" [35] – originate from the loss or dysfunction of ion channels. Some well-known channelopathies affect the central nervous system (episodic ataxias, familial hemiplegic migraine and inherited epilepsies), skeletal muscles (myotonias, periodic paralyses, malignant hyperthermia) and the heart (long-QT syndromes, idiopathic ventricular fibrillation). In terms of effective therapies, ion channels represent excellent targets for drugs. Worldwide, the total revenue of drugs used to treat channelopathies amounts to six billion US$. Moreover, accompanied by the remarkable progress in molecular biology, the number of identified hereditary ion channel diseases is increasing constantly. Therefore, the screening for novel ion-modulating compounds that can be used as potential therapeutic drugs creates an enormous demand for novel HTS tools.

The patch-clamp technique, which was introduced by Neher and Sakmann in 1976 [36], is a traditional and outstanding method for the functional investigation of ion channel behavior on the cellular surface. Depending on the configuration, patch-clamping allows the direct recording of multiple or single ion channels. However, since the conventional patch-clamp technique is very slow, very laborious, and less cost-effective, this method is rather a method for basic research than for pharmaceutical testing, as the latter asks for HTS tools. At present, some promising chip-based patch-clamp arrays are available commercially, including PatchXpress and SealChip™ from Axon Industries, CYTOPATCH™ from Cytocentrics AG Reutlingen, and QPatch™ from Sophion Bioscience. When using planar chip devices, their fabrication can be achieved by using conventional semiconductor technologies, allowing individual chip designs and a cost-effective production. For the proper recording on a planar chip, the positioning of cells and the tight contact between cell and device is essential for a successful patch-clamping (giga seal formation). To overcome this problem, small micro-openings are generated, where cells can be placed and held by applying an appropriate negative pressure, as realized by the CYTOPATCH™ chip (Fig. 3.9). Here, one site of the CYTOPATCH™ chip is able to perform patch-clamp recordings on 200 cells per day.

Another chip-based automated patch clamp system has been developed by Molecular Devices Corp., and purchased by Axon Industries. This enables the simultaneous measurement of 16 mammalian cells, which potentiates the screening of 1000 components a week. The Qpatch™ system from Sophion Bioscience provides a so-called Qplate that is microfabricated by conventional photolithography techniques and consists of 16 individually controlled recording sites. An elaborated microfluidic channel system that is implemented in the silicon chip allows the controlled trapping of cells at the patch-clamping micro-opening site. In order to achieve a higher signal-to-noise ratio, the membrane at the patch aperture is disturbed by applying an acute and short suction pulse, which in turn transfers the patch-clamping from "on-cell" into "whole-cell" configuration. The first generation of automated chip-based patch-clamping systems (as described above) is well suited to distinct high-throughput approaches in pharmaceutical companies. In this context, it is possible to use these techniques at much earlier stages of functional drug discovery and drug testing processes and, therefore, to minimize development and preclinical evaluation costs. However, as the quality of chip- and non-chip-based automated patch-clamp platforms cannot yet be compared with the traditional patch-clamp technique, ongoing and further improvements of automated patch-clamping are necessary. Nevertheless, a number of studies have shown how automated patch-clamping can be improved

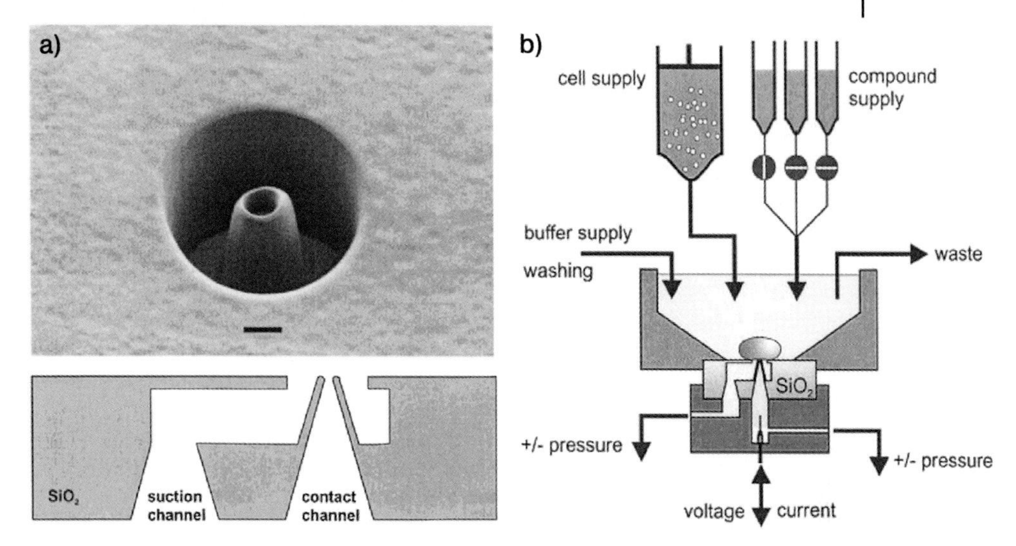

Fig. 3.9 The CYTOPATCH™ chip is characterized by two concentric openings a) formed by a focused ion beam in a 10 μm-thick silicon dioxide layer (SEM image; scale bar 2 μm). Positioning and recording are carried out independently by a suction channel and contact channel. (b) A schematic drawing of automated cell-by-cell patch-clamping using the CYTOPATCH™ chip technique. (From Multi Channel Systems, Reutlingen, Germany).

and optimized towards a real high-throughput application, especially in terms of the traditional patch-clamp quality [37–44].

3.3.3 Extracellular Recording of Electrically Excitable Cells: Multiple Site Recording of Field Potentials by MEAs

In general, for extracellular recording of field potentials or low-frequency potentials from electrogenic cells, glass and silicon-based MMEAs with embedded passive metal electrodes [45, 46] or integrated FETs are commonly used [7, 29, 47]. Although, in general, intracellular recording techniques are more sensitive than extracellular recordings, the latter – when carried out with MEAs – have several remarkable advantages. For example, cells cultured on substrate-integrated electrodes allow the noninvasive and simultaneous recording of extracellular potentials on multiple recording sites [48]. Based on these noninvasive recordings, the same cells can be repeatedly monitored, allowing true long-term experiments in parallel. Over the years, a number of companies (Multi Channel Systems, Reutlingen, Germany; Ayanda Biosytems, Lausanne, Switzerland; Alpha MED Sciences (Panasonic), Osaka, Japan; UNT Center for Network Neuroscience, Denton, USA) have offered customer-friendly and fully equipped systems that can be used without time-consuming staff training. A large number of publications relating to various applications can be found on the homepages of these companies.

In general, cell- or tissue-based MEAs are placed in a heatable amplifier which allows an adequate signal amplification and simultaneous stimulation and recording of multiple microelectrodes. For data acquisition and processing, the MEA and the amplifier are connected to an analogue/digital (A/D) converter and computer. Typically, MEAs can be used for almost all electrically active cells derived either from heart, muscle, or neural tissues. Although several studies using neuronal cells and tissues in combination with MEAs have revealed promising applications for the pharmaceutical industry [2, 3, 5, 8, 9, 46, 49–55], here we focus only on the use of cardiomyocytes for extracellular recordings.

In contrast to neuronal cell cultures, heart muscle cells show several advantages for HTS as they possess a spontaneous electric activity, are stable in culture even over long periods of time, and provide feasible and reliable extracellular recordings. For instance, Robitzki and co-workers used MEA-based extracellular recordings to identify extremely low concentrations of positive chronotropic compounds (range of 10^{-11} M), as shown for example by the application of angiotensin II to rat cardiomyocytes [56]. Moreover, by improving their functional cardiomyocyte-based sensor, these authors were able to detect low concentrations of autoimmune antibodies in sera from pregnant women suffering from pre-eclampsia [1]. The existence of these autoimmune antibodies has been demonstrated by an automated counting of extracellular field potentials after application of pre-eclamptic sera. Due to the binding of antibodies to the angiotensin II type 1 receptors, an increase in contraction frequency was observed.

With a view to its use as a diagnostic and/or drug screening tool, this MEA platform may need to be implemented in an automated work station for liquid handling, chip positioning, cell growth, and maintenance. The work station allows the screening of 10 to 20 chips per hour, with each chip being divided into four application chambers, each containing six to ten electrodes that can be recorded simultaneously. Thus, between 40 and 80 compounds per hour – that is, 960 to 1920 per day – can be screened. Since the work station is placed in a CO_2 incubator, long-term experiments are possible by repeated measurement of the same chips under cell culture conditions (over a period of 28 days). The ability of cardiomyocyte-based MEAs to detect chronotropic and arrhythmic effects of drugs and compounds has been demonstrated in several previous reports.

MEAs have also been shown applicable for the screening of QT prolongation [6, 9, 57, 58]. For example, Multi Channel Systems GmbH, Reutlingen, Germany has developed a 96-well QTplate™ for HTS of drug-induced QT prolongation in primary cardiomyocytes (Fig. 3.10). Each well of the QT-plate consists of a round recording electrode and an octagonal reference electrode; this allows the simultaneous extracellular recording of field potentials from all 96 wells. A combination of the QT-Screen system (amplifier and recording unit) with an automated liquid-handling system permits the daily screening of approximately 6000 data points or compounds [57, 58]. Based on the low fabrication costs, the average cost of each data point is US$ 0.20.

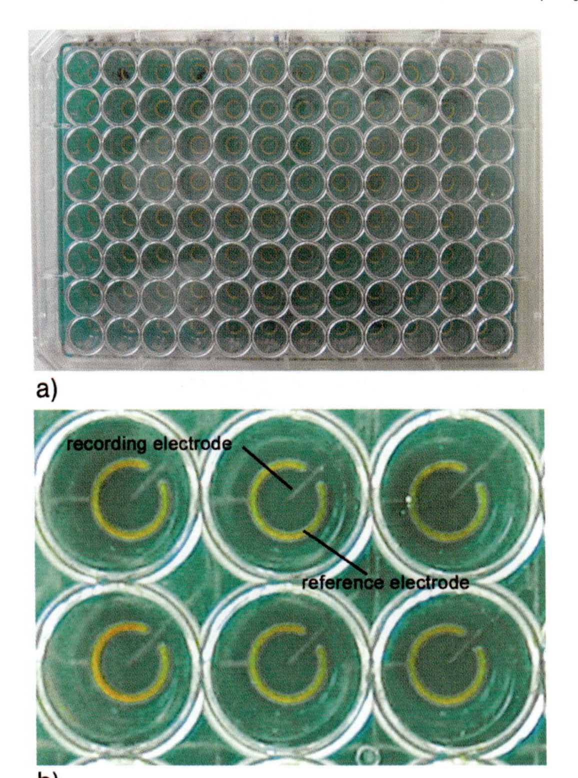

Fig. 3.10 QTplate™ from Multi Channel Systems enables HTS of QT prolongation of cardiomyocytes in a 96-well plate (a). For real-time recording of extracellular field potentials, microplates are connected to an amplifier board and computer. Cells can be cultured directly on the recording and reference electrodes implemented on the bottom of each well (b).

Another application of MEAs is the quantification of neurotoxins and drugs on neurons and/or neuronal networks. Nerve cells must express electrical activity depending on the molecular expression pattern as a part of their physiological function. Any interference with these patterns of electrophysiology – created, for example, by toxic agents or drugs – can generate alterations or malfunctions that might be classified as a reaction and could be monitored in real-time by microelectrodes. Therefore, spontaneously active neuronal monolayer networks *in vitro* can be cultured on thin film microelectrode arrays as a screening platform to determine the activities of drugs and other compounds under test. Such a screening module has been developed as an experimental platform by Gramowski et al. [59] to measure the acute neurotoxicological effects of trimethyltin (TMT). Murine spinal cord and auditory cortex both exhibited dose-dependent changes of their electrophysiological activity after treatment with TMT. However, microscopic inspection showed no acute cytotoxic effects on the neurons, such as beading of axons, retraction of dendrites, shrinking or swelling of somata, up to ranges which were 10-fold higher than doses of TMT that interfered with spontaneous activity. An ongoing effort to validate a cellular test system based on cultured networks by a simultaneous multichannel chart recording based on sputtered indium tin oxide (ITO) plates, spin-insulated with polysiloxane and electrolytically gold-plated at the electrode tips, has also been reported [59].

The measurement of different cellular parameters under noninvasive and real-time conditions on a single chip has been realized by Wolf et al., with the fabrication of a multiparametric sensor chip [16]. Here, tumor cells were cultured directly onto the sensor chip that comprises interdigital electrode structures (IDES) to record electric current-associated changes in cellular membrane properties, ion-sensitive effect transistors (ISFETs) to analyze the extracellular pH of the culture medium, and an amperometric oxygen electrode to detect oxygen consumption. Although the chip has been designed primarily for monitoring the efficiency of chemotherapeutic drugs, it can also be applied to almost all other cell types.

Bionas GmbH (Rostock, Germany) has developed a similar multifunctional system to monitor various cellular parameters on a single device. Different chip-integrated sensors allow HCS of cell metabolism-dependent parameters such as acidification, respiration, cell adhesion, and membrane integrity. Based on these properties, this cell-based sensor represents a promising tool for functional drug and cytotoxicity screening.

3.4 Concluding Remarks: Secondary Screening for Safety and Cost-Effective Drug Testing and Discovery

During the past few decades, an amazing number of innovative cell-based sensors has been developed, and it has not been possible to consider all of these in this chapter. The common focus here is the combination of living cells with inanimate matter to produce an electronic and noninvasive data read-out of cellular parameters under real-time conditions. The appropriate adaptation of these manifold biosensors to automated fluid-handling and application systems opens new possibilities for feasible and reliable HTS or HCS approaches. A hallmark of chip-based sensors is that they combine the properties of both primary and secondary screening tools; in other words, biosensor arrays are ultra-fast-working systems that may produce several thousand data points each day. In this way, not only the activity of compounds is identified, but information is also provided about functional cellular changes that are normally obtained via cost-intensive secondary screening processes. Thus, it is clear that cell-based sensors are suited to reducing costs during the early and late phases of drug discovery and screening processes. Moreover, they can be used in the preclinical phase for improved drug safety screening, especially with regard to drug-induced side effects, as recently claimed by the Food and Drug Administration in the United States (FDA) and the European Agency for the Evaluation of Medical Products (EMEA) in the European Union.

References

1 Rothermel, A., Kurz, R., Ruffer, M., Weigel, W., Jahnke, H. G., Sedello, A., Stepan, H., Faber, R., Schulze-Forster, K., and Robitzki, A. Cells on a chip – the use of electric properties for highly sensitive monitoring of blood-derived factors involved in angiotensin II type 1 receptor signalling. *Cell Physiol. Biochem.* **2005**, *16*(1–3): 51–58.

2 Heuschkel, M. O., Fejtl, M., Raggenbass, M., Bertrand, D., and Renaud, P. A three-dimensional multi-electrode array for multi-site stimulation and recording in acute brain slices. *J. Neurosci. Methods* **2002**, *114*(2): 135–148.

3 Hutzler, M. and Fromherz, P. Silicon chip with capacitors and transistors for interfacing organotypic brain slice of rat hippocampus. *Eur. J. Neurosci.* **2004**, *19*(8): 2231–2238.

4 Reppel, M., Boettinger, C., and Hescheler, J. Beta-adrenergic and muscarinic modulation of human embryonic stem cell-derived cardiomyocytes. *Cell Physiol. Biochem.* **2004**, *14*(4–6): 187–196.

5 Segev, R., Goodhouse, J., Puchalla J., and Berry, M. J. II. Recording spikes from a large fraction of the ganglion cells in a retinal patch. *Nat. Neurosci.* **2004**, *7*(10): 1154–1161.

6 Halbach, M., Egert, U., Hescheler, J., and Banach, K. Estimation of action potential changes from field potential recordings in multicellular mouse cardiac myocyte cultures. *Cell Physiol. Biochem.* **2003**, *13*(5): 271–284.

7 Ingebrandt, S., Yeung, C. K., Staab, W., Zetterer, T., and Offenhausser, A. Backside contacted field effect transistor array for extracellular signal recording. *Biosens. Bioelectron.* **2003**, *18*(4): 429–435.

8 Pancrazio, J. J., Gray, S. A., Shubin, Y. S., Kulagina, N., Cuttino, D. S., Shaffer, K. M., Eisemann, K., Curran, A., Zim, B., Gross, G. W., and O'Shaughnessy, T. J. A portable microelectrode array recording system incorporating cultured neuronal networks for neurotoxin detection. *Biosens. Bioelectron.* **2003**, *18*(11): 1339–1347.

9 Stett, A., Egert, U., Guenther, E., Hofmann, F., Meyer, T., Nisch, W., and Haemmerle, H. Biological application of microelectrode arrays in drug discovery and basic research. *Ann. Biomol. Chem.* **2003**, *377*(3): 486–495.

10 Bartholomä, P., Impidjati, Reininger-Mack, A., Zhang, Z., Thielecke, H., and Robitzki, A. A more aggressive breast cancer spheroid model coupled to an electronic capillary sensor system for a high-content screening of cytotoxic agents in cancer therapy: 3-dimensional *in vitro* tumor spheroids as a screening model. *J. Biomol. Screen.* **2005**, *10*(7): 705–714.

11 Ciambrone, G. J., Liu, V. F., Lin, D. C., McGuinness, R. P., Leung, G. K., and Pitchford, S. Cellular dielectric spectroscopy: a powerful new approach to label-free cellular analysis. *J. Biomol. Screen.* **2004**, 9(6): 467–480.

12 Hug, T. S. Biophysical methods for monitoring cell-substrate interactions in drug discovery. *Assays Drug Dev. Technol.* **2003**, *1*(3): 479–488.

13 Thielecke, H. Capillary chip-based characterisation of small tissue samples. *Med. Device Technol.* **2003**, *14*(9): 18–20.

14 Reininger-Mack, A., Thielecke, H., and Robitzki, A. A. 3D-biohybrid systems: applications in drug screening. *Trends Biotechnol.* **2002**, *20*(2): 56–61.

15 Mestres-Ventura, P., Morguet, A., Schofer, A., Laue, M., and Schmidt, W. Application of silicon sensor technologies to tumor tissue *in vitro*: detection of metabolic correlates of chemosensitivity. *Methods Mol. Med.* **2005**, *111*: 109–125.

16 Otto, A. M., Brischwein, M., Motrescu, E., and Wolf, B. Analysis of drug action on tumor cell metabolism using electronic sensor chips. *Arch. Pharm.* **2004**, *337*(12): 682–686.

17 Gimsa, J., and Wachner, D. A unified resistor-capacitor model for impedance, dielectrophoresis, electrorotation, and induced transmembrane potential. *Biophys. J.* **1998**, *75*(2): 1107–1116.

18 Foster, K. R. and Schwan, H. P. Dielectric properties of tissues and biological materials: a critical review. *Crit. Rev. Biomed. Eng.* **1989**, *17*(1): 25–104.

19 Schwan, H. P. Electrical properties of tissue and cell suspensions. *Adv. Biol. Med. Phys.* **1957**, *5*: 147–209.

20 Giaever, I. and Keese, C. R. Monitoring fibroblast behavior in tissue culture with an applied electric field. *Proc. Natl. Acad. Sci. USA* **1984**, *81*(12): 3761–3764.

21 Arndt, S., Seebach, J., Psathaki, K., Galla, H. J., and Wegener, J. Bioelectrical impedance assay to monitor changes in cell shape during apoptosis. *Biosens. Bioelectron.* **2004**, *19*(6): 583–594.

22 Thielecke, H., Mack, A., and Robitzki, A. A multicellular spheroid-based sensor for anti-cancer therapeutics. *Biosens. Bioelectron.* **2001**, *16*(4–5): 261–269.

23 Thielecke, H., Mack, A., and Robitzki, A. Biohybrid microarrays – impedimetric biosensors with 3D *in vitro* tissues for toxicological and biomedical screening. *Fresenius J. Anal. Chem.* **2001**, *369*(1): 23–29.

24 Hoheisel, D., Nitz, T., Franke, H., Wegener, J., Hakvoort, A., Tilling, T., and Galla, H. J. Hydrocortisone reinforces the blood-brain barrier properties in a serum free cell culture system. *Biochem. Biophys. Res. Commun.* **1998**, *247*(2): 312–315.

25 Janshoff, A., Wegener, J., Sieber, M., and Galla, H. J. Double-mode impedance analysis of epithelial cell monolayers cultured on shear wave resonators. *Eur. Biophys. J.* **1996**, *25*(2): 93–103.

26 Wegener, J., Sieber, M., and Galla, H. J. Impedance analysis of epithelial and endothelial cell monolayers cultured on gold surfaces. *J. Biochem. Biophys. Methods* **1996**, *32*(3): 151–170.

27 Tilling, T., Korte, D., Hoheisel, D., and Galla, H. J. Basement membrane proteins influence brain capillary endothelial barrier function in vitro. *J. Neurochem.* **1998**, *71*(3): 1151–1157.

28 Wegener, J., Abrams, D., Willenbrink, W., Galla, H. J., and Janshoff, A. Automated multi-well device to measure transepithelial electrical resistances under physiological conditions. *Biotechniques* **2004**, *37*(4): 590, 592–594, 596–597.

29 Fromherz, P., Kiessling, V., Kottig, K., and Zeck. G. Membrane transistor with giant lipid vesicle touching a silicon chip. *Appl. Phys.* **1999**, *A69*, 571–576.

30 Fromherz, P. and Arden, W. Transmission of chemical signal by sequential energy and electron transfer in pigment membrane on semiconductor. *Ber. Bunsenges Phys. Chem.* **1980**, *84*, 1045–1050.

31 Tam, L. K. and McConnell, H. M. Supported phospholipid bilayers. *Biophys. J.* **1985**, *47*(1): 105–113.

32 Kalb, E., Frey, S., and Tamm, L. K. Formation of supported planar bilayers by fusion of vesicles to supported phospholipid monolayers. *Biochim. Biophys. Acta* **1992**, *1103*(2): 307–316.

33 Sackmann, E. Supported membranes: scientific and practical applications. *Science* **1996**, *271*(5245): 43–48.

34 Gritsch, S., Nollert, P., Jähnig, F., and Sackmann, E. Impedance spectroscopy of porin and gramicidin pores reconstituted into supported lipid bilayers on indium-tin-oxide electrodes. *Langmuir* **1998**, *14*(11): 3118–3125.

35 Lehmann-Horn, F. and Jurkat-Rott, K. Voltage-gated ion channels and hereditary disease. *Physiol. Rev.* **1999**, *79*(4): 1317–1372.

36 Neher, E. and Sakmann, B. Single-channel currents recorded from membrane of denervated frog muscle fibres. *Nature* **1976**, *260*(5554): 799–802.

37 Lepple-Wienhues, A., Ferlinz, K., Seeger, A., and Schafer, A. Flip the tip: an automated, high quality, cost-effective patch clamp screen. *Receptors Channels* **2003**, *9*(1): 13–17.

38 Trumbull, J. D., Maslana, E. S., McKenna, D. G., Nemcek, T. A., Niforatos, W., Pan, J. Y., Parihar, A. S., Shieh, C. C., Wilkins, J. A., Briggs, C. A., and Bertrand, D. High throughput electrophysiology using a fully automated, multiplexed recording system. *Receptors Channels* **2003**, *9*(1): 19–28.

39 Asmild, M., Oswald, N., Krzywkowski, K. M., Friis, S., Jacobsen, R. B., Reuter, D., Taboryski, R., Kutchinsky, J., Vestergaard, R. K., Schroder, R. L., Sorensen, C. B., Bech, M., Korsgaard, M. P., and Willumsen, N. J. Upscaling and automation of electrophysiology: toward high throughput screening in ion channel drug discovery. *Receptors Channels* **2003**, *9*(1): 49–58.

40 Tao, H., Santa Ana, D., Guia, A., Huang, M., Ligutti, J., Walker, G., Sithiphong, K., Chan, F., Guoliang, T., Zozulya, Z., Saya, S., Phimmachack, R., Sie, C., Yuan, J., Wu, L., Xu, J., and Ghetti, A. Automated tight seal electrophysiology for assessing the potential hERG liability of pharmaceutical compounds. *Assay Drug Dev. Technol.* **2004**, *2*(5): 497–506.

41 Shieh, C. C. Automated high-throughput patch-clamp techniques. *Drug Discov. Today* **2004**, *9*(13): 551–552.

42 Zheng, W., Spencer, R. H., and Kiss, L. High throughput assay technologies for ion channel drug discovery. *Assay Drug Dev. Technol.* **2004**, *2*(5): 543–552.

43 Guo, L. and Guthrie, H. Automated electrophysiology in the preclinical evaluation of drugs for potential QT prolongation. *J. Pharmacol. Toxicol. Methods* **2005**, *52*(1): 123–135.

44 Ionescu-Zanetti, C., Shaw, R. M., Seo, J., Jan, Y. N., Jan, L. Y., and Lee, L. P. Mammalian electrophysiology on a microfluidic platform. *Proc. Natl. Acad. Sci. USA* **2005**, *102*(26): 9112–9117.

45 Thomas, C. A. Jr, Springer, P. A., Loeb, G. E., Berwald-Netter, Y., and Okun, L. M. A miniature microelectrode array to monitor the bioelectric activity of cultured cells. *Exp. Cell Res.* **1972**, *74*(1): 61–66.

46 Gross, G. W., Rieske, E., Kreutzberg, G. W. and Meyer, A. A new fixed-array multi-microelectrode system designed for long-term monitoring of extracellular single unit neuronal activity *in vitro*. *Neurosci. Lett.* **1977**, *6*(2–3): 101–105.

47 Offenhausser, A. and Knoll, W. Cell-transistor hybrid systems and their potential applications. *Trends Biotechnol.* **2001**, *19*(2): 62–66.

48 Connolly, P., Clark, P., Curtis, A. S., Dow, J. A., and Wilkinson, C. D. An extracellular microelectrode array for monitoring electrogenic cells in culture. *Biosens. Bioelectron.* **1990**, *5*(3): 223–234.

49 Kaul, R. A., Syed, N. I., and Fromherz, P. Neuron-semiconductor chip with chemical synapse between identified neurons. *Phys. Rev. Lett.* **2004**, *92*(3): 038102.

50 Pancrazio, J. J., Bey, P. P. Jr, Loloee, A., Manne, S., Chao, H. C., Howard, L. L., Gosney, W. M., Borkholder, D. A., Kovacs, G. T., Manos, P., Cuttino, D. S., and Stenger, D. A. Description and demonstration of a CMOS amplifier-based-system with measurement and stimulation capability for bioelectrical signal transduction. *Biosens. Bioelectron.* **1998**, *13*(9): 971–979.

51 Fromherz, P., Offenhausser, A., Vetter, T., and Weis, J. A neuron-silicon junction: a Retzius cell of the leech on an insulated-gate. *Science* **1991**, *252*(5010): 1290–1293.

52 Sproessler, C., Denyer, M., Britland, S., Knoll, W., and Offenhausser, A. Electrical recordings from rat cardiac muscle cells using field-effect transistors. *Phys. Rev. E. Stat. Phys. Plasmas Fluids Relat. Interdiscip. Topics* **1999**, *60*(2 PtB): 2171–2176.

53 Breckenridge, L. J., Wilson, R. J., Connolly, P., Curtis, A. S., Dow, J. A., Blackshaw, S. E., and Wilkinson, C. D. Advantages of using microfabricated extracellular electrodes for in vitro neuronal recording. *J. Neurosci. Res.* **1995**, *42*(2): 266–276.

54 Pine, J. Recording action potentials from cultured neurons with extracellular microcircuit electrodes. *J. Neurosci. Methods* **1980**, *2*(1): 19–31.

55 Grumet, A. E., Wyatt, J. L. Jr, and Rizzo, J. F. III. Multi-electrode stimulation and recording in the isolated retina. *J. Neurosci. Methods* **2000**, *101*(1): 31–42.

56 Kurz, R., Rothermel, A., Rüffer, M., Urban, C., Jahnke, H.-G., Weigel, W., and Robitzki, A. A functional cardiomyocyte based biosensor for pre-diagnostic monitoring: an angiotensin II study. Proceedings, IFMBE, Medical & Biological Engineering & Computing, **2004**, pp. 1727–1983.

57 Meyer, T., Leisgen, C., Gonser, B., and Gunther, E. QT-screen: high-throughput cardiac safety pharmacology by extracellular electrophysiology on primary cardiac myocytes. *Assay Drug Dev. Technol.* **2004**, *2*(5): 507–514.

58 Meyer, T., Boven, K. H., Gunther, E., and Fejtl, M. Micro-electrode arrays in cardiac safety pharmacology: a novel tool to study QT interval prolongation. *Drug Safety* **2004**, *27*(11): 763–772.

59 Gramowski, A., Schiffmann, D., and Gross, G. W. Quantification of acute neurotoxic effects of trimethyltin using neuronal networks cultured on microelectrode arrays. *Neurotoxicology* **2000**, *21*(3): 331–342.

60 Ivorra, A., Genesca, M., Sola, A., Palacios, L., Villa, R., Hotter, G., and Aguilo, J. Bioimpedance dispersion width as a parameter to monitor living tissues. *Physiol. Meas.* **2005**, *26*(2): 165–173.

4
Novel *In-Vitro* Exposure Techniques for Toxicity Testing and Biomonitoring of Airborne Contaminants

Amanda Hayes, Shahnaz Bakand, and Chris Winder

4.1
Introduction

Exposure to occupational and environmental airborne contaminants is a major contributor to human health problems. The inhalation of gases, vapors, solid and liquid aerosols, and also mixtures of these, can cause a wide range of adverse health effects, ranging from simple irritation to systemic diseases [1–4]. Despite significant achievements in the risk assessment of chemicals, the toxicological database, particularly for airborne contaminants, remains limited. Therefore, as a part of preventive strategies that can assist in reducing the effects of toxic chemicals, it is critical to develop new alternative approaches that are both informative and time-/cost-efficient to identify the potential hazards of airborne chemicals.

4.2
The Inhalation of Air Contaminants

Air contaminants are exogenous substances in indoor or outdoor air, and include both particulates and gaseous contaminants that may cause adverse health effects in humans or animals. In addition, they may affect plant life and impact on the global environment by changing the atmosphere of the Earth [5]. Various physical, chemical and dynamic processes may generate air pollution leading to the emission of gases, particulates, or mixtures of these, into the atmosphere [6]. Whilst major attempts have been made to reduce emissions from both stationary and mobile sources, millions of people today face excessive air pollution in both occupational and urban environments [7, 8]. Many industrial and commercial activities release toxic contaminants in gas, vapor or particulate forms, and occupational and environmental health problems can potentially arise where such releases are not controlled properly. Moreover, air contaminants today are not limited to the urban environment, nor to the industrial workplaces, but may

Drug Testing In Vitro: Breakthroughs and Trends in Cell Culture Technology
Edited by Uwe Marx and Volker Sandig

ISBN: 978-3-527-31488-1

Table 4.1 Types of airborne contaminants (from [3, 9, 10]).

Type	Properties
Gas	These are substances that exist in a gaseous state at room temperature and pressure. Only by the combined effects of increased pressure and decreased temperature, can gas be changed to a liquid or perhaps solid state. Processes that involve high temperature, such as welding operations and exhaust from engines, can potentially generate toxic gases such as oxides of carbon, nitrogen or sulfur.
Vapor	This is the gaseous phase of a substance that ordinarily is in a liquid or solid state at room temperature and pressure. Vapor can convert to a liquid state either by increasing pressure and/or reducing temperature. Several occupational practices may produce toxic vapors, such as charging and mixing liquids, painting, spraying, and dry cleaning or any other activities which involve volatile solvents or chemicals.
Dust	This is a small solid particle, usually produced by different mechanical processes such as grinding, cutting, sawing, crushing, screening, or sieving. Dust particles may originate from organic materials. Dust particles have a nonspherical shape, with a wide range of sized from a few nanometers (nanoparticles) to larger particles of > 100 μm.
Fiber	This is an elongated or long solid particle with an aspect ratio (length:width) more than 3 : 1. There are two types of fibers: natural (e.g., asbestos), and synthetic or man-made (e.g., glass) fiber. Asbestos is the most important natural fiber, as it can induce asbestosis and lung cancer in workers who have experienced heavy exposure.
Fume	This is a solid aerosol produced by combustion, sublimation or condensation of vaporized materials. Metallic fumes usually form in air due to the oxidation of metallic vapors. Fume particles are spherical and extremely fine, usually < 0.1 μm diameter, though the size can increase by aggregation or flocculation as the fume ages. High-temperature operations such as arc-welding, torch-cutting and metal smelting can generate extremely fine metal oxide fumes.
Smoke	This is a complex compound that involves solid and liquid aerosols, gases and vapors that usually result from the incomplete combustion of organic materials. For example, tobacco smoke contains thousands of chemical substances, most of which are toxic or carcinogenic. Although primary smoke particles are between 0.01 and 1 μm diameter, they can aggregate and produce extremely larger particles known as soot.
Mist	This is a suspended spherical liquid droplet formed by mechanical dispersing of a bulk liquid such as spraying and atomizing. Mist droplets have their parent liquid properties with a wide range of sizes, ranging from a few to more than 100 μm. All processes involving high-pressure liquids which can potentially generate mists, such as paint-spraying, need to be adequately controlled. Other examples of mists include oil mists in cutting and grinding operations, acid mists in electroplating, and acid or alkali mists from pickling operations.
Fog	This is a suspended spherical liquid droplet aerosol (similar to mist), but it is formed by the condensation of vapor phase on particle nuclei of the air. The size of fog droplets (1–10 μm) is less than that of mists.

be common indoor air contaminants present in office workplaces, schools, and hospitals.

Based on their physical properties, airborne contaminants can be classified as two main types:

- Gases and vapors or dissolved air contaminants.
- Aerosols or suspended air contaminants.

The term "aerosol" refers to both liquid droplets and solid particles suspended in the air such as dust, fiber, smoke, mist, and fog. These terms can be defined as shown in Table 4.1.

Three main routes of exposure to chemicals are inhalation, dermal absorption, and ingestion [11]. Inhalation is considered to be the most important means by which humans are exposed to airborne chemicals. The severity of toxic effects of inhaled chemicals is influenced by several factors such as the type of air contaminant, airborne concentration, size of airborne chemical (for particles), solubility in tissue fluids, reactivity with tissue compounds, blood–gas partition coefficient (for gases and vapors), frequency and duration of exposure, interactions with other air toxicants, and individual immunological status [9–12]. The site of deposition/action of inhaled toxicants will determine, to a great extent, the ultimate response of the respiratory tract to inhaled chemicals. After inhalation, airborne contaminants may be deposited in different regions of the respiratory tract including the nasopharyngeal, tracheobronchial and pulmonary regions.

The major physiological function of the respiratory tract is the transfer of oxygen to the blood and removal of carbon dioxide as a metabolic waste product. The human respiratory tract has a very large surface area of approximately 140 m^2, and a high daily exchange volume in excess of 10 m^3 [13–16]. In addition, the membrane between air and blood in the gas exchange region is extremely thin (ca. 0.4–2.5 μm) [17, 18]. As well as olfactory, gas exchange and blood oxygenation functions, the respiratory system has evolved to deal with xenobiotics and airborne materials such as those usually occurring in the air environment [17]. However, this system cannot always deal adequately with the wide range of airborne contaminants that may occur in urban, and particularly occupational, environments [10]. As a result, the respiratory system is both a site of toxicity for pulmonary toxicants, and a pathway for inhaled chemicals to reach other organs distant from the lungs and to elicit their toxic effects at these extrapulmonary sites. Responses to inhaled toxicants range from immediate reactions to long-term chronic effects, and from specific impacts on single tissue to generalized, systemic effects [9, 12].

4.3 Toxicological Assessment

Toxicology is the science of poisons or, to be more precise, it is the study of the adverse effects of xenobiotics on biological systems [19, 20]. Toxicology is an ever-developing science, and modern toxicologists consider cellular and molecular responses as the earliest indicators of exposure to exogenous agents. Toxicological developments are due both to theoretical expansion and technical improvements of different branches of science, notably biological sciences, chemistry, mathematics, and physics. Toxicology is a very broad science that can be classified into several branches relating to discipline, application and function.

Toxicological information about chemicals is, in many cases, still limited [21–24]. While data obtained from human experiences would be most useful in assessing the toxic effects of chemicals, human data are not always available for developing safety evaluations on chemicals and airborne contaminants. Moreover, as a result of many unfortunate human experiences, such as those with pharmaceutical agents such as diethylstilbestrol and thalidomide, or occupational and environmental contaminants such as lead and polychlorinated biphenyls (PCBs), it is now understood that the risks of chemicals, new products and technologies need to be assessed before adverse human experience occurs [25–27].

Therefore, as part of a preventive strategy, it is critical to develop new approaches that are both informative and time-/cost-efficient in order to identify the potential hazards in the absence of widespread human exposures [19]. In general, no single method can cover the complexity of general toxicity in humans [28]. However, toxicity data can be obtained from several sources, including *in-vivo* and *in-vitro* toxicological studies, epidemiological studies, quantitative structure–activity relationships (QSAR), and physiologically based toxicokinetic (PBTK) studies.

Toxicology has made a major contribution in providing chemical toxicity information for many years. From a methodology perspective, toxicological test methods range from conventional whole-animal (*in-vivo*) usage to modern cell culture (*in-vitro*) techniques. Conventional methods of toxicological assessment are based on whole-animal (*in-vivo*) studies; indeed, animal toxicity studies began in 1927 when the LD_{50} (lethal dose 50%) test was introduced by Trevan in the USA [29].

With regard to time of exposure, toxicity studies range from single exposure to short-term and long-term repeated exposure studies. Usually, single-exposure toxicity tests are performed to determine the acute toxicity of chemicals. These toxicity tests often involve the administration of relatively large amounts of chemicals in order to measure an appropriate endpoint such as lethality, organ damage, or cell death. Short-term repeated exposure studies are performed by repeated administration of a chemical in order to determine both subacute toxicity information and to establish doses for longer-term studies. Long-term repeated exposure studies are performed to determine the long-term toxicity of chemicals; these tests often involve the administration of low chemical doses over extended periods in order to establish the cumulative toxicity of the materials.

Conventional animal toxicity tests such as the LD_{50}, have been criticized due to the heavy reliance on animal data. The Organisation for Economic and Cooperative Development (OECD) has modified the LD_{50} acute oral toxicity test (401) in order to reduce the number of animals used for each test substance [30]. Similarly, in the area of inhalation toxicology, the OECD has also proposed new test guidelines including an Acute Inhalation Toxicity-Fixed Dose Procedure (433) and an Acute Inhalation Toxicity-Acute Toxic Class (ATC) Method (436), both of which are in draft form [31]. These guidelines will replace conventional guidelines including: Acute Inhalation Toxicity (403); Repeated Dose Inhalation Toxicity 28/14-Day (412); and Subchronic Inhalation Toxicity 90-Day (413).

However, the prediction of biological activities of toxic compounds in humans by placing reliance on animal data poses some degree of uncertainty due to inter-species differences between animals and humans [32–34].

4.4 *In-Vitro* Toxicological Studies

The focus of toxicology has shifted somewhat since the mid-1980s, from whole-animal toxicity tests to alternative *in-vitro* toxicity methods [19, 34]. The application of cell culture techniques in toxicological studies is referred to as *in-vitro* toxicology, which describes a field of study that applies technology using isolated organs, tissues and cell culture to investigate the toxic effects of chemicals [35].

The development of *in-vitro* toxicity test methods has been influenced by a number of factors [36]. Animal welfare issues are one of the important social concerns that have impacted on the recent shift towards alternatives in toxicity testing. Another social issue is the increasing public interest on the safety of chemicals and new products. Each year, thousands of new cosmetics, pharmaceuticals, pesticides and consumer products are introduced into the marketplace. Considering there are approximately 80 000 chemicals in commerce [37], as well as an extremely large number of chemical mixtures, the *in-vivo* testing of this number of chemicals would require a large number of expensive, time-consuming – and in some cases nonhumane – tests on animal species [38]. The need to determine the potential toxic effects of this large number of chemicals has provoked the search for rapid, sensitive, and specific test methods.

4.5 Applications of *In-Vitro* Test Methods

In-vitro toxicity methods were first developed to study the mechanisms of actions of chemical substances at the molecular and cellular levels [39]. Increasingly, *in-vitro* methods have been applied in many fields such as cancer biology, drug discovery and toxicology. Genotoxicity was the first field of toxicology in which *in-vitro* test methods have been used for toxicity testing to identify the mutagenic characteristics of chemicals [40, 41]. At present, *in-vitro* methods cover a broad range of techniques and models, and a standardized battery of *in-vitro* tests can be used to assess acute local and systemic toxicity. Cytotoxicity testing, reproductive toxicity, mutagenicity, irritancy testing, immunology and target organ toxicity are the main areas of *in-vitro* toxicology [32].

Meanwhile, the application of *in-vitro* methods in toxicology is also faced by a number of difficulties. One important shortcoming of this approach is the lack of complexity [42]. Although cells in culture represent the elementary living systems, as a very simplified system, they cannot represent the complexity of the entire organism. Hence, an *in-vitro* system cannot replicate exactly the biodynamics of the whole human body due to the lack of possible mitigating systems (e.g., hormones, nervous system and immunity), and the lack of biotransformation and excretion pathways for their elimination *in vitro* [35]. The challenge of how to relate *in-vitro* concentrations which produce cellular toxicity *in vitro* to equivalent *in-vivo* dosages may be improved by the development and application of predictive tools such as PBTK models [39]. The basic toxicology knowledge of absorption, distribution, metabolism and elimination (ADME) of chemicals is essential for the development of appropriate toxicokinetic models.

The next limitation of these methods is related to chronic toxicity testing. Although some studies have shown promise [43], *in-vitro* test systems are to date mainly focused on acute toxicity testing rather than on short-term or long-term repeated dose toxicity investigations [44]. More knowledge of the mechanisms of toxicity is still needed before *in-vitro* methods could broadly be implemented for repeated-dose toxicity testing [45]. The physico-chemical properties of test chemicals, including low water solubility or high vapor pressure, may also cause technical problems during the course of *in-vitro* tests [46–48].

Nevertheless, *in-vitro* toxicity test systems offer new advantages when compared to traditional *in-vivo* approaches [39, 42, 44, 49–52]. *In-vitro* test methods eliminate the interspecies extrapolation by using human cells and tissues, which may in turn help to generate more representative data for human toxicological risk assessments. These methods are comparatively simpler, faster and hence less time-consuming and more cost-efficient [50]. Moreover, they facilitate the application of biochemical, cellular and molecular biology techniques to study the underlying mechanism of toxic action [51, 53]. *In-vitro* models for organ toxicity evaluation which have been developed for many organs and tissues (such as nephrotoxic, neurotoxic and hematotoxic models) allow the study of components of potential target organs and target systems [52]. Despite the complex organization of cells, *in-vitro* methods are also being developed for respiratory toxicity assessment using relevant airway cells, lung cells or tissues and for the implementation of target-specific endpoints [17, 54].

4.6
In-Vitro Toxicity Endpoints

In order to assess the potential toxicity of chemical substances, several *in-vitro* tests have been developed by measuring different biological endpoints, which are summarized in Table 4.2.

Over recent years, research investigations into apoptosis have provided extensive knowledge of the mechanisms involved in cell death, and this in turn has led to the development of mechanistically based endpoints. Many morphological and biological changes that may occur at the cellular membrane, nucleus, specific

Table 4.2 Common biological endpoints assessed by *in-vitro* toxicity tests (from [28, 42, 50, 52, 56–59]).

Biological endpoint	Detection method
Cell morphology	Cell size and shape Cell–cell contacts Nuclear number, size, shape and inclusions Nuclear or cytoplasmic vacuolation
Cell viability	Trypan blue dye exclusion Diacetyl fluorescein uptake Cell counting Replating efficiency
Cell metabolism	Mitochondrial integrity (MTT, MTS and XTT tetrazolium salt assays) Lysosome and Golgi body activity (Neutral red uptake) Cofactor depletion (e.g., ATP content)
Membrane leakage	Loss of enzymes (e.g., LDH), ions or cofactors (e.g., Ca^{2+}, K^{+}, NADPH) Leakage of pre-labeled markers (e.g., ^{51}Cr or fluorescein)
Cell proliferation	Cell counting Total protein content (e.g., methylene blue, Coomassie blue, kenacid blue) DNA content (e.g., Hoechst 33342) Colony formation
Cell adhesion	Attachment to culture surface Detachment from culture surface Cell–cell adhesion
Radioisotope incorporation	Thymidine incorporation into DNA Uridine incorporation into RNA Amino acids incorporation into proteins

proteases and DNA level can be used as biological endpoints for measuring apoptosis [52, 59]. Moreover, the rapid progress in genomic, transcriptomic (gene expression) and proteomic technologies has created a unique powerful tool in toxicological investigations [44].

4.7 *In-Vitro* Toxicity Testing of Air Contaminants

The study of the toxic effects of inhaled chemicals is typically more challenging due to the technology required to generate and characterize the test atmospheres, and to develop effective and reproducible techniques for the exposure of cell cultures to airborne contaminants. The generation and characterization of known concentrations of air contaminants and reproducible exposure conditions requires different equipment and techniques (Fig. 4.1). For example, inhalation exposure systems involve several efficient and precise subsystems, including: a conditioned air supply system; a suitable gas or aerosol generator for the test chemical; an

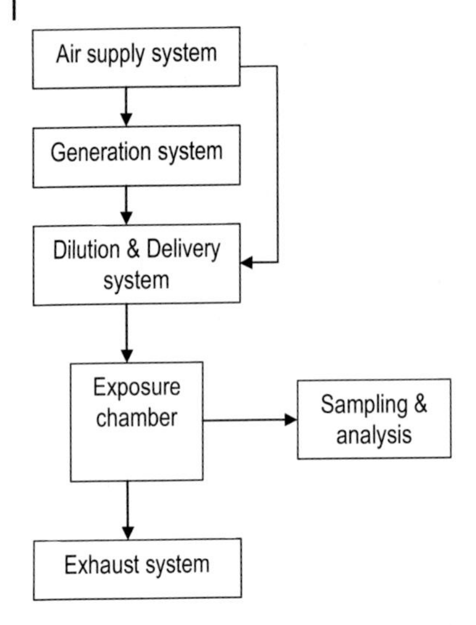

Fig. 4.1 The components of a test atmosphere generation system (adopted from [62]).

atmosphere dilution and delivery system; an exposure chamber; a real-time monitoring, sampling and analytical system; a filter or scrubbing system; and an exhaust system [15, 60, 61].

A practical approach for *in-vitro* respiratory toxicity testing has been proposed by the European Centre for the Validation of Alternative Methods (ECVAM) [17]. This systematic approach is initiated with the consultation of existing literature, evaluating the physico-chemical characteristics of test chemicals, and predicting potential toxic effects based on SARs. The physico-chemical characteristics of chemicals such as molecular structure, solubility, vapor pressure, pH sensitivity, electrophilicity and chemical reactivity are important properties that may provide critical information for hazard identification and toxicity prediction [24, 34].

Initial *in-vitro* tests should be conducted to identify likely target cells and the toxic potency of test chemicals. Based on the obtained result, *in-vitro* tests may be followed by a second phase using the following cells: nasal olfactory cells, airway epithelial cells, type II cells, alveolar macrophages, vascular endothelial cells, fibroblasts, and mesothelial cells [17]. Moreover, the advantages and limitations of using different cell types and endpoints for toxicity measurement of air contaminants have been discussed [17]. Whilst over ten main cell types have been identified in the epithelium of the respiratory tract, for the assessment of respiratory toxicity it is important to utilize specific cell types with appropriate metabolizing activities. Therefore, freshly isolated cells maintained in suitable culture media that can mimic biotransformation activities and cellular functions comparable to the *in-vivo* environment are preferred to long-term cultures or cell lines that may differentiate, lose their organ-specific functions, and lack the enzyme systems required for biotransformation [17, 28]. It has been suggested that the endpoints used should be selected based on knowledge of the toxic effects of test

Table 4.3 Indirect and direct *in-vitro* exposure techniques developed for the study of the toxicity of air contaminants.

Exposure technique	Exposure achievement procedure
Indirect methods	
Exposure to the test chemical itself	Cells are exposed to test chemicals solubilized or suspended in culture media.
Exposure to collected air samples	Cells are exposed to air samples collected by filtration or impingement methods.
Direct methods	
Submerged exposure condition	Test gas is introduced to cell suspension under submerged conditions using impinger or vacuum test tubes.
Intermittent exposure	Cells are periodically exposed to gaseous compound and culture medium at regular intervals using variation of techniques: rocker platforms, rolling bottles.
Continuous direct exposure at the air–liquid interface	Cells are continuously exposed to airborne contaminants during the exposure time, usually on their apical side, while being nourished from their basolateral side using; collagen-coated or porous membranes permeable to culture media.

chemicals, and should always include cell viability testing in at least two different cell types. A better understanding of mechanisms involved in respiratory toxicity, as well as the development of standardized and reproducible *in-vitro* exposure and delivery systems which simulate inhalation exposure *in vivo*, were encouraged by the ECVAM workshop [17].

To evaluate the potential applications of *in-vitro* methods for studying respiratory toxicity, more recent models developed for the toxicity testing of airborne contaminants have been reviewed [62, 63]. The toxic effects of air contaminants have been studied using several indirect and direct *in-vitro* exposure techniques (Table 4.3).

4.7.1
Indirect Methods

Most of these studies – and especially those conducted on particulates – are limited to the exposure of cells to test chemicals that are either solubilized or suspended in the culture medium [64–71].Although this exposure condition may be adequate for soluble test materials, this may not follow the *in-vivo* exposure pattern of airborne aerosols, particularly for insoluble aerosols, due to unexpected alternation of their compositions and particle–medium or particle–cell interactions [72]. Such techniques of exposure may also ignore size, which is crucial in toxicity testing of inhaled particles.

Some research groups have employed sampling aerosols by filtration techniques, followed by an investigation of the effects of suspended and extracted particles, including those conducted on atmospheric aerosols [73–77] or cigarette smoke condensates [78–81]. The cytotoxicity of roadside airborne particulates was studied in rodent and human lung fibroblasts using the filtration technique [74]. Cytotoxicity was investigated using cell proliferation, (3-(4,5-dimethylthiazol-2-yl)-5-(3-carboxymethoxyphenyl)-2-(4-sulfophenyl)-2H-tetrazolium) (MTS) and lactate dehydrogenase (LDH) *in-vitro* assays. Airborne particulates were sampled on glass-fiber filters using a high-volume sampler, after which the filters were sonicated using benzene-ethanol solvents; a crude extract was then obtained by evaporating the solvents to dryness. This extract was further fractionated by acid-base partitioning, and all extracts were dissolved in dimethyl sulfoxide (DMSO) for cytotoxicity assays [74].

Although filtration offers an advantage for the on-site toxicity assessment of aerosols, the technique usually requires sample preparation (e.g., extraction) in order to isolate the components of interest from a sample matrix. Ultimately, solubilization or suspension in the culture medium will potentially increase experimental errors and further toxicity interactions.

An indirect exposure technique has been developed in which samples of airborne formaldehyde were collected in serum-free culture media using an impingement method [82]. Cytotoxicity was investigated after treating human lung-derived cells with collected air samples. The objective of this study was to develop an *in-vitro* sampling and exposure technique that could be used for toxicity testing of soluble airborne contaminants, but with the potential for on-site applications. An average of 96.8% was calculated for the collection efficiency of airborne formaldehyde in serum-free culture media, which signified the potential application of this method for sampling airborne formaldehyde or other soluble airborne contaminants. The use of serum-free culture media as a collection solution for soluble airborne contaminants proved to be a simple technique, with no specific sample preparation or extraction steps required; moreover, this prevented any potential toxic interactions of the test chemical with other toxic organic solvents during preparation.

4.7.2 Direct Methods

Several direct *in-vitro* models have also been developed to deal with gas-phase exposure of airborne contaminants, using different exposure techniques. The different features of these techniques have been discussed in terms of their relevance, advantages, and limitations [83, 84]. In principle, these methods include the exposure of cells under submerged conditions, intermittent exposure procedures and, more recently, direct exposure techniques at the air–liquid interface.

Exposure by bubbling test gaseous compounds through cells suspended in media can easily be achieved using variations of standard laboratory processes (Fig. 4.2). One such example was an investigation of the *in-vitro* toxic effects of ozone on human hematic mononucleated cells (HHMC), with exposure to ozone being effected by single injection of the gas into cell suspensions in vacuum test

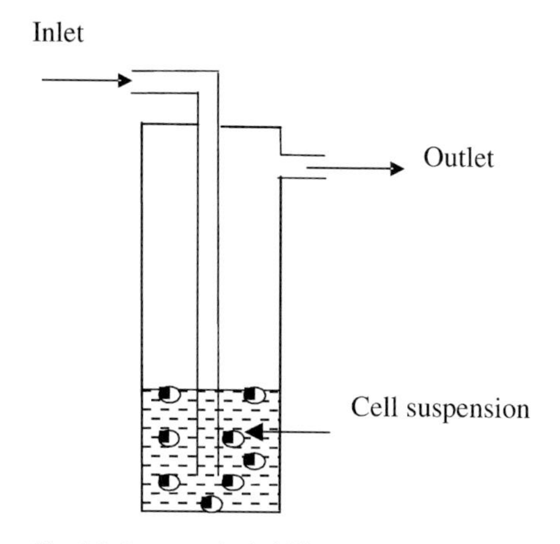

Fig. 4.2 Exposure by bubbling test gaseous compounds through cells suspended in media.

tubes [85]. Exposure patterns *in vivo* may not be closely simulated by this method, however, as only a very small interface between the test gas and the target cells can be provided when using the submerged exposure technique.

Variations of laboratory techniques have been developed that allow the intermittent exposure of cultures to gaseous contaminants. Cell culture dishes held on chambers or platforms rounded, stacked or tilted at certain angles, were exposed periodically to gaseous compounds [86–88]. Cell culture flasks were also tilted at regular intervals to expose the cell cultures to volatile anesthetics [89], while rolling culture bottles on roller drums were set up for *in-vitro* gas exposure [38, 90]. Lung slices were alternatively fed by culture medium and exposed to diesel exhausts by rotating the culture vial on the internal wall of a flow-through chamber [91]. Tissue culture flasks on a rocking platform were used to expose the cells to mainstream cigarette smoke, followed by intermittent immersion in culture media [92].

A micro-roller bottle system has been developed for cytotoxicity screening of volatile compounds in which primary hepatocytes are attached to a collagen-coated nylon mesh (Fig. 4.3). The primary hepatocytes were exposed to volatile compounds injected into the roller bottle, which was placed on a roller apparatus in an incubator at 37 °C, with the hepatocytes being exposed alternately to the medium and the test atmosphere. Medium samples were then taken to measure cellular LDH and aspartate aminotransferase activities [38]. Compared to submerged exposure, the intermittent exposure technique provides a larger interface between the gaseous compound and the target cells. Nevertheless, under such exposure conditions, cells are always covered by an intervening layer of medium that may influence both the accuracy and reproducibility of the results.

During the 1990s, the technology became available that would allow cells to be cultured on permeable porous membranes in commercially available transwell or snapwell inserts (Fig. 4.4). Once the cells have become established on the membrane, the upper layer of culture medium can be removed, and the cells exposed directly to air contaminants. In a direct exposure technique, at the

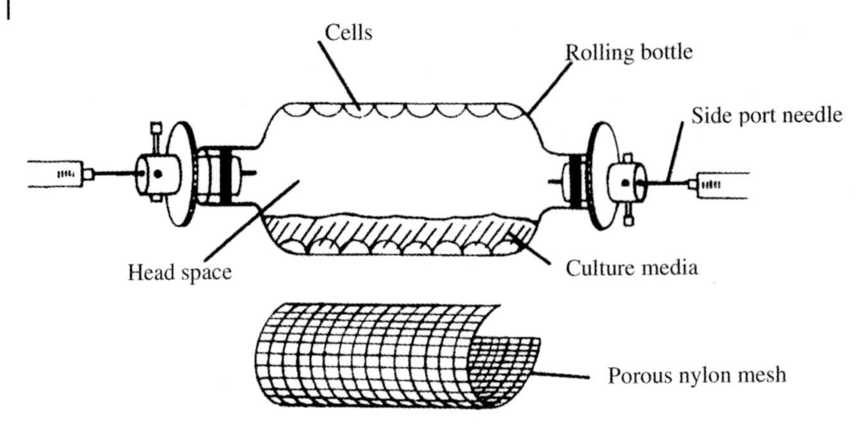

Fig. 4.3 A micro roller bottle system (modified from [39]).

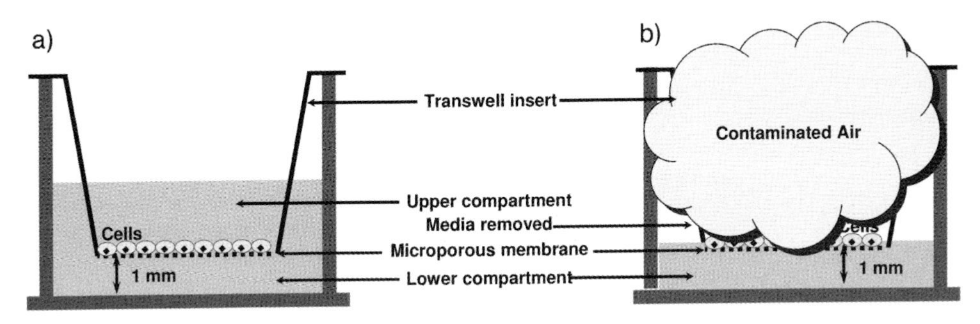

Fig. 4.4 The culture of human cells on porous membranes.
(a) Culture of cells on microporous membranes.
(b) Exposure of cells to airborne contaminants following removal of the media in the upper compartment.

air–liquid interface the target cells can be exposed to airborne contaminants continuously on their apical side, while receiving nourishment via their basolateral side. The direct exposure of cells to airborne contaminants was initially achieved by growing cells on collagen-coated membrane located on special platforms [93], and more recently on porous membranes in transwell inserts [94–96], or snap well inserts [97, 98]. Both static and dynamic direct exposure methods were established for exposure purposes.

Although exposure to volatile chemicals represents a significant contribution to human health problems, the toxicity testing of volatile compounds has always faced major technological problems [43, 47, 48, 99]. Apart from high volatility, many volatile organic compounds (VOCs) are less water-soluble, or insoluble in water, and these physico-chemical properties may lead to technical challenges during the course of *in-vitro* experiments. Static direct exposure methods have been developed for the toxicity assessment of VOCs, in which test atmospheres of selected chemicals were generated in sealed glass chambers with known volumes [97].

Human cells, including A549-pulmonary type II-like cell lines, HepG2-hepatoma cell lines and skin fibroblasts, were exposed to airborne toxicants at different concentrations directly at the air–liquid interface. Post-exposure cytotoxicity was investigated using the tetrazolium salt (MTS) and neutral red uptake (NRU) *in-vitro* assays. Using the static direct exposure method enabled the establishment of airborne IC_{50} (50% inhibitory concentration) values for selected VOCs such as xylene (IC_{50} = 5350–8200 ppm) and toluene (IC_{50} = 10 500–16 600 ppm) after 1 h exposure. Indeed, the static direct exposure method proved to be both practical and reproducible for *in-vitro* inhalation studies with volatile chemicals.

A typical experimental set-up for dynamic direct exposure at the air–liquid interface requires the use of appropriate exposure chambers. Standard tissue culture incubators were used for exposure purposes [100], while dynamic delivery and direct exposure of human cells to airborne contaminants was achieved with specific exposure chambers [101] or horizontal diffusion chamber systems [97]. The toxic effects of single airborne chemicals such as ozone and nitrogen dioxide [97, 102], and complex mixtures such as diesel motor exhaust [94], cigarette smoke [96] and combustion products [103, 104], were studied using cultured human lung cells on porous membranes permeable to culture media. The dynamic direct exposure technique at the air–liquid interface offers a reproducible contact between chemically and physically unmodified airborne contaminants and target cells and, from a technically standpoint, more closely reflects inhalation exposure *in vivo* [83, 95–97].

4.8 Conclusions

Analyses of air pollution provide evidence that occupational and environmental exposure to airborne contaminants is significantly associated with human health risks, ranging from bronchial reactivity to morbidity and mortality due to acute intense or long-term, low-level repeated exposures [1, 2, 4, 8]. Although an extensive background database from toxicological studies has been developed on animal models, most toxicity data have been obtained from oral and dermal exposures rather than inhalation exposure [22, 105]. Although inhalation studies are technologically complicated and require unique equipment and resources [37], recent studies have shown that *in-vitro* methods may also have significant potential for assessing the toxicity of airborne contaminants.

An optimal *in-vitro* exposure system for studying cellular interactions upon chemical airborne exposures must meet several requirements [62, 83, 98, 106]. Such requirements include the direct exposure of target cells to unmodified airborne chemicals without an intervening layer of media, the constant nourishment of cells during the exposure time, the duplication of *in-vivo* parameters without system-related toxicity, an uncomplicated exposure system, and a reasonable price. Studies have confirmed that both static and dynamic direct exposure techniques at the air–liquid interface have the potential for more extensive use in studying the toxic effects of airborne contaminants under more representative physiological conditions [62, 96, 98].

A diversified battery of *in-vitro* test methods measuring different cytotoxic endpoints (see Table 4.2), and a multiple human cell-based system may potentially provide a better understanding of mechanisms involved in the toxicity testing of chemicals. However, in order to develop new insights into mechanisms inducing acute cytotoxicity, the implementation of toxicity endpoints such as proinflammatory cytokine production, reactive oxygen species formation, lipid peroxidation, and mitochondrial function are recommended. In order to extend observations to lower doses, further morphological and biological changes that might occur at the cellular membrane, nucleus, specific proteases and DNA level must be considered which may increase the potential for on-site toxicity evaluations.

Dynamic direct exposure techniques that can be operated independently from the cell culture incubator offer an advantage for on-site toxicity assessments. Considering the multitude of airborne chemicals that usually occur in real environments, dynamic direct exposure techniques at the air–liquid interface can be used for the comprehensive toxicity assessment of airborne contaminants including gases, vapors, solid/liquid aerosols, and complex atmospheres. However, studying the toxic effects of airborne particulates requires sampling inlets with specific characteristics and design for the accurate and homogeneous distribution of test atmospheres close to the target cells. In the future, a range of *in-vitro* bioassays, in conjunction with direct-exposure techniques, may provide an advanced technology for the toxicity testing and biomonitoring of airborne contaminants.

References

1 Klaassen, C. D. (Ed.) (2001). *Casarett and Doull's Toxicology: the Basic Science of Poisons*, 6th edition. McGraw-Hill, New York.

2 Greenberg, M. I., Hamilton, R. J., Phillips, S. D., McCluskey, G. J. (Eds.) (2003). *Occupational, Industrial, and Environmental Toxicology*, 2nd edition. Mosby, Inc., Philadelphia, Pennsylvania.

3 Johnson, D. L., Swift, D. L. (1997). Sampling and sizing particles. In: Dinardi, S. R. (Ed.), *The Occupational Environment – Its evaluation and Control*. American Industrial Hygiene Association, Virginia, pp. 243–262.

4 Winder, C., Stacey, N. H. (Eds.) (2004). *Occupational Toxicology*. 2nd edition. CRC Press, Boca Raton.

5 Raabe, O. G. (1999). Respiratory exposure to air pollutants. In: *Air Pollutants and the Respiratory Tract*. Swift, D. L., Foster, W. M., (Eds.), Marcel Dekker, Inc., New York, pp. 39–73.

6 Lioy, P J., Zhang, J. (1999). Air pollution. In: Swift, D. L., Foster, W. M. (Eds.), *Air Pollutants and the Respiratory Tract*. Marcel Dekker, Inc., New York, Basel, pp. 1–38.

7 Costa, D. L. (2001). Air pollution. In: Klaassen, C. D. (Ed.), *Casarett and Doull's Toxicology: the Basic Science of Poisons*. 6th edition. McGraw-Hill, New York, pp. 979–1012.
8 Chauhan, A. J., Johnston, S. L. (2003). Air pollution and infection in respiratory illness. *Br. Med. Bull.* 68: 95–112.
9 David, A., Wagner, G. R. (1998). Respiratory system. In: Stellman, J. M. (Ed.), *Encyclopaedia of Occupational Health and Safety*, 4th edition. International Labour Office, Geneva, pp. 10.1–10.7.
10 Winder, C. (2004). Toxicology of gases, vapours and particulates. In: Winder, C., Stacey, N. H. (Eds.), *Occupational Toxicology*, 2nd edition. CRC Press, Boca Raton, pp. 399–424.
11 Rozman, K. K., Klaassen, C. D. (2001). Absorption, distribution and excretion of toxicants. In: Klaassen, C. D. (Ed.), *Casarett and Doull's Toxicology: the Basic Science of Poisons*, 6th edition. McGraw-Hill, New York, pp. 105–132.
12 Winder, C. (2004). Occupational respiratory diseases. In: Winder, C., Stacey, N. H. (Eds.), *Occupational Toxicology*, 2nd edition. CRC Press, Boca Raton, pp. 71–114.
13 Beckett, W. S. (1999). Detecting respiratory tract responses to air pollutants. In: Swift, D. L., Foster, W. M. (Eds.), *Air Pollutants and the Respiratory Tract*. Marcel Dekker, New York, pp. 105–118.
14 Boulet, L. P., Bowie, D. (1999). Acute occupational respiratory diseases. In: Mapp, C. E. (Ed.), *Occupational Lung Disorders*. European Respiratory Society, Huddersfield, UK, pp. 320–346.
15 Hext, P. M. (2000). Inhalation toxicology. In: Ballantyne, B., Marrs, T. C., Syversen, T. (Eds.), *General and Applied Toxicology, Volume 1*, 2nd edition. Macmillan, London, pp. 587–601.
16 Witschi, H. P., Last, J. A. (2001). Toxic responses of the respiratory system. In: Klaassen, C. D. (Ed.), *Casarett and Doull's Toxicology: the Basic Science of Poisons*, 6th edition. McGraw-Hill, New York, pp. 515–534.
17 Lambre, C. R., Auftherheide, M., Bolton, R. E., Fubini, B., Haagsman, H. P., Hext, P. M., Jorissen, M., Landry, Y., Morin, J. P., Nemery, B., Nettesheim, P., Pauluhn, J., Richards, R. J., Vickers, A. E. M., Wu, R. (1996). *In vitro* tests for respiratory toxicity: The report and recommendations of ECVAM Workshop 18. *Altern. Lab. Anim.* 24: 671–681.
18 Brouder, J., Tardif, R. (1998). Absorption. In: Wexler, P. (Ed.), *Encyclopedia of Toxicology*, Vol. 1. Academic Press, San Diego, pp. 1–7.
19 Silbergeld, E. K. (1998). Toxicology. In: Stellman, J. M. (Ed.), *Encyclopaedia of Occupational Health and Safety*, 4th edition. International Labour Office, Geneva, pp. 33.1–33.74.
20 Gallo, M. A. (2001). History and scope of toxicology. In: Klaassen, C. D. (Ed.), *Casarett and Doull's Toxicology: the Basic Science of Poisons*, 6th edition. McGraw-Hill, New York, pp. 3–10.
21 National Toxicology Program (1984). *Toxicology Testing Strategies to Determine Needs and Priorities*. National Toxicology Program, National Research Council, Washington.

22 Agrawal, M. R., Winder, C. (1996). The frequency and occurrence of LD_{50} values for materials in the workplace. *J. Appl. Toxicol.* 16: 407–422.

23 EPA (1998). *Chemical Hazard Availability Study.* US Environmental Protection Agency, Office of Pollution Prevention and Toxics, Washington, DC.

24 Faustman, E. M., Omenn, G. S. (2001). Risk Assessment. In: Klaassen, C. D. (Ed.), *Casarett and Doull's Toxicology: the Basic Science of Poisons,* 6th edition. McGraw-Hill, New York, pp. 83–104.

25 McClellan, R. O. (1999). Health risk assessments and regulatory considerations for air pollutants. In: Swift, D. L., Foster, W. M. (Eds.), *Air Pollutants and the Respiratory Tract.* Marcel Dekker, Inc., New York, pp. 289–338.

26 Thorne, P. S. (2001). Occupational toxicology. In: Klaassen, C. D. (Ed.), *Casarett and Doull's Toxicology: the Basic Science of Poisons,* 6th edition. McGraw-Hill, New York, pp. 1123–1140.

27 Greenberg, M. I., Phillips, S. D. (2003). A brief history of occupational, industrial and environmental toxicology. In: Greenberg, M. I., Hamilton, R. J. and Phillips, S. D. (Eds.), *Occupational, Industrial and Environmental Toxicology,* 2nd edition. Mosby, Philadelphia, pp. 2–5.

28 Barile, F. A. (1994). *Introduction to* in vitro *Cytotoxicity, Mechanisms and Methods.* CRC Press, Boca Raton.

29 Trevan, J. W. (1927). Error of determination of toxicity. *Proc. Royal Soc. London,* Ser. B 101: 483–514.

30 OECD. (2001). *Guidance Document on Acute Oral Toxicity Testing.* OECD Environment, Health and Safety Publications Series on Testing and Assessment, No. 24. Organisation for Economic and Cooperative Development, Paris.

31 OECD. (2004). Chemicals Testing: OECD Guidelines for the Testing of Chemicals. Section 4: Health Effects. Organisation for Economic and Cooperative Development, Paris.

32 O'Hare, S., Atterwill, C. K. (1995). *Methods in Molecular Biology: In vitro Toxicity Testing Protocols.* Humana Press, New Jersey.

33 Blaauboer, B. J. (2002). The applicability of *in vitro*-derived data in hazard identification and characterisation of chemicals. *Environ. Toxicol. Pharmacol.* 11: 213–225.

34 Gad, S. C. (2000). *In Vitro Toxicology.* Taylor and Francis, New York.

35 Hayes, A. J., Markovic, B. (1999). Alternative to animal testing for determining the safety of cosmetics. *Cosmetics, Aerosols Toiletries Austr.* 12: 24–30.

36 Purchase, I. C. H., Botham, P. A., Bruner, L. H., Flint, O. P., Frazier, J. M., Stokes, W. S. (1998). Workshop overview: Scientific and regulatory challenges for the reduction, refinement and replacement of animals in toxicity testing. *Toxicol. Sci.* 43: 86–101.

37 NTP (2001). *The National Toxicology Program Annual Plan Fiscal Year 2001.* US Department of Health and Human Services, Public Health Service, NIH publication No. 02-5092, Washington.

38 Del Raso, N. J. (1992). *In vitro* methods for assessing chemical or drug toxicity and metabolism in primary hepatocytes. In: Watson, R. R. (Ed.), *In Vitro Methods of Toxicology*. CRC Press, Boca Raton.

39 Frazier, J. M. (1992). In Vitro *Toxicity Testing Applications to Safety Evaluation*. Marcel Dekker, Inc., New York.

40 Ames, B. N., Durston, W. E., Yamasaki, E., Lee, F. D. (1973). Carcinogens are mutagens. Simple test system combining liver homogenates for activation and bacteria for detection. *Proc. Natl. Acad. Sci. USA* 70: 2281–2285.

41 Ames, B. N., McCann, J., Yamasaki, E. (1975). Methods for detecting carcinogens and mutagens with the *Salmonella*/mammalian-microsome mutagenicity test. *Mutat. Res.* 31: 347–363.

42 Zucco, F., De Angelis, I., Stammati, A. (1998). Cellular models for *in vitro* toxicity testing. In: Clynes, M. (Ed.), *Animal Cell Culture Techniques, Springer Laboratory Manual*. Springer-Verlag, Berlin, pp. 395–422.

43 Scheers, E. M., Ekwall, B., Dierickx, P. J. (2001). *In vitro* long-term cytotoxicity testing of 27 MEIC chemicals on Hep G2 cells and comparison with acute human toxicity data. *Toxicol. In Vitro* 15: 153–161.

44 Eisenbrand, G., Pool-Zobel, B. P., Baker, V., Balls, M., Blaauboer, B. J., Boobis, A., Carere, A., Kevekordes, S., Lhuguenot, J. C., Pieters, R., Kleiner, J. (2002). Methods of *in vitro* toxicology. *Food Chem. Toxicol.* 40: 193–236.

45 Lambre, C. R., Auftherheide, M., Bolton, R. E., Fubini, B., Haagsman, H. P., Hext, P. M., Jorissen, M., Landry, Y., Morin, J. P., Nemery, B., Nettesheim, P., Pauluhn, J., Richards, R. J., Vickers, A. E. M., Wu, R. (1996). *In vitro* tests for respiratory toxicity: The report and recommendations of ECVAM Workshop 18. *Altern. Lab. Anim.* 24: 671–681.

46 Stark, D. M., Shopsis, C., Borenfreund, E., Babich, H. (1986). Progress and problems in evaluating and validating alternative assays in toxicology. *Food Chem. Toxicol.* 24: 449–455.

47 Frazier, J. M., Bradlaw, J. A. (1989). Technical problems associated with *in vitro* toxicity testing systems. A report of the CAAT technical workshop of May 17–18, 1989. The Johns Hopkins Center for Alternatives to Animal Testing, Baltimore, MD.

48 Ciapetti, G., Granchi, D., Verri, E., Savarino, L., Stea, S., Savioli, F., Gori, A., Pizzoferrato, A. (1998). False positive in cytotoxicity testing due to unexpectedly volatile compounds. *Biomed. Mater. Res.* 39: 286–291.

49 Balls, M., Fentem, J. H. (1992). The use of basal cytotoxicity and target organ toxicity tests in hazard identification and risk assessment. *Altern. Lab. Anim.* 20: 368–388.

50 Anderson, D., Russell, T. (1995). *The Status of Alternative Methods in Toxicology*. The Royal Society of Chemistry, Cambridge, UK.

51 Holme, J. A., Dybing, E. (2002). The use of *in vitro* methods for hazard characterisation of chemicals. *Toxicol. Lett.* 127 (1–3): 135–141.

52 Zucco, F., De Angelis, I., Testai, E., Stammati, A. (2004). Toxicology investigations with cell culture systems: 20 years after. *Toxicol. In Vitro* 18: 153–163.

53 Blaauboer, B. J. (2002). The applicability of *in vitro*-derived data in hazard identification and characterisation of chemicals. *Environ. Toxicol. Pharmacol.* 11: 213–225.

54 ICCVAM (2001). *Report of the International Workshop on in Vitro Methods for Assessing Acute Systemic Toxicity.* Interagency Coordinating Committee on the Validation of Alternative Methods, NIH Publication No. 01-4499.

55 Marchant, G. E. (2005). Genetics and the future of environmental policy. In: Olson, R., Rejeski, D. (Eds.), *Environmentalism and the Technologies of Tomorrow, Shaping the Next Industrial Revolution.* Island Press, Washington, pp. 61–70.

56 Balls, M., Clothier, R. (1992). Cytotoxicity assays for intrinsic toxicity and irritancy. In: Watson, R. R. (Ed.), *In Vitro Methods of Toxicology.* CRC Press, Boca Raton, pp. 37–52.

57 Doyle, A., Griffiths, J. B. (Eds.) (2000). *Cell and Tissue Culture for Medical Research.* John Wiley, West Sussex, England.

58 Freshney, I. R. (2000). *Culture of Animal Cells – A Manual of Basic Techniques,* 4th edition. John Wiley & Sons, Inc., New York.

59 Wilson, A. P. (2000). Cytotoxicity and viability assays. In: Masters, J. R. W. (Ed.), *Animal Cell Culture,* 3rd edition. Oxford University Press, New York, pp. 175–219.

60 Phalen, R. F. (Ed.) (1997). *Methods in Inhalation Toxicology.* CRC Press, Boca Raton.

61 Valentine, R., Kennedy, J. G. L. (2001). Inhalation toxicology. In: Wallace Hayes, A. (Ed.), *Principles and Methods of Toxicology,* 4th edition. Taylor and Francis, Philadelphia, USA, pp. 1085–1143.

62 Bakand, S., Winder, C., Khalil, C., Hayes, A. (2005). Toxicity assessment of industrial chemicals and airborne contaminants: Transition from *in vivo* to *in vitro* test methods: A review. *Inhalation Toxicol.* 17: 775–787.

63 Allen, C. B. (2006). *In vitro* models for lung toxicology. In: Gardner, D. E. (Ed.), *Toxicology of the Lung,* 4th edition. Taylor and Francis, Boca Raton, pp. 107–150.

64 Nadeau, D., Vincent, R., Kumarathasan, P., Brook, J., Dufresne, A. (1995). Cytotoxicity of ambient air particles to rat lung macrophages: comparison of cellular and functional assays. *Toxicol. In Vitro* 10: 161–172.

65 Governa, M., Valentino, M., Amati, M., Visona, I., Botta, G. C., Marcer, G., Gemignani, C. (1997). Biological effects of contaminated silicon carbide particles from a work station in a plant producing abrasives. *Toxicol. In Vitro* 11: 201–207.

66 Goegan, P., Vincent, R., Kumarathasan, P., Brook, J. (1998). Sequential *in vitro* effects of airborne particles in lung macrophages and reporter Cat-gene cell lines. *Toxicol. In Vitro* 12: 25–37.

67 Baeza-Squiban, A. B., Bonvallot, V., Boland, S., Marano, F. (1999). Diesel exhaust particles increase NF-KB DNA binding activity and C-Fos proto-oncogene expression in human bronchial epithelial cells. *Toxicol. In Vitro* 13: 817–822.

68 Becker, S., Soukup, J. M., Gallagher, J. E. (2002). Differential particulate air pollution induced oxidant stress in human granulocytes, monocytes and alveolar macrophages. *Toxicol. In Vitro* 16: 209–218.

69 Takano, Y., Taguchi, T., Suzuki, I., Balis, J. U., Kazunari, Y. (2002). Cytotoxicity of heavy metals on primary cultured alveolar type II cells. *Environ. Res.* 89: 138–145.

70 Riley, M. R., Boesewetter, D. E., Kim, A. M., Sirvent, F. P. (2003). Effects of metals Cu, Fe, Ni, V, and Zn on rat lung epithelial cells. *Toxicology* 190: 171–184.

71 Okeson, C. D., Riley, M. R., Riley-Saxton, E. (2004). *In vitro* alveolar cytotoxicity of soluble components of airborne particulate matter: effects of serum on toxicity of transition metals. *Toxicol. In Vitro* 18: 673–680.

72 Diabate, S., Mulhopt, S., Paur, H. R., Krug, H. F. (2002). Pro-inflammatory effects in lung cells after exposure to fly ash aerosol via the atmosphere or the liquid phase. *Ann. Occup. Hygiene* 46 (suppl 1): 382–385.

73 Hamers, T., Van Schaardenburg, M. D., Felzel, E. C., Murk, A. J., Koeman, J. H. (2000). The application of reporter gene assay for the determination of the toxic potency of diffuse air pollution. *Sci. Total Environ.* 262: 159–174.

74 Yamaguchi, T., Yamazaki, H. (2001). Cytotoxicity of airborne particulates sampled roadside in rodent and human lung fibroblasts. *J. Health Sci.* 47: 272–277.

75 Alfaro Moreno, E., Martinez, L., Garcia Cuellar, C., Bonner, J. C., Murray, J. C., Rosas, I., de Leon Rosales, S. P., Osornio Vargas, A. R. (2002). Biologic effects induced *in vitro* by PM_{10} from three different zones of Mexico City. *Environ. Health Perspect.* 110: 715–720.

76 Glowala, M., Mazurek, A., Piddubnyak, V., Fiszer-Kierzkowska, A., Michalska, J., Krawczyk, Z. (2002). HSP70 overexpression increases resistance of V79 cells to cytotoxicity of airborne pollutants, but does not protect the mitotic spindle against damage caused by airborne toxins. *Toxicology* 170: 211–219.

77 Baulig, A., Sourdeval, M., Meyer, M., Marano, F., Baeza-Squiban, A. (2003). Biological effects of atmospheric particles on human bronchial epithelial cells. Comparison with diesel exhaust particles. *Toxicol. In Vitro* 17: 567–573.

78 Bombick, D. W., Putnam, K., Doolittle, D. J. (1998). Comparative cytotoxicity studies of smoke condensates from different types of cigarettes and tobaccos. *Toxicol. In Vitro* 12: 241–249.

79 Bombick, B. R., Murli, H., Avalos, J. T., Bombick, D. W., Morgan, W. T., Putnam, K. P., Doolittle, D. J. (1998). Chemical and biological studies of a new Cigarette that primarily heats tobacco; Part 2: *In vitro* toxicology of mainstream smoke condensate. *Food Chem. Toxicol.* 36: 183–190.

80 McKarns, S. C., Bombic, D. W., Morton, M. J., Doolittle, D. J. (2000). Gap junction intercellular communication and cytotoxicity in normal human cells after exposure to smoke condensates from cigarettes that burn or primarily heat tobacco. *Toxicol. In Vitro* 14: 41–51.

81 Putnam, K. P., Bombic, D. W., Doolittle, D. J. (2002). Evaluation of eight *in vitro* assays for assessing the cytotoxicity of cigarette smoke condensate. *Toxicol. In Vitro* 16: 599–607.

82 Bakand, S., Hayes, A., Winder, C., Khalil, C., Markovic, B. (2005). *In vitro* cytotoxicity testing of airborne formaldehyde collected in serum-free culture media. *Toxicol. Indust. Health* 21: 147–154.

83 Ritter, D., Knebel, J. W., Aufderheide, M. (2001). *In vitro* exposure of isolated cells to native gaseous compounds. Development and validation of an optimized system for human lung cells. *Exp. Toxicol. Pathol.* 53: 373–386.

84 Aufderheide, M. (2005). Direct exposure methods for testing native atmospheres. *Exp. Toxicol. Pathol.* 57 (Suppl. 1): 213–226.

85 Cardile, V., Russo, J. A., Casella, F., Renis, M., Bindoni, M. (1995). Effects of ozone on some biological activities of cells *in vitro. Cell Biol. Toxicol.* 11: 11–21.

86 van der Zee, J., Dubbelman, T. M. A. R., Raap, T. K., Van Steveninck, J. (1987). Toxic effects of ozone on murine L929 fibroblasts. Enzyme inactivation and glutathione depletion. *Biochem. J.* 242: 707–712.

87 Blanquart, C., Giuliani, I., Houcine, O., Jeulin, C., Guennou, C., Marano, F. (1995). *In vitro* exposure of rabbit trachelium to SO_2: effects on morphology and ciliarity beating. *Toxicol. In Vitro* 9: 123–132.

88 Rusznak, C., Devalia, J. L., Sapsford, R. J., Davies, R. J. (1996). Ozone-induced mediator release from human bronchial epithelial cells *in vitro* and the influence of nedocromil sodium. *Eur. Respir. J.* 9: 2298–2305.

89 Muckter, H., Zwing, M., Bader, S., Marx, T., Doklea, E., Liebl, B., Fichtl, B., Georgieff, M. (1998). A novel apparatus for the exposure of cultured cells to volatile agents. *J. Pharmacol. Toxicol. Methods,* 40: 63–69.

90 Shiraishi, F., Hashimoto, S., Bandow, H. (1986). Induction of sister-chromatid exchanges in Chinese hamster V79 cells by exposure to the photochemical reaction products of toluene plus NO_2 in the gas phase. *Mutat. Res.* 173: 135–139.

91 Morin, J. P., Fouquet, F., Monteil, C., Leprieur, E., Vaz, E., Dionnet, F. (1999). Development of a new *in vitro* system for continuous *in vitro* exposure of lung tissue to complex atmospheres: application to diesel exhaust toxicology. *Cell Biol. Toxicol.* 15: 143–152.

92 Bombick, D. W., Ayres, P. H., Putnam, K., Bombick, B. R., Doolittle, D. J. (1998). Chemical and biological studies of a new cigarette that primarily heats tobacco; Part 3: *In vitro* toxicity of whole smoke. *Food Chem. Toxicol.* 36: 191–197.

93 Chen, L. C., Fang, C. P., Qu, Q. S., Fine, J. M., Schlesinger, R. B. (1993). A novel system for the *in vitro* exposure of pulmonary cells to acid sulfate aerosols. *Fundam. Appl. Toxicol.* 20: 170–176.

94 Knebel, J. W., Ritter, D., Aufderheide, M. (2002). Exposure of human lung cells to native diesel motor exhaust-development of an optimized *in vitro* test strategy. *Toxicol. In Vitro* 16: 185–192.

95 Diabate, S., Mulhopt, S., Paur, H. R., Krug, H. F. (2002). Pro-inflammatory effects in lung cells after exposure to fly ash aerosol via the atmosphere or the liquid phase. *Ann. Occup. Hygiene* 46 (Suppl. 1): 382–385.

96 Aufderheide, M., Knebel, J. W., Ritter, D. (2003). Novel approaches for studying pulmonary toxicity *in vitro*. *Toxicol. Lett.* 140–141: 205–211.

97 Bakand, S., Winder, C., Khalil, C., Hayes, A. (2005). A novel *in vitro* exposure technique for toxicity assessment of volatile organic compounds. *J. Environ. Monitor.* 8: 100–105.

98 Bakand, S., Winder, C., Khalil, C., Hayes, A. (2006). An experimental *in vitro* model for dynamic direct exposures of human cells to airborne contaminants. *Toxicol. Lett.* 165: 1–10.

99 Dierickx, P. J. (2003). Evidence for delayed cytotoxicity effects following exposure of rat hepatoma-derived Fa32 cells: Implications for predicting human acute toxicity. *Toxicol. In Vitro* 17: 797–801.

100 Lang, D. S., Jorres, R. A., Mucke, M., Siegfried, W., Magnussen, H. (1998). Interactions between human bronchoepithelial cells and lung fibroblasts after ozone exposure *in vitro*. *Toxicol. Lett.* 96–97: 13–24.

101 Aufderheide, M., Mohr, U. (2000). CULTEX – an alternative technique for cultivation and exposure of cells of the respiratory tract to the airborne pollutants at the air/liquid interface. *Exp. Toxicol. Pathol.* 52: 265–270.

102 Knebel, J. W., Ritter, D., Aufderheide, M. (1998). Development of an *in vitro* system for studying effects of native and photochemically transformed gaseous compounds using an air/liquid culture technique. *Toxicol. Lett.* 96–97: 1–11.

103 Lestari, F., Markovic, B., Green, A. R., Chattopadhyay, G., Hayes, A. J. (2006). Comparative assessment of three *in vitro* exposure methods for combustion toxicity. *J. Appl. Toxicol.* 26: 99–114.

104 Lestari, F., Green, A. R., Chattopadhyay, G., Hayes, A. J. (2006). An alternative method for fire smoke toxicity assessment using human lung cells. *Fire Safety J.* (in press).

105 Miller, G. C., Klonne, D. R. (1997). Occupational Exposure Limits. In: Dinardi, S. R. (Ed.), *The Occupational Environment – Its Evaluation and Control*. American Industrial Hygienists Association, Virginia, pp. 21–42.

106 Tu, B., Wallin, A., Moldeus, P., Cotgreave, I. A. (1995). Cytotoxicity of NO_2 gas to cultured human and murine cells in an inverted monolayer exposure system. *Toxicology* 96: 7–18.

Part II
Primary Tissues and Cell Lines in Drug Screening/Testing

Drug Testing In Vitro: Breakthroughs and Trends in Cell Culture Technology
Edited by Uwe Marx and Volker Sandig

ISBN: 978-3-527-31488-1

5
Drug Screening Using Cell Lines: Cell Supply, High-Throughput and High-Content Assays

Christa Burger, Oliver Pöschke, and Mirek R. Jurzak

5.1
Introduction

Progress made in the areas of molecular biology, combinatorial chemistry and laboratory automation during the past few decades has transformed the process of drug discovery. Whilst the sequencing of the human genome has fuelled our imagination on the availability of novel drug targets, combinatorial and parallel chemistry technologies have rendered large libraries available to drug companies which can be screened efficiently only by using sophisticated and reliable laboratory automation. A plethora of detection technologies has been developed which allows for the rapid screening of compounds on both small and economical scales. Hence, during the 1990s, as dedicated units became established within pharmaceutical companies, the discipline of high-throughput screening (HTS) was born.

The processes and technological developments in the area of cellular screening form the focus of this chapter. Cellular assays provide some clear technical advantages, for example when the isolation of a biochemical target is difficult or cannot be up-scaled, as is often found with membrane proteins. In addition, there are conceptual advantages when certain targets, for example nuclear receptors, undergo complex intracellular interactions which cannot be adequately reproduced in a simple biochemical set-up. It must be emphasized however that, as has been discussed widely in earlier debates in the field, biochemical and cellular screening approaches still have their intrinsic "pros" and "cons" [1, 2]. Although the natural environment of the cell will present the biochemical targets under biologically relevant conditions, the cellular screening of intracellular targets bears the disadvantage that penetration of the cellular membrane might distort the chemical structure–activity relationship and thus render the series analysis of hits and chemical lead optimization difficult. On the other hand, screening directly for compounds that already are known to have relevant cellular activity might simplify the hit selection process and shorten lead optimization cycles.

Drug Testing In Vitro: Breakthroughs and Trends in Cell Culture Technology
Edited by Uwe Marx and Volker Sandig

ISBN: 978-3-527-31488-1

5.2
Cell Lines for HTS

5.2.1
Selection of the Most Suitable Cell Line

A selected cell line for HTS (Fig. 5.1) must express the desired function in the correct signal context with a stable and reproducible signal window.

The different cell types from which to choose have different advantages: primary cells best resemble the physiological situation [3], but cultivation and available cell numbers are critical issues, even when using techniques such as conditional immortalization [4] or stem cell differentiation.

A large panel of easy-to-cultivate immortalized cell lines is available with cell history, growth behavior and functional information [5–8]. These cells are ideal for functional screens as proliferation assays.

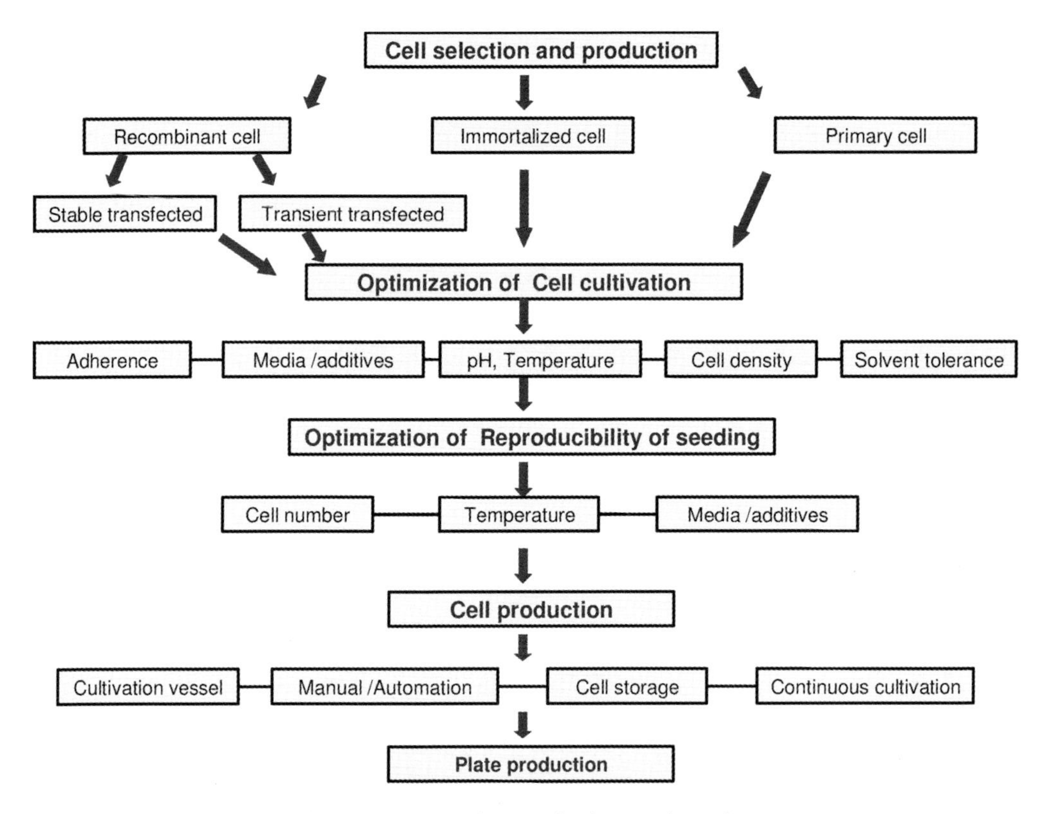

Fig. 5.1 From cell selection to plate production. The diagram shows the different work packages with the most important parameters for analysis and optimization.

Table 5.1 The generation of recombinant cell lines.

Stable cell lines	Reproducible material with selected expression level	
	Frequently used plasmid delivery methods:	• Nonviral: electroporation (mainly suspension cells) • Polyethylenimine (PEI), calcium phosphate, liposomal reagents • Retroviruses (e.g., lentiviruses)
	Plasmid integration methods:	• Random gene expression ± amplification; high producer • Targeted expression as Flip-in system; fast cell generation • Random gene activation; work-intensive but no intellectual property
Transient bulk transfection	Reproducible if portions of one batch are produced	
	Toxic proteins can be expressed, no cell clone selection	
	Timeframe can be shorter as generation of stable clone	
	Delivery methods:	• Developed method for few cells as 293 EBNA with inexpensive reagents as PEI and calcium phosphate • Infection with adenoviruses and Semliki Forest virus
Transient transfection of every well	Reproducibility often a problem without automation	
	On-time delivery for HTS is not easy without automation	

Recombinant cell lines overexpressing the target are often used for screening, because cell lines with an optimal growth behavior and signal window can be selected (Table 5.1). In order to obtain reproducible cell material for HTS, the favorite procedure is to generate a stable target-expressing cell line. This allows adjustment of the expression level by using different promoter elements and/or amplification with the methotrexate or glutamine synthase [9, 10] system combined with clone selection. Several transfection methods are available [11, 12] as well as infection with retroviral vectors [13, 14].

Targeting expression systems such as the Flip-in system (Invitrogen) [15] can help to shorten cell line development time to between one and three months, and if the target is restricted by intellectual property or the generation of a recombinant cell is complex, then the random gene activation strategy can be considered [16]. Before HTS can be carried out, a master cell bank must be prepared and the stability of the functional target expression should be tested for a period of more than four weeks.

In addition to stable cell lines, identical cell material can be produced by transient transfection of one large cell batch, followed by freezing the transfected cells in portions. Directly before HTS, the frozen cells are seeded into assay plates. For an optimal large-scale transient transfection, special standards must be fulfilled for the cell line, the expression plasmid, the medium, and the transfection reagent

[12, 17, 18]. By using inexpensive reagents such as PEI [19] or calcium phosphate [20, 21], cells could be reproducible transfected in suspension culture [21–23], roller culture or cell factories (Nunc). An alternative is to infect the cells with Semliki Forest virus (SFV) vectors [24] or adenoviruses (QBiogen) with a broad host range.

The transient transfection of every single well of an assay plate requires a long development time, is work-intensive, and tends to be less reproducible, though it can be improved by using automated cell cultivation and plating systems [25].

5.2.2 Optimizing Cell Cultivation

In general, in order to be cost-effective, an HTS screen must provide a robust and reproducible signal and an adequate throughput of about 20 to 100 000 samples per day. For cell-based assays, both the cell line itself and the cell production process must be selected to optimize stability and the signal window. The ideal cell for HTS is fast growing, has good adherence, a stable cellular phenotype, a broad tolerance against solvents, and is insensitive against culture modifications. All of these parameters must to be tested for each cell line and each assay, using HTS equipment and conditions early during cell development (see Fig. 5.1).

5.2.2.1 Adherence

Most HTS are plate-based assays involving one washing or separation step, as well as several pipetting steps. Good adherence of the cell line used is necessary, as the uncontrolled loss of cells is a frequent reason for low assay reproducibility. The degree of negative charges on the cell at physiological pH seems to be one important factor for good adherence. In addition, the presence of bivalent cations (Mg^{2+}, Ca^{2+}) and the presence of extracellular proteins determine the adherence of cells [26]. Examples of good adherent cell lines in standard culture plasticware are Chinese hamster ovary (CHO), baby hamster kidney (BHK), COS and NIH-3T3 cells. Poor adherence can be improved by coating the wells with charged substrates such as poly-L- or -D-lysine, or with components of the cellular matrix. Examples include fibronectin or collagen for fibroblasts, and laminin for epithelial cells; alternatively, a special plastic such as CellBind (Corning) can be used. In our experience, coating is necessary for cells such as HEK293.

5.2.2.2 pH and Temperature

Most cells grow best at a pH between 6.8 to 7.6, and the pH of most commercial media is 7.2 to 7.4. Media based on Earl's salt composition are buffered by bicarbonate/carbonic acid, and the pH is regulated by constant CO_2 delivery in an incubator [11, 27]. After 30 min at atmospheric pressure, the pH rises by about 0.1 to 0.2 pH units; after 4 h, the pH is raised by about 0.3 to 0.5 pH units. Media with Hank's salt composition are buffered by phosphoric acid and should be used at normal atmosphere. The addition of 10–20 mM HEPES to the media with a good buffering capacity between pH 7.2 and 7.4 helps to keep the pH constant.

Depending upon the available instruments, screen conditions minimizing pH variations must be selected.

The ideal temperature for mammalian cells is between 36 and 37.5 °C. Short periods at room temperature are usually not critical, but the functional activity of channels [28] and the metabolic activity of the cell can be changed under these conditions.

5.2.2.3 Media and Additives

The medium is not only a food source but also represents the environment in which the cells live. Medium selection for the screen is dependent on the cell type and on the assay. The medium composition must support the desired cell function needed, and provide an optimal energy source as well as all important nutrients such as vitamins, growth factors, minerals, and essential lipids that the cells need [29, 30]. A media for cell production can differ from one used for plate generation, as the latter often contains antibiotics and HEPES. The optimal amount of fetal calf serum (FCS) and ingredients such as glucose and glutamine can be different for the production phase compared to the plating medium. For fluorescence read-outs, as well as some reporter assays, the presence of phenol red in the medium enhances the background. All medium components such as FCS, growth factors, hormones, vitamins, ions, and nucleotides should be evaluated for possible interference with the assay signal. FCS contains different amounts of stress-protecting factors such as vitamin E and ATP, as well as calcium, which modulates cell function and influences cell adhesion, proliferation [31], and differentiation [32].

Charges of media and additives must be checked for functionality for each assay, and should not be changed during HTS and development.

5.2.2.4 Solvent Tolerance

Most compound libraries are dissolved in dimethyl sulfoxide (DMSO), the tolerance of which is dependent on the cell line and the duration of exposure. The solubility of a compound can also determine its useable concentration in the screen. Often-used cell lines such as CHO, BHK, NIH-3T3 and HEK293 show no growth inhibition and functional difference in the presence of up to 1% DMSO or ethanol [25, 28], but primary cells are more sensitive in general.

5.2.2.5 Cell Density

The optimal cell density must be determined. For transport and receptor function assays, a good confluent cell layer enhancing the target density is often advantageous, and cell numbers in the range of 8000 to 12 000 per 384 wells should be considered. The influence of cell number and variations on the signal must also be monitored. Functions such as sensitivity to drugs, transcriptional activity, permeability, cells in mitosis or the differentiation state often change with cell density. For these assays, semi-confluent cultures are often preferable.

Surface proteins (e.g., receptors) can be sensitive to proteases such as trypsin. Receptor function must be carefully checked if cells are to be used at different time-points after seeding. It is recommended that cells are harvested using nonenzymatic solutions such as cell dissociation solution (Sigma) or HyQutase (Perbio).

For image analysis, and for localization assays, a good separation of the single cells is needed, and aggregates must be avoided. Larger cells (e.g., U2-OS cells) with a clearly separated, round nucleus and minimal morphological variation are ideal.

5.2.3 Optimizing the Reproducibility of Seeding

To identify an active compound from a large library of compounds in a single test, the assay must provide a robust and reproducible signal. For a cellular HTS, the cell material over the complete run must be robust for passage-to-passage, operator-to-operator, plate-to-plate and for well-to-well parameters. The Z′-value [33] over the assay should be > 0.5.

5.2.3.1 Signal Shift

A signal shift during the day of the screen is one frequently faced problem. In screens with long plate-handling or incubation periods, plates prepared during 1 to 4 h are used for more than 24 h. Cell plates used later after seeding can have higher cell numbers or different medium compositions compared to those plates used earlier. In addition to growth arrest and hypothermia, the seeding of plates for one day with two to four different cell concentrations can help to reduce the shift effect. An automated plate seeding procedure reflecting the time frame of the assay procedure is ideal.

5.2.3.2 Edge Effect

Edge effects are another frequent phenomena in cellular screens, and involve a number of different factors. Thermal gradients across the plates have been described as one main reason [34], as this leads to irregular patterns of cell adhesion and distribution in the well. Coating of the plates to improve cell adhesion, adding a pH stabilizer such as HEPES, and not using the peripheral wells can reduce the problem. Adjusting the media volume may also help, with a depth of not more than 2–5 mm being recommended for good gaseous diffusion [10, 35]. Incubation of the newly seeded plates for 1–2 h at room temperature minimizes the thermal gradient during cell settlement and results in more even cell distribution and adhesion, as shown for CHO cells [34] and (by the present authors) for BHK cells.

5.2.4 Cell Production and Plate Delivery

5.2.4.1 The Amount of Cells Needed

In order to run a total of five hundred 384-well plates each week, 1.9×10^7 cells are needed when using 100 cells per well, but 1.9×10^9 will be needed when using 10 000 cells per well. For most assays, between 1000 and 10 000 cells per well are needed for reliable results, as using cell numbers below 100 can lead to statistical problems. Typically, $1–4 \times 10^7$ cells are harvested from one T175 flask which, for an average cellular HTS, means handling 48 to 95 T175 flasks, or 16 to 48 triple flasks or roller bottles, each week.

With a "multidrop machine" (Labsystems) or an "Aliquot" (Genetix), approximately one hundred 384-well plates can be prepared in 4–6 h under a sterile bench semi-manually.

5.2.4.2 **Cell Storage**

Unlike "ordinary" assay reagents (e.g., enzymes), living cells are prone to change during their time in culture, in response to environmental changes, and they require between 4 and 24 h after plating to adhere properly. Differences in cell density, metabolism and function one has to take into a consideration if using cells as reagents. Cells often function optimally at defined time points after plating or confluence [26].

For the optimal use of investment and throughput, an HTS should be performed for five to seven days each week, and consequently cell plates with functional identical cells must be available during the entire week. Preparing cell plates every day is a relatively safe solution to this problem, but it requires expensive investment in automation systems. Equipment for automated cell cultivation and plate preparation is available from The Automation Partnership (Cellmate, TSelect) and RTS Life Science International (acCellerator).

For several assays it is possible to adjust cell numbers and media conditions in order to provide functionally comparable cells for one to four days after seeding. Using this concept, cells could be delivered for HTS throughout the week, by preparing cell plates on two or three days. For other assays, however, an identical signal could not be produced at different times after plate seeding. In particular, the signals of reporter assays are known to be sensitive to changes in cell number [36].

The use of growth-arrested cells for screening provides another possibility of enhancing flexibility and reproducibility. Kunapuli et al. used CHO cells which had been frozen after mitomycin C-induced division arrest in a calcium flux assay, and reported a stable signal-to-noise ratio for four days after plating [37]. Similarly, by using mitomycin C treatment, Fursov et al. improved the robustness of a reporter assay to a Z′-value of 0.79, compared to 0.35 for normal cultured cells [36].

Differentiated cells reflect naturally growth-arrested cells by keeping their functional phenotype stable for several days in culture. Differentiated cells are more difficult to prepare, but this is acceptable if they provide reliable cell material for screening, as do Caco-2 cells [4] and differentiated 3T3-L1 adipocytes.

Maintaining plated cells under hypothermic conditions (4–8 °C) offers another possibility of enhancing flexibility in cell delivery. By using special solutions (e.g., Hypo-Thermosol; BioLife), normal coronary artery smooth cells, hepatic cells and skeletal muscle cells can be kept for between 2 and 7 days at 4–8 °C [38]. This effect is due to the solutions balancing the altered cellular ion concentrations and counteracting cell swelling at low temperature.

5.3 Conventional Cellular Screening Assays

5.3.1 General HTS Assay Prerequisites

A certain level of automation is a prerequisite for an efficient HTS laboratory. Initially, higher productivity and freedom from performing repetitive tasks has been a major motivator to introduce laboratory automation [39]. However, as the equipment became more sophisticated, miniaturization of the assay volume and increased parallelization became important in order to reduce screening costs and to respond to increasing pressure on program cycle times. In general, the microplate has established itself as the major assay platform, starting with the 96-well plate initially used for diagnostics, and being the first major step towards parallelization. The pressure to reduce the costs of screening, combined with technical improvements in pipetting equipment and plate manufacture, has led to increased use of higher-density plates with 384 and 1536 wells. Despite the clear advantage in reagent savings, the use of smaller volumes is more demanding on the equipment used for liquid handling and detection. Thus, the reduction of assay volumes can lead to compromised assay parameters, especially due to stronger evaporation effects and changed surface-to-volume ratios.

Although the number of screened wells per unit time is an easily measurable and objective performance parameter, this does not directly define the desired output, which is the efficient identification of novel lead structures. As a consequence, further emphasis must be given to novel screening strategies and physiologically relevant screening approaches. Advanced screening approaches using cellular read-outs are one of means by which this goal may be achieved.

5.3.2 Evaluation of Assay Quality

Before an expensive, large-scale screening campaign is started, certain quality parameters must be met during the development of the assay. One of the basic parameters is to outline the controls used to define assay performance, and to calculate activities. Whereas uninhibited samples typically are used to serve as the positive control, the negative or blank control is often more difficult to define. In the best case, a pharmacologically relevant standard can be used to define the level of maximal effect or the level of nonspecific signal blank (NSB). During assay development, it must be determined if there is sufficient faith that these control values are not too "artificial" to be reached by a pharmacologically active compound.

In some cases, the positive controls might require the addition of an agonistic compound to stimulate the system. Assays searching for agonistic activities might have accordingly reversed controls – for example, the blank value might represent the unperturbed control.

Assay performance and its suitability for HTS is usually calculated using the mean (M) and standard deviation (SD) of such control wells. The screening window coefficient is the most commonly used statistical parameters to describe the quality of an assay [33]:

$$z' = 1 - \frac{3\,\mathrm{SD}_{\text{positive controls}} + 3\,\mathrm{SD}_{\text{NSB controls}}}{\mathrm{M}_{\text{positive controls}} - \mathrm{M}_{\text{NSB controls}}}$$

z' is a statistical tool that reflects the assay's dynamic range and the variability associated with the measurements. In recent years, the z'-value has been accepted as the most relevant parameter describing the assay robustness, and z'-values above 0.5 are considered as sufficient for screening campaigns. However, all of these parameters describe to a certain level only the robustness of the assay, and not whether the assay can detect the correct pharmacology and/or is capable of detecting hits with the desired sensitivity. In addition to the difficult quest for the correct pharmacology of novel targets, target-unrelated effects have been recognized as a major challenge for HTS operation [40]. Such undesired assay interference leads to so-called "false positive" or negative results, and these can create a significant burden on resources during the hit validation phase. Phenomena such as the "inner filter effect", which is caused by colored compounds, the quenching of fluorescence by various mechanisms, autofluorescence of compounds, light scattering resulting from particles, and photo bleaching are common mechanisms of assay interference.

Many compound-related assay interferences are concentration-dependent, and this creates a barrier for the selection of high compound concentrations for screening. Another major disturbing effect is caused by compound precipitation; hence, exceeding a compound's solubility limits in HTS is not recommended but, due to the variety within the compound collection, this cannot always be avoided [41]. However, the most important parameter – the relevance of the screening approach for the pathophysiological *in vivo* situation is the most difficult to appreciate a priori. One strategy to improve the level of confidence in the physiological nature of the screening is to present the target of choice within its cellular environment.

5.3.3 ELISA-Based Assays

The enzyme-linked immunosorbent assay (ELISA) is one of best-established and most often-used screening technologies since the start of the HTS era. Today, ELISA technology is widely used in the field of diagnostics, and can be easily transferred to a microtiter format. Although a variety of different assay protocols exist, all of them depend upon the availability of an antibody with a specificity for the antigen of interest. Binding of the primary antibody to the antigen can be detected by a secondary antibody which recognizes species-specific sequence differences in the FC portion of the primary antibody. These secondary antibodies are often coupled with an enzyme such as alkaline phosphatase or horseradish peroxidase. When provided with appropriate substrates, these enzymes can produce colored, chemifluorescent or chemiluminescent products, which can easily be quantified [42]. The amount of colored product is, therefore, within certain limits, proportional to the amount of antigen in the well.

In combination with cellular screens, ELISA-type technologies can be used, for example, to detect and measure antigens produced by the cells. Examples of such applications are the detection of cell-surface [43] or viral antigens [44]. A novel variant of ELISA technology is the antibody-mediated quantification of phosphorylated kinase substrates [45]. These *P*hosphospecific *A*ntibody *C*ell-based *E*LISA (PACE) screens are used to monitor the activity of kinase pathways within cells, which is of considerable value as kinases have become a major focus of drug discovery programs.

As an example, Versteeg et al. have developed an ELISA test to detect p42/p44 MAP kinase activation, p38 mitogen-activated protein kinase (MAPK), phosphokinase B and cAMP-response-element-binding (CREB) protein [46]. A PACE-based screening not only allows for large-scale analysis of signal transduction within a cellular environment, but also determines if a compound can sufficiently penetrate the cell membrane. In some PACE applications the secondary antibody used can be labeled with a fluorophore, which allows localization of the signal specifically to cells, or even to subcellular compartments.

ELISAs are typical examples of nonhomogeneous assay technologies, and the many washing steps require huge amounts of washing buffers. Therefore, they are not favored by most HTS laboratories as the many necessary steps slow down the assay protocol and reduce throughput. Despite the difficult screening logistics, however, ELISA technologies are well established and provide robust and sensitive measurements, and homogeneous variants of the detection technology have been described [45].

5.3.4
Radiometric Cellular Assays

Receptor binding or transporter binding assays using radioligands are classical biochemical methods used to characterize and screen these membrane-bound proteins. Although often performed on membrane preparations, these assays can in principle also be performed on whole cells. In its most simple form, the cells are seeded to grow on plates and the medium is exchanged for the appropriate binding buffer when the cells have reached the desired density. The compounds to be screened are added with the radioligand and, after incubation, the bound radioactivity is determined after a number of washing steps and the addition of scintillant. This type of assay is classically performed on ligand-gated transport proteins such as the dopamine transporter [47, 48], but they can also be used for receptor binding or enzymatic studies [49–52]. In the latter case, it should be kept in mind that, due to receptor-internalization, a fraction of the bound radioactivity might not be in equilibrium as it is trapped inside the cell [53]. In addition, a nonspecific trapping of radioactivity within cellular organelles has been described, and this must be considered for certain types of radioligand [54]. As described for the ELISA technology, due to the necessary washing steps these types of assays must be characterized as nonhomogeneous. However, they can be transformed into homogeneous technology when receptor–ligand complexes are detected by scintillation proximity [55, 56]. In contrast to the liquid scintillation system used for filter binding assays, in scintillation proximity only the receptor-bound β-radiation-emitting ligand is able to generate light emission, which can be monitored using a scintillation counter. Depending on the isotope used, the decay generates β-particles, which have a limited travel distance with sufficient energy to excite the scintillant. For whole-cell assays, Scintiplate™ (PerkinElmer) or Cytostar™ (GE Helthcare) plates can be used for a pseudohomogeneous spatial separation of the bound/nonbound signals. In such plates, the solid scintillant is incorporated into the plastic; the plates exhibit clear-bottomed wells for microscopic control of cell growth, and are cell-culture-treated for better cell attachment. Cytostar-based assays have been described for bile acid uptake in transfected HEK-293 cells [57], ionotropic glutamate receptors [58, 59], and thymidine incorporation [60]. The similar Scintiplates have been used for radioligand binding studies on whole cells [61].

5.3.5
Reporter Gene Assays

Initially, reporter gene assays have been one of the few really scalable cellular assays, which can be run in high-density plates [62, 63]. These assays monitor the transcriptional regulation of a promoter of the gene of interest linked to the coding region of a reporter gene. Using this principle, the activity of a cellular target and its subsequent signal transduction pathway will be linked to the expression of a readily detectable protein. Such proteins will mostly be enzymes, for example firefly and renilla luciferase, β-galactosidase, secreted alkaline phosphatase, chloramphenicol acetyltransferase, or β-lactamase. By coupling the response to the expression of an enzyme, a highly amplified – and therefore sensitive – signal is obtained [64, 65]. Alternatively, green fluorescent protein (GFP) can be used as signal [66]. Many drug targets will influence some transcriptional events in the cells, and are therefore amendable for reporter gene assays. Growth receptor pathways will, by definition, change transcriptional regulation, which is also achieved by nuclear receptors which are ligand-regulated transcription factors. G-protein coupled receptors (GPCRs) signaling can be monitored using cAMP-responsive element binding (CREB) or serum-responsive element (SRE) -activated promoter activity [67]. By using different constructs, a wide range of cellular events can be coupled to different read-outs via various substrates. Due to the broad application of the assay principle, only review articles are cited here. As a general disadvantage, reporter gene assays might suffer from interference of compounds acting distally to the target. Therefore, their results must be verified in a number of control assays aimed at filtering the non-target-related or cytotoxic effects [68].

5.3.6
Second Messenger Assays

To be more specific and closer to the molecular target of interest, receptor-targeted screens have been developed which directly monitor second messengers such as cAMP or changes in intracellular Ca^{2+}-concentration. These assays have been well established to investigate the functional effects of compounds targeting GPCRs. A number of different detection technologies exist for cAMP-detection which are mostly based on an immunochemical detection of the cyclic nucleotide (for a review, see [69, 70]). In principle, measurements of cAMP concentration in cells are mostly performed using competitive immunoassays. Here, cAMP-specific antibodies bind either labeled cAMP analogues, or unlabeled, free cAMP generated by the cells. In HTS applications, the read-out will usually be homogeneous and can use (for example) Flashplate™, AlphaScreen, LANCE (PerkinElmer), or HTRF (Cisbio) technologies. The activation of a Gαq-coupled GPCR leads to phospholipase C stimulation, turnover of inositol phosphates (IPs), and a subsequent increase in intracellular Ca^{2+}-concentration. Whereas early studies used laborious radioactive methods for $^{45}Ca^{2+}$-uptake and $[^{3}H]$-IP_3 turnover, methods based on the fluorescence of cell-permeable Ca^{2+}-chelators such as Fura2-AM, Fluo3-AM or Ca-mediated bioluminescence of aequorin, are widely used today [71, 72]. Due to the transient nature of the signal, these assays were not HTS-compatible until the introduction of the Fluorescence Imaging Plate Reader (FLIPR®), produced by Molecular Devices [73]. This instrument has not only stimulated cellular approaches in HTS, but also initiated the development of new generations of instruments designed specifically to measure cellular fluorescence or luminescence signals [74]. Today, these instruments can be integrated into robotic systems [71], and can perform assays in 1536-well format [75]. Recently, a HTRF-based IP1 immunoassay was described which can measure IP-signaling using an endpoint assay [76]. The above-mentioned instruments have addressed the need of functional assays for the important target class of GPCRs.

5.3.7
Ion Channel Assays

Especially for voltage-gated ion channels HTS technologies only recently became available. Initially, only radiometric binding or ion flux assays were used, but later fluorescent ion-chelating or membrane potential dyes such as DiBac4 were utilized for read-outs on FLIPR-like cellular imagers [77]. An alternative has been to use FRET-voltage sensor probes on readers capable of dual emission reading, such as VIPR™ and CellLux™ [78]. However, these dyes are prone to fluorescent assay artifacts and can only slowly follow changes in the membrane potential; thus, they do not provide the necessary temporal resolution of the signal. The reference technology for measuring ion currents across cellular membranes is patch-clamping, and automated systems such as the IonWorks® or IonWorks® Quattro (Molecular Devices) [79] have recently been introduced to the market.

These systems use premanufactured 384-well plates, in contrast to micropipettes in conventional electrophysiology, and can reach throughputs of up to 2500 data points per day [80]. In particular, for the measurement of the fast currents of voltage-gated channels, automated patch-clamp technologies will open new possibilities of directly addressing this target class in rapid screening with high fidelity [81].

5.4 The Definition of High-Content Screening

The term "high-content screening" (HCS) describes the analysis of drug activities in cell-based assays. It is a technology platform designed to define the temporal and spatial activities of genes, proteins, and other cellular constituents in living cells [82]. The measured events can include general effects (e.g., cellular morphology, apoptosis, migration), metabolic stimulation or inhibition, and specific effects on discrete target proteins. From a biology-orientated perspective, the term "phenotypic screening" is also often used to describe complex cell-based compound screening [83]. HCS is not a novel technique, but combines cellular manipulation, data acquisition and data processing within an integrated process.

5.4.1 Instrumentation for HCS

The first reader to be developed especially for HCS was launched in 1997. The Array Scan® (Cellomics, Pittsburgh, PA, USA) was the first system which integrated microscopy, digital imaging and data processing within a single platform [82].

Today, individual instruments available commercially can be classified by their optics (inverted microscope versus confocal microscope), light sources (laser versus mercury-xenon lamps) and their capabilities of performing live cell/kinetic measurements (Table 5.2).

The ArrayScan (Cellomics, Pittsburgh, PA, USA) and DISCOVERY-1 (Molecular Devices, Sunnyvale, CA, USA) contain optics based on an inverted microscope.

Confocal microscopes, used in the INCELL Analyzer 3000 (GE Healthcare) and the Opera (Evotec, Hamburg, Germany) have the capability of delivering high-resolution images.

Another HCS platform which is not microscope-based is the Acumen Explorer (TTP Labtech). This is a laser-scanning fluorescence microplate cytometer, which uses a scanning laser and optics with fluorescence filters to direct light emissions onto four photomultiplier detectors.

All HCS platforms described allow the performance of cellular assays in dense plate formats (usually 96- or 384-well plates). The Opera and the INCELL Analyzer 3000 are assembled with an internal pipetting device, which provides the capability of performing rapid kinetic measurements, even under environmental control using the INCELL Analyzer 3000.

Table 5.2 A summary of high-content screening (HCS) reader characteristics.

	ArrayScan VTI	DISCOVERY-1	INCELL Analyzer 3000	Opera	Acumen explorer
Optics	Inverted microscope/ CCD camera	Inverted microscope/ CCD camera	Confocal microscope/ CCD camera	Confocal microscope/ CCD camera	Photo-multiplier
Light source	UV, Mercury-Xenon lamp	Arc lamp	Laser	Laser	Laser
Objective magnifications	5–40×	5–40×	40×	10–60×	NA
Optics	Inverted microscope/ CCD camera	Inverted microscope/ CCD camera	Confocal microscope/ CCD camera	Confocal microscope/ CCD camera	Photo-multiplier
Light source	UV, Mercury-Xenon lamp	Arc lamp	Laser	Laser	Laser
Plate format (well)	96, 384	96–1536	96, 384	96–2000	96–1536
Internal environmental control	No	No	Yes	No	No
Internal pipetting	No	No	Yes	Yes	No

NA = not applicable.

The data processing is accomplished by the systems' software, which is often developed by the suppliers to provide their customers with an integrated solution for their HCS needs. A recently launched software package, entitled Cellenger (Definiens AG, Munich, Germany), not only delivers ready-to-use assays and the ability to generate novel algorithms, but also provides a database solution for handling the large data sets obtained during HCS campaigns.

5.4.2 Reagents (Fluorescent Probes) for HCS

The selection and application of optimal fluorescent probes to measure cellular features is crucial to successful HCS. Two different approaches are available to image spatial and/or temporal changes of cellular textures. First, low-molecular-weight fluorogenic dyes are coupled chemically to proteins in order to visualize the presence of the latter within cells. Second, fluorogenic reporter proteins, such as GFP, are genetically fused to the desired cellular target.

5.4.2.1 Low-Molecular-Weight Fluorophores

One of the major benefits of HCS is the ability to perform multiplexed assays where different cellular characteristics are simultaneously measured using multicolor fluorophores containing narrow emission spectra.

These fluorophores should have the optical properties of a bright fluorescence signal, combined with a large Stoke's shift and a high molar absorption [84]. Additional practical considerations when selecting fluorophores are their desired insensitivity to fixation procedures in end-point assays, their photostability, resistance against photobleaching [85], and solubility in aqueous buffers. One practical issue is the chemical labeling of the protein with dyes, as conjugation at high molar ratios of the fluorophore with the protein can lead to fluorescence quenching, presumably due to dye–dye interactions [86].

In general, the fluorophores are chemically conjugated to the protein of interest, or the protein is indirectly visualized via a fluorophore tagged antibody. Such fluorescent dyes with emission maxima ranging over a broad range are assembled by the indocyanine dyes Cy- (Amersham Biosciences, Piscataway, NJ, USA) and the Alexa fluor dyes (Molecular Probes, Eugene, OR, USA).

The long-wavelength dyes (> 550 nm excitation wavelength) are especially favored in multicolor assays because they fluoresce at wavelengths longer than the usual sources of cellular autofluorescence, and the background fluorescence of the dyes is generally low [87]. Both dyes are similar with respect to absorption and emission maxima, Stoke's shift, and extinction coefficient. However, the Cy-dyes were significantly less resistant to photobleaching than the Alexa dyes, and formed aggregates when coupled to proteins at high degrees of labeling [88].

A further step with regard to the use of optimal fluorophores for cellular imaging was taken with the development of fluorescent semiconductor nanocrystals – so-called Quantum dots – for biological applications. These have a broadband absorption spectrum, which makes them ideal for multicolor detection because only a single excitation source is needed. Quantum dots tend to be brighter than commonly used dyes because of the compounded effects of extinction coefficients, which are a magnitude larger than those of most other dyes. The Quantum dots are also more resistant to photobleaching than other dyes [89].

5.4.2.2 **Genetically Encoded Reporter for Fluorescence Detection**

One drawback of proteins, when chemically labeled with low-molecular-weight fluorescent dyes, is their intracellular visualization, because labeled proteins cannot enter living cells without penetrating the cell membrane.

The identification and cloning [90] of GFP from *Aequoria victoria* has revolutionized the use of intracellular protein sensoring. GFP can be genetically fused to the target protein, and the fusion protein visualized after expression within the cells. GFP contains an internal chromophore that is responsible for the emission of green light, and is generated upon cyclization and oxidation of a Ser-Tyr-Gly sequence at positions 65–67 within the protein [91].

The color spectrum of fluorescent proteins was expanded with the identification of fluorescent proteins from *Anthozoa* species. These proteins share the same structure as GFP from *Aequoria victoria*, but two of them have totally different spectral characteristics, emitting on yellow and red wavelengths [92].

The use of GFP has certain limitations, notably that the protein has a relatively high molecular mass of 248 amino acids, and this can lead to aggregation and subsequent quenching of the fluorescence. One approach to overcome these limitations is to label the fusion proteins with synthetic fluorophores in living cells. The target protein of interest is genetically tagged with a receptor, which binds its synthetic ligand, labeled with a fluorescent dye, inside the cells. This technique was first introduced with the use of a tetracysteine peptide as the receptor and biarsenical fluorophores as the ligands. This ligand is cell-permeable and nonfluorescent until it binds, with high affinity and specificity, to the tetracystein peptide of the fusion protein [93].

Another example of intracellular fluorophore labeling of fusion proteins is the use of O^6-alkylguanine-DNA alkyltransferase (AGT). This enzyme binds the substrate O^6-benzylguanosine, which is then derivatized with fluorescein, thereby enabling intracellular visualization of the tagged fusion protein [94].

5.4.3
Assays and Target-Based Applications of HCS

These assays/applications can be grouped into two classes, namely target-orientated or mechanism-orientated. Target-based approaches seek the activity or localization of a defined protein within the cells, whereas mechanism-based approaches measure general cellular parameters such as intercellular communication or multi-faceted phenomena, such as apoptosis.

5.4.3.1 GPCRs

Today, GPCRs are becoming increasingly accepted as excellent targets in HCS because of the availability of appropriate assay techniques for drug screening.

The majority of known GPCRs desensitize and internalize into the cytoplasm upon ligand binding to the receptor. Desensitization (receptor aggregation on the cell surface) is accompanied by receptor internalization via clathrin-coated pits into an endosomal compartment [95]. This internalization phenomenon was quantitatively measured using a fusion protein consisting of the parathyroid hormone receptor (PTH) and GFP in mammalian cells. Upon PTH binding, the receptor-GFP internalization, visualized as punctuate spots within the cells, was both time- and dose-dependent, and shown to be selective for the ligand PTH [96]. Pharmacological characterization of the PTH receptor-GFP construct revealed no differences in K_d and V_{max} values, as well as in a functional Ca^{2+} release assay, between the wild-type PTH receptor and the recombinant receptor-GFP construct expressed in cells, thus validating the internalization assay for this receptor [97].

Another generic assay format for HCS is to measure the translocation of GPCR-interacting proteins. Beta-arrestins interact with GPCRs which have been phosphorylated in response to agonists. This interaction is central to the internalization of GPCRs via clathrin-coated pits, owing to the ability of beta-arrestins to interact with clathrin adapter proteins [95]. Measurement of the internalized GPCR is accomplished by using a beta-arrestin2 fused to GFP as a biosensor which binds with high affinity to ligand-occupied phosphorylated

GPCRs [98]. The universality of this approach has been demonstrated for many GPCRs, whether they couple to different G proteins (Gs, GQ/11 or Gi/o), are classified by ligand (biogenic amines, peptides or lipids), or by the receptor sequences (class A rhodopsin-like, class B secretin-like family or class C metabotropic-like subfamily) [99].

As GPCRs are sequestered into acidic compartments during the internalization process, the use of a receptor-tagged dye, Cypher 5, which is fluorescent only at acidic pH, greatly expands the options for GPCR assays in HCS. On internalization into cells, the dye is protonated and becomes highly fluorescent in the acidic endosomal compartments. The dye can be bound to the receptor via dye-labeled antibodies or ligands. This approach has been used successfully to monitor the agonist-mediated internalization of GPCRs that couple to Gs-, Gi- and GQ/11-mediated signaling pathways [100].

To date, HCS approaches have been used in secondary screening approaches for lead identification or optimization. However, one of the first applications of HCS in primary screening was conducted at Hoffmann-La Roche Inc., USA. A primary cell-based HTS was carried out to identify agonists of a novel orphan GPCR using an assay based on receptor internalization measuring GFP-tagged beta-arrestin. A library of 800 000 compounds was screened, and approximately 800 hits were initially identified. Follow-up studies were carried out to filter out fluorescent compounds and compounds inducing morphological changes. Some compounds were identified that induced beta-arrestin relocalization independently of GPCR binding, possibly acting as kinase inhibitors [101].

5.4.3.2 **Kinases**

Over the past few years, apart from GPCRs, kinases have been the most important drug targets in the pharmaceutical industry. In fact, the activation of kinase-mediated signaling pathways has emerged as an area that is particularly suitable for HCS. The activation of many signaling proteins involves either phosphorylation on specific tyrosine, threonine or serine residues, and/or intracellular trafficking of proteins within the pathways. These phenomena can be detected by immunostaining using phospho-specific antibodies or, in the case of protein relocalization, GFP-tagged proteins.

One example of a HCS approach was recently carried out at Merck KgaA [102], whereby a cellular assay was set up which determined the autophosphorylation of a membrane-bound receptor tyrosine kinase (RTK). Activation of the RTK with ligand induced RTK autophosphorylation which was detected with a phospho-specific antibody, while a second anti-primary antibody was labeled with Alexa 488 (Molecular Probes). Cellular counterstaining was carried out with the red fluorescent dye Syto 64, which enabled the toxic compounds to be identified. Using this approach, the activation (autophosphorylation) status of the RTK could be measured independently of whether the cells were treated with a kinase inhibitor, or not. The assay was run on the Acumen Explorer; although this platform is a laser-scanning cytometer without microscopic optics, the resolution was comparable to that obtained with a fluorescence microscope (Fig. 5.2).

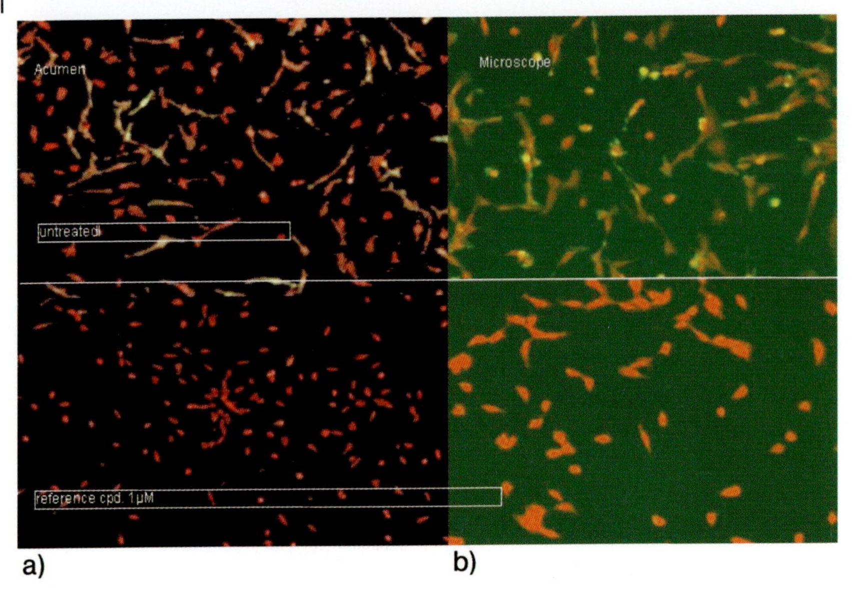

Fig. 5.2 High-content screening (HCS) assay for receptor tyrosine kinase (RTK) activation. Green objects reflect cells with phosphorylated RTK, red objects the total cell number. Screenshots are shown from the reader (a) and microscopic pictures (b).

The activation of a signaling pathway by a specific receptor can also be measured using a multiplexed assay for HCS. Activation of the MAPK by the epidermal growth factor (EGF) receptor was quantitated using the nuclear localization of phosphorylated ERK by immunofluorescence, together with internalization of the receptor ligand EGF, which was labeled with Texas Red. This assay allowed the identification of compounds able to block activation of the MAPK pathway by a specific relevant oncology target [103].

Using a protein translocation as readout, a HCS screen for inhibitors of p38 kinase was performed at Eli Lilly & Co, USA. The assay was based on the inhibition of anisomycin-induced nuclear import of MAPKAP-K2, fused to GFP. Among 32 000 compounds tested, 48 were identified with an $IC_{50} < 10$ µM. Two of the identified compounds were found to be novel p38 inhibitors [101].

HCS can significantly accelerate lead optimization, resulting in the faster selection of a clinical candidate. This was demonstrated using a multiplexed assay by the Oncology group at Astra Zeneca, who replaced three different individual assays for aurora kinase by a single multiplexed assay that simultaneously measured histone H3 phosphorylation, nuclear area change (a marker for failed cell division), and cell count [101].

5.4.3.3 Other Drug Targets

HCS can also be applied to the use of primary human cells, as shown in a report demonstrating the assay development and compound screening for the identification of activators of the Wnt/Frizzled (Fzd) pathway. Using primary human bone cells (preosteoblasts) as a model, activation of the Wnt/Fzd pathway was determined by immunofluorescent staining of translocation of the transcription factor beta-catenin. A library screen of 51 000 compounds yielded a hit rate of 1.4%, which was reduced to 0.6% after the removal of false-positive compounds, whether these were autofluorescent or cytotoxic. A major benefits of HCS was demonstrated by this approach, in that the assay requires only a limited cell number per well, thus eliminating the bottleneck in cell supply [104].

As HCS measures compound activities in a cellular context, it should be – to some extent – predictive with regard to compound activities *in vivo*. This was exemplified in a report where inhibitors of a MAPK phosphatase were found. Using a HCS assay based on phosphorylation of the kinase Erk, a compound with *in-vivo* activity was identified which was shown not to be active in a biochemical screening assay [105].

By using a translocation assay which measures the cytoplasm to nuclear localization of a GFP-tagged forkhead transcription factor, a signaling pathway inhibitor was identified which demonstrated anti-tumorous activity in a breast cancer xenograft model [101].

5.4.4 HCS Applications Targeting Generic Cellular Parameters and Morphology

In contrast to target-based approaches, assays which recapture generic cellular parameters deliver in-depth results with regard to the mode of action of screening compounds.

An elegant HCS approach targeting cellular communication was recently performed at Aventis Pharmaceuticals, USA, whereby an assay was set up to identify compounds that block gap junctions between cells. Cells were labeled with the dye calcein and dispensed with unlabeled cells into 384-well plates. The assay was run as a live cell imaging assay in order to prevent dye leakage which occurs when cells are fixed.

The primary screen was run with 486 000 test compounds, from which 1515 primary hits were identified; among these were 53 compounds that produced concentration–response curves. As the read time per plate was 43 min, in order to prevent passive leakage of the calcein dye from the cells, the consequent throughput was limited to only 13 assay plates per day. This low throughput was partially compensated by the application of four test compounds per well.

This screening campaign highlighted one of the benefits of HCS, namely that it also permits HTS for those targets not amenable to primary screening by using more conventional assays such as radioactive metabolite transfer or dye transfer after microinjection [106].

Cellular assays with multiparametric readouts are desired when a multifaceted cellular phenomenon such as apoptosis is studied. Such an assay was set up to analyze three parameters that are cellular hallmarks of apoptosis, namely caspase-3 activation, nuclear condensation, and mitochondrial membrane potential. Using two tumor cell lines (HeLa and u937 cells), standard cytotoxic agents were applied and the three apoptosis markers measured simultaneously using an antibody against activated caspase-3, the Hoechst 33342 dye for nuclear staining, and chloromethyl-X-rosmaine to measure mitochondrial membrane potential.

The value of this assay was demonstrated by the actions of the applied drugs, which induced caspase-3 activation and nuclear condensation, but reduced the mitochondrial membrane potential. This multiplexed HCS approach combined biochemical (caspase-3 activation) and morphological (nuclear condensation) information in one assay, thus eliminating the need for different, laborious, single-point assays [107].

5.5 Outlook

The examples of HCS assays referred to in Section 5.4 for compound screening highlight the benefits of cellular over conventional assays. First, they can provide access to targets not amenable to screening in conventional assays. Second, multiplexed assays enable the direct identification of pharmacologically active compounds which can modulate the physiology or morphology of a cell.

These advantages must be reflected knowing that most successful drugs descend from natural products selected in intact biological systems, and many drugs show several biological activities, for example nonsteroidal anti-inflammatory agents, cyclosporine, and antihistamines [108]. Even for the most popular drugs, not all of the targets and signaling pathways influenced in one cell are realized, much less the effects on the whole organism.

The minimization of biological complexity by screening isolated targets or individual pathways causes most emergent properties and all network responses of a cell to be missed. Hence, the future use of complex cell systems in combination with testing of several targets will allow the collection of information covering several disease parameters and pathways by selecting for cell-permeable, nontoxic drug candidates. Here, the studies of Kunkel and colleagues have provided a hint of the possibilities that lie ahead [109], when they assayed the effect of 33 kinase inhibitors on 51 protein readouts in four inflammatory cell systems, and ultimately identified an unexpectedly high number of active drugs [110].

Today, a series of new, more physiologically relevant cell culture methods have become available: tumor cells have been grown to three-dimensional spheroids and tested for their response against cytotoxic agents [111], while embryonic stem (ES) cells [112] allow the effects of compounds to be monitored in complex cell systems.

The microscope has long been recognized as an important instrument in cell biology research, yet today its combination with fluorescence techniques and automation has revealed it to be a very powerful and very sensitive tool. Now, the use of automated fluorescence microscopy allows a small number of cells expressing a low number of biomolecules to be screened. Indeed, the potential of this technique has been well documented by Mitchison's group, in testing 90 compounds for mitotic arrest, differentiation, toxicity, and key signaling pathways [113, 114].

In future, the main challenges for complex cellular screens are threefold, namely the identification of drug target(s), the determination of target-based structure–activity relationships of the compounds, and regulatory issues. Moreover, the management and presentation of the growing amount of data, as well as the integration of analyzed cellular parameters into a physiological context, represent a major challenge for cellular bioinformatics.

References

1 S. P. Manly, *J. Biomol. Screen.* **1997**, *2*, 197–199.
2 K. Moore, S. Rees, *J. Biomol. Screen.* **2001**, *6*, 69–74.
3 F. Gaunitz, K. Heise, *Assay Drug Dev. Technol.* **2003**, *1*, 469–477.
4 C. Horrocks, R. Halse, R. Suzuki, P. R. Shepherd, *Curr. Opin. Drug Discov. Devel.* **2003**, *6*, 570–575.
5 ATCC, http://www.lgcpromochem.com/atcc/ **2005**.
6 ECACC, http://www.ecacc.org.uk/ **2005**.
7 DSMZ, http://www.dsmz.de/human_and_animal_cell_lines/main.php?menu_id=2 **2005**.
8 NIH 60 cell screen program, http://dtp.nci.nih.gov/branches/btb/ivclsp.html **2005**.
9 A. Hussain, D. Lewis, M. Yu, P. W. Melera, *Gene* **1992**, *112*, 179–188.
10 P. G. Sanders, A. Hussein, L. Coggins, R. Wilson, *Dev. Biol. Stand.* **1987**, *66*, 55–63.
11 T. Lindl, *Zell- und Gewebekultur*. Spektrum Akademischer Verlag, Heidelberg, **2000**.
12 M. E. Al-Rubeai, *Cell Engineering, Volume 2: Transient Expression*, **2000**.
13 A. D. Miller, G. J. Rosman, *BioTechniques* **1989**, *7*, 980–986, 989.
14 S. L. Kozak, D. Kabat, *J. Virol.* **1990**, *64*, 3500–3508.
15 N. L. Craig, *Annu. Rev. Genet.* **1988**, *22*, 77–105.
16 J. Song, C. Doucette, D. Hanniford, K. Hunady, N. Wang, B. Sherf, J. J. Harrington, K. R. Brunden, A. Stricker-Krongrad, *Assay Drug Dev. Technol.* **2005**, *3*, 309–318.
17 F. Wurm, A. Bernard, *Curr. Opin. Biotechnol.* **1999**, *10*, 156–159.
18 E.-J. Schlaeger, K. Christensen, *Cytotechnology* **1999**, *30*, 71–83.
19 O. Boussif, T. Delair, C. Brua, L. Veron, A. Pavirani, H. V. Kolbe, *Bioconjug. Chem.* **1999**, *10*, 877–883.
20 C. Chen, H. Okayama, *Mol. Cell Biol.* **1987**, *7*, 2745–2752.
21 S. Geisse, M. Jordan, F. M. Wurm, *Methods Mol. Biol.* **2005**, *308*, 87–98.
22 Y. Durocher, S. Perret, A. Kamen, *Nucleic Acids Res.* **2002**, *30*, E9.
23 E. J. Schlaeger, K. Christensen, *Cytotechnology* **2005**, *30*, 71–83.

24 K. Lundstrom, A. Michel, H. Blasey, A. R. Bernard, R. Hovius, H. Vogel, A. Surprenant, *J. Recept. Signal. Transduct. Res.* **1997**, *17*, 115–126.
25 Y. M. Mishina, C. J. Wilson, L. Bruett, J. J. Smith, C. Stoop-Myer, S. Jong, L. P. Amaral, R. Pedersen, S. K. Lyman, V. E. Myer, B. L. Kreider, C. M. Thompson, *J. Biomol. Screen.* **2004**, *9*, 196–207.
26 M. D. Evans, J. G. Steele, *J. Biomed. Mater. Res.* **1998**, *40*, 621–630.
27 J. S. Bonifacio, et al. *Short Protocols in Cell Biology*. Wiley, **2004**.
28 P. A. Johnston, *Drug Discov. Today* **2002**, *7*, 353–363.
29 C. Scott, *BioProcess International* **2005**, *3*, 16–27.
30 D. W. Jayme, K. E. Blackman, *Adv. Biotechnol. Processes* **1985**, *5*, 1–30.
31 L. Munaron, S. Antoniotti, D. Lovisolo, *J. Cell Mol. Med.* **2004**, *8*, 161–168.
32 T. Korkiamaki, H. Yla-Outinen, P. Leinonen, J. Koivunen, J. Peltonen, *Arch. Dermatol. Res.* **2005**, *296*, 465–472.
33 J. H. Zhang, T. D. Chung, K. R. Oldenburg, *J. Biomol. Screen.* **1999**, *4*, 67–73.
34 B. K. Lundholt, K. M. Scudder, L. Pagliaro, *J. Biomol. Screen.* **2003**, *8*, 566–570.
35 I. R. Freshney. *Culture of Animal Cells, A Manual of Basic Techniques.* New York, Wiley-Liss, Inc., New York, **1994**.
36 N. Fursov, M. Cong, M. Federici, M. Platchek, P. Haytko, R. Tacke, T. Livelli, Z. Zhong, *Assay Drug Dev. Technol.* **2005**, *3*, 7–15.
37 P. Kunapuli, W. Zheng, M. Weber, K. Solly, R. Mull, M. Platchek, M. Cong, Z. Zhong, B. Strulovici, *Assay Drug Dev. Technol.* **2005**, *3*, 17–26.
38 L. Moce-Llivina, J. Jofre, *Cytotechnology* **2005**, *46*, 57–61.
39 J. P. Elands, R. Seethala, P. B. Fernandes. *Handbook of Drug Screening.* Marcel Dekker Inc., New York, **2001**, pp. 477–492.
40 J. Comley, *Drug Discovery World* **2003**, Summer, 91–97.
41 S. L. McGovern, E. Caselli, N. Grigorieff, B. K. Shoichet, *J. Med. Chem.* **2002**, *45*, 1712–1722.
42 J. R. Crowther. *Methods in Molecular Biology, The ELISA Guide-book.* Humana Press Inc., New Jersey, **1995**.
43 G. A. Bishop, J. Hwang, *Biotechniques* **1992**, *12*, 326–330.
44 A. Myc, M. J. Anderson, J. R. Baker, Jr., *J. Virol. Methods* **1999**, *77*, 165–177.
45 L. K. Minor, *Curr. Opin. Drug Discov. Devel.* **2003**, *6*, 760–765.
46 H. H. Versteeg, E. Nijhuis, G. R. van den Brink, M. Evertzen, G. N. Pynaert, S. J. van Deventer, P. J. Coffer, M. P. Peppelenbosch, *Biochem. J.* **2000**, *350* (Pt 3), 717–722.
47 Z. Lin, W. Wang, T. Kopajtic, R. S. Revay, G. R. Uhl, *Mol. Pharmacol.* **1999**, *56*, 434–447.
48 K. Y. Little, L. W. Elmer, H. Zhong, J. O. Scheys, L. Zhang, *Mol. Pharmacol.* **2002**, *61*, 436–445.
49 L. W. Fitzgerald, J. P. Patterson, D. S. Conklin, R. Horlick, B. L. Largent, *J. Pharmacol. Exp. Ther.* **1998**, *287*, 448–456.
50 M. M. Voigt, L. G. Davis, J. H. Wyche, *J. Neurochem.* **1984**, *43*, 1106–1113.
51 M. Tamura, Y. Wanaka, E. J. Landon, T. Inagami, *Hypertension* **1999**, *33*, 626–632.

52 R. B. Lobell, J. P. Davide, N. E. Kohl, H. D. Burns, W. S. Eng, R. E. Gibson, *J. Biomol. Screen.* **2003**, *8*, 430–438.
53 D. A. Jans, P. Jans, H. Luzius, F. Fahrenholz, *Mol. Cell. Endocrinol.* **1991**, *81*, 165–174.
54 J. M. Maloteaux, A. Gossuin, C. Waterkeyn, P. M. Laduron, *Biochem. Pharmacol.* **1983**, *32*, 2543–2548.
55 N. Bosworth, P. Towers, *Nature* **1989**, *341*, 167–168.
56 N. D. Cook, *Drug Discov. Today* **1996**, *1*, 287–294.
57 H. Bonge, S. Hallen, J. Fryklund, J. E. Sjostrom, *Anal. Biochem.* **2000**, *282*, 94–101.
58 L. Smith, et al. *Epilepsia* **2000**, *42*, 48–51.
59 A. Cushing, M. J. Price-Jones, R. Graves, A. J. Harris, K. T. Hughes, D. Bleakman, D. Lodge, *J. Neurosci. Methods* **1999**, *90*, 33–36.
60 R. Graves, R. Davies, G. Brophy, G. O'Beirne, N. Cook, *Anal. Biochem.* **1997**, *248*, 251–257.
61 S. S. Jaakola, *Anal. Biochem.* **2002**, *307*, 280–286.
62 A. Dhundale, C. Goddard, *J. Biomol. Screen.* **1996**, *1*, 115–118.
63 C. M. Suto, D. M. Ignar, *J. Biomol. Screen.* **1997**, *2*, 7–9.
64 A. M. Maffia, III, I. Kariv, K. R. Oldenburg, *J. Biomol. Screen.* **1999**, *4*, 137–142.
65 L. R. Fitzgerald, I. J. Mannan, G. M. Dytko, H. L. Wu, P. Nambi, *Anal. Biochem.* **1999**, *275*, 54–61.
66 J. C. March, G. Rao, W. E. Bentley, *Appl. Microbiol. Biotechnol.* **2003**, *62*, 303–315.
67 K. Kotarsky, N. E. Nilsson, B. Olde, C. Owman, *Pharmacol. Toxicol.* **2003**, *93*, 249–258.
68 P. A. Johnston, *Drug Discov. Today* **2002**, *7*, 353–363.
69 C. Williams, *Nature Rev. Drug Discov.* **2004**, *3*, 125–135.
70 D. Gabriel, M. Vernier, M. J. Pfeifer, B. Dasen, L. Tenaillon, R. Bouhelal, *Assay Drug Dev. Technol.* **2003**, *1*, 291–303.
71 C. Chambers, F. Smith, C. Williams, S. Marcos, Z. H. Liu, P. Hayter, G. Ciaramella, W. Keighley, P. Gribbon, A. Sewing, *Comb. Chem. High Throughput Screen.* **2003**, *6*, 355–362.
72 E. Le Poul, S. Hisada, Y. Mizuguchi, V. J. Dupriez, E. Burgeon, M. Detheux, *J. Biomol. Screen.* **2002**, *7*, 57–65.
73 K. S. Schroeder, *J. Biomol. Screen.* **1996**, *1*, 75–80.
74 J. Comley, *Drug Discovery World* **2004**, *Winter*, 49–60.
75 P. Hodder, R. Mull, J. Cassaday, K. Berry, B. Strulovici, *J. Biomol. Screen.* **2004**, *9*, 417–426.
76 www.htrf-assays.com **2005**.
77 W. Tang, J. Kang, X. Wu, D. Rampe, L. Wang, H. Shen, Z. Li, D. Dunnington, T. Garyantes, *J. Biomol. Screen.* **2001**, *6*, 325–331.
78 J. Xu, X. Wang, B. Ensign, M. Li, L. Wu, A. Guia, J. Xu, *Drug Discov. Today* **2001**, *6*, 1278–1287.
79 K. S. Schroeder, *J. Biomol. Screen.* **1996**, *1*, 75–80.

80 A. Southan, *Drug Discovery World* **2005**, *Summer*, 17–23.
81 W. Zheng, R. H. Spencer, L. Kiss, *Assay Drug Dev. Technol.* **2004**, *2*, 543–552.
82 K. A. Giuliano, R. L. Debiasio, R. T. Dunlay, A. Gough, J. M. Volosky, J. Zock, G. N. Pavlakis, D. L. Taylor, *J. Biomol. Screen.* **1997**, *2*, 249–259.
83 P. A. Clemons, *Curr. Opin. Chem. Biol.* **2004**, *8*, 334–338.
84 B. Herman, K. Jacobson, *Optical Microscopy for Biology* **1990**, 143–157.
85 L. Song, C. A. Varma, J. W. Verhoeven, H. J. Tanke, *Biophys. J.* **1996**, *70*, 2959–2968.
86 J. B. Randolph, A. S. Waggoner, *Nucleic Acids Res.* **1997**, *25*, 2923–2929.
87 C. Cullander, *J. Microsc.* **1994**, *176* (Pt 3), 281–286.
88 J. E. Berlier, A. Rothe, G. Buller, J. Bradford, D. R. Gray, B. J. Filanoski, W. G. Telford, S. Yue, J. Liu, C. Y. Cheung, W. Chang, J. D. Hirsch, J. M. Beechem, R. P. Haugland, *J. Histochem. Cytochem.* **2003**, *51*, 1699–1712.
89 X. Michalet, F. F. Pinaud, L. A. Bentolila, J. M. Tsay, S. Doose, J. J. Li, G. Sundaresan, A. M. Wu, S. S. Gambhir, S. Weiss, *Science* **2005**, *307*, 538–544.
90 D. C. Prasher, V. K. Eckenrode, W. W. Ward, F. G. Prendergast, M. J. Cormier, *Gene* **1992**, *111*, 229–233.
91 R. Heim, D. C. Prasher, R. Y. Tsien, *Proc. Natl. Acad. Sci. USA* **1994**, *91*, 12501–12504.
92 M. V. Matz, A. F. Fradkov, Y. A. Labas, A. P. Savitsky, A. G. Zaraisky, M. L. Markelov, S. A. Lukyanov, *Nat. Biotechnol.* **1999**, *17*, 969–973.
93 B. A. Griffin, S. R. Adams, R. Y. Tsien, *Science* **1998**, *281*, 269–272.
94 A. Keppler, S. Gendreizig, T. Gronemeyer, H. Pick, H. Vogel, K. Johnsson, *Nat. Biotechnol.* **2003**, *21*, 86–89.
95 S. S. Ferguson, *Pharmacol. Rev.* **2001**, *53*, 1–24.
96 B. R. Conway, L. K. Minor, J. Z. Xu, J. W. Gunnet, R. DeBiasio, M. R. D'Andrea, R. Rubin, R. DeBiasio, K. Giuliano, L. DeBiasio, K. T. Demarest, *J. Biomol. Screen.* **1999**, *4*, 75–86.
97 B. R. Conway, L. K. Minor, J. Z. Xu, M. R. D'Andrea, R. N. Ghosh, K. T. Demarest, *J. Cell Physiol.* **2001**, *189*, 341–355.
98 L. S. Barak, S. S. Ferguson, J. Zhang, M. G. Caron, *J. Biol. Chem.* **1997**, *272*, 27497–27500.
99 R. H. Oakley, C. C. Hudson, R. D. Cruickshank, D. M. Meyers, R. E. Payne, Jr., S. M. Rhem, C. R. Loomis, *Assay Drug Dev. Technol.* **2002**, *1*, 21–30.
100 E. J. Adie, M. J. Francis, J. Davies, L. Smith, A. Marenghi, C. Hather, K. Hadingham, N. P. Michael, G. Milligan, S. Game, *Assay Drug Dev. Technol.* **2003**, *1*, 251–259.
101 O. Rausch, *IDrugs* **2005**, *8*, 197–199.
102 O. Pöschke, Poster abstract, SBS meeting, Orlando, Florida, **2005**.
103 R. N. Ghosh, L. Grove, O. Lapets, *Assay Drug Dev. Technol.* **2004**, *2*, 473–481.

104 K. M. Borchert, R. J. Galvin, C. A. Frolik, L. V. Hale, D. L. Halladay, R. J. Gonyier, O. J. Trask, D. R. Nickischer, K. A. Houck, *Assay Drug Dev. Technol.* **2005**, *3*, 133–141.
105 A. Vogt, K. A. Cooley, M. Brisson, M. G. Tarpley, P. Wipf, J. S. Lazo, *Chem. Biol.* **2003**, *10*, 733–742.
106 Z. Li, Y. Yan, E. A. Powers, X. Ying, K. Janjua, T. Garyantes, B. Baron, *J. Biomol. Screen.* **2003**, *8*, 489–499.
107 H. Lovborg, P. Nygren, R. Larsson, *Mol. Cancer Ther.* **2004**, *3*, 521–526.
108 H. Kubinyi, *Nat. Rev. Drug Discov.* **2003**, *2*, 665–668.
109 E. J. Kunkel, I. Plavec, D. Nguyen, J. Melrose, E. S. Rosler, L. T. Kao, Y. Wang, E. Hytopoulos, A. C. Bishop, R. Bateman, K. M. Shokat, E. C. Butcher, E. L. Berg, *Assay Drug Dev. Technol.* **2004**, *2*, 431–441.
110 E. C. Butcher, E. L. Berg, E. J. Kunkel, *Nat. Biotechnol.* **2004**, *22*, 1253–1259.
111 P. Bartholoma, Impidjati, A. Reininger-Mack, Z. Zhang, H. Thielecke, A. Robitzki, *J. Biomol. Screen.* **2005**.
112 G. Keller, *Genes Dev.* **2005**, *19*, 1129–1155.
113 Z. E. Perlman, T. J. Mitchison, T. U. Mayer, *Chembiochem.* **2005**, *6*, 145–151.
114 T. J. Mitchison, *Chembiochem.* **2005**, *6*, 33–39.

6
Cell Lines and Primary Tissues for *In-Vitro* Evaluation of Vaccine Efficacy

Anthony Meager

6.1
Introduction

Vaccines are normally considered to be a class of biological medicines. The vaccine materials are produced by living biological systems or biotechnological processes, and usually also tested in biological systems. Preventive or prophylactic vaccines are designed to induce sterilizing immunity against pathogenic organisms, and thus reduce the disease burden in humans and domestic animals. In general, these vaccines deliver the antigenic components of microorganisms and viruses in a manner that simulates to some degree the natural infection; this in turn stimulates the immune system to develop neutralizing antibodies and protective T-cell responses that persist in the long term. The degree to which individual vaccines are effective for their intended purpose depends on a number of factors, including antigenic structure, live-attenuated versus killed vaccine, adjuvant effects, mode of delivery, age of recipient, and disease target. Efficacy is thus often difficult to predict or to measure accurately, especially at the nonclinical testing phase of vaccine development, and usually remains uncertain until substantial evidence for it accrues from clinical trials. Nevertheless, it is crucial that experimental testing ahead of clinical usage is carried out in order to assess both potential efficacy and safety. Suitable animal models are frequently the means of choice for determining efficacy and safety, but the great social debate on the use of animals in scientific research, and legal and economic pressure to reduce usage at all costs, has emphasized the need for appropriate, valid *in-vitro* testing methods.

While it is evident that some animal tests are difficult to justify and/or have serious shortcomings with regard to their prediction of efficacy in humans, it is very important that any *in-vitro* tests developed to replace or support them are scientifically validated, and do not themselves suffer shortcomings. For instance, for certain single-protein vaccines (e.g., hepatitis B virus vaccine) the amount of antigen (hepatitis B virus surface antigen {HBsAg}) may be determined by immunoassays [1]. The antigen concentration is then assumed to be a direct measure of the amount of immunogen. However, variable production methods of

Drug Testing In Vitro: Breakthroughs and Trends in Cell Culture Technology
Edited by Uwe Marx and Volker Sandig

ISBN: 978-3-527-31488-1

hepatitis B virus vaccine can lead to unexpected changes to the amount of antigen recorded in the immunoassay, without concomitant changes in immunogenicity of the vaccine [1, 2]. These cell-free tests appear also to be of little use for evaluating the efficacy of more complex vaccines, such as inactivated- and live-attenuated-viral vaccines. For example, several collaborative studies of the potency/efficacy of inactivated polio vaccines have attempted to relate the amount of antigen to immunogenicity in animal models and in humans, but have failed to identify a single antigen measurement method that can be applied to all of the different vaccines available [3, 4]. As alternatives, *in-vitro* tests in cell lines or primary tissues can sometimes yield informative data on potential efficacy. Here, the aim is to measure vaccine potency (or some other surrogate parameter) that can be linked to vaccine efficacy or correlates of protection.

There are three main scenarios where such testing might be appropriately carried out:

- The use of mostly attenuated live viral vaccines and viral/plasmid vector vaccines expressing a foreign (heterologous) antigen(s). Suitable mammalian cell lines are infected/transfected with the vaccine, and the amount of the intended antigen(s) measured at selected time points.
- The testing of blood leukocytes from animals/humans following immunization with the vaccine. Estimations of T-cell activation, for example, interferon-gamma (IFN-γ) production, the generation of specific cytotoxic T lymphocytes (CTL), following antigenic stimulation *in vitro* are among the immunological tests carried out to assess whether a vaccine is effective, or not.
- A form of *in-vitro* immunization where lymphoid cells/tissues/organoids are exposed to the vaccine to stimulate quantifiable immune responses; this is more for consideration in the future.

This aim of this chapter is to consider the relevance and validity of *in-vitro* testing methods in evaluating the potential efficacy of prophylactic vaccines. Certain aspects of such testing also apply to therapeutic vaccines – vaccines administered after infection by pathogen or during disease progression, for example, the so-called "cancer vaccines" – but these will be considered only briefly. Specific examples are used to illustrate vaccination regimes and assay methods. Since one of the main current foci of vaccine development is in the human immunodeficiency virus type 1 (HIV-1) field, and because the types of cell-based assays used to evaluate potential preventative HIV-1 vaccines are applicable to several novel vaccines against other pathogens, this clinical research area will be used to provide and illustrate assay methods and discussion.

6.2 Measurement of Antigen Expression

In the case of live-attenuated viral vaccines, viral antigens are expressed in infected cells with limited virus replication, and without the pathogenic effects of the parental live virus. The viral antigens may be detected and measured by using specific polyclonal antisera or monoclonal antibodies (MAbs) for immunostaining of fixed cells or Western blotting of extracted proteins. Such techniques were less frequently used on the now well-established viral vaccines such as live-attenuated polio vaccine, but are increasingly applied to novel vaccines based on live recombinant viral vectors encoding and expressing heterologous antigens. For example, poxviruses have large DNA genomes with a high capacity for the insertion of heterologous DNA sequences encoding and expressing single/multiple heterologous antigen(s). In particular, vaccinia virus and avian poxviruses (canarypox and fowlpox) are considered as promising candidates for the development of vaccines for use in humans and in the veterinary field [5–8]. Attenuated strains of vaccinia virus – for example modified vaccinia virus Ankara (MVA) and NYVAC, canarypox virus (e.g., ALVAC) and fowlpox virus (e.g., FP9 and TROVAC) – have been isolated that replicate in chicken embryo fibroblasts (CEF), but not in human cells, and these are now being widely developed as recombinant viral vector vaccines [6, 9–11, 13]. For example, recombinant poxviruses encoding various antigens of disease-causing viruses, including influenza virus, parainfluenza virus, measles virus, respiratory syncytial virus, dengue virus, Japanese encephalitis virus, human immunodeficiency virus-1 (HIV-1) and *P. falciparum* (malaria) antigens, have been characterized *in vitro* [10–14, 94]. Despite abortive replication in human cells, the heterologous genes are still expressed; the heterologous antigens are produced and have been found to stimulate immune responses. Examples are the development of antigen-specific T cells and antibodies in animal model systems and in humans.

Since mutations (including deletions) within recombinant MVA or ALVAC vector vaccine DNA can occur at high passage levels, it is considered essential to monitor the consistency of heterologous antigen expression to ensure the batch-to-batch stability of potency [15]. Such monitoring requires the development of validated potency assays to determine expression levels, and concurrently the development of stable, well-characterized reference materials to permit comparison of results among assays and different vaccine batches. For ALVAC-HIV, a recombinant ALVAC viral vector that expresses the HIV-1 p24 and gp120 antigens, potency assays have been based on: (1) HIV-1 gene expression in CEF (immunoplaque assay); (2) HIV-1 gene expression in the human HeLa cell line (*F*luorescence *A*ctivated *C*ell *S*orter, FACS, analysis); and (3) immunoreactivity of HIV-1 p24 and gp120 following the lysis of infected HeLa cells.

In the immunoplaque assay, *double staining* is used to measure expression levels of the HIV-1 antigen and the ALVAC vector, respectively, in order to demonstrate a consistent ratio between antigen and vector expression levels [15]. In CEF, antigen expression levels are however 100%. In contrast, only 70% of infected HeLa cells appear to express HIV-1 antigens, suggesting that only limited amounts of these are produced during early gene expression in HeLa cells, which are nonpermissive for ALVAC replication [14, 15]. This finding suggests that these tests are more important for establishing the consistency of potency of vaccine batches; they would not by themselves provide any certain predictive measure of vaccine efficacy in humans. It is evident that the level of expression of heterologous antigen is dependent on the type of cell infected, and could be quite variable. In addition, it remains unclear as to how important correct post-translational modification, such as glycosylation, is for antigen presentation and immunogenicity.

Other viruses also show promise for the development of recombinant viral vector vaccines; these include human and animal adenoviruses, and certain alphaviruses.

Foreign DNA may be readily incorporated into adenoviruses by replacement of the viral early E1 sequences with a heterologous gene construct [16, 17]. The resulting recombinant adenoviral vector can be propagated in "special" E1-complementing cell lines, but is otherwise replication-deficient in human cells. Nevertheless, as with recombinant poxviral vectors, the heterologous gene is expressed in human cells to provide measurable amounts of the intended antigen. Recombinant adenoviral vectors have been widely used in the gene therapy field (for monogenic disorders, malignancies and cardiovascular disease) [16, 17], but are now increasingly being developed as prophylactic vaccines, particularly against HIV-1. For example, recombinant adenovirus type 5 vectors expressing consensus genes encoding HIV-1 clade B *gag* or trivalent *gag/pol/nef* are being tested as HIV-1 vaccines in clinical trials to assess safety and efficacy [95, 99]. As in the case of recombinant poxviral vectors, levels of antigen expression may be measured *in vitro* following infection of suitable cell lines. Again, it is not possible to predict immunogenicity/efficacy *in vivo* simply by reference to antigen expression levels *in vitro*.

Alphavirus "replicon" vectors based on the RNA viruses Semliki Forest virus (SFV), Sindbis virus (SINV) and Venezuelan Equine Encephalitis virus (VEE), are being proposed as potential preventative vaccines [15, 18–21], especially against HIV-1 and influenza viruses, but are not as advanced in their development as recombinant poxviral or adenoviral vector vaccines. They are produced by replacing the alphaviral structural genes with a heterologous gene insert, which renders the recombinant replicon vectors propagation-incompetent. The RNA replicons may be packaged into virus-like particles, or alternatively they can be converted into their corresponding functional DNA counterparts that directly transcribe the RNA replicons *in situ*. The alphavirus DNA replicon is considered to be essentially similar in physical properties to "conventional" plasmid DNA vaccines. However, in contrast to nonreplicating plasmid DNA, translation of the replicon RNA produces the alphavirus replicase complex, which catalyzes cytoplasmic self-amplification of the replicon RNA and, advantageously, leads to high-level synthesis of the heterologous antigen [18]. Such replicon RNA self-amplification causes cell death within a few days and is thus self-limiting. Following vaccination with DNA replicons, it is therefore expected that the expression of an antigen-coding gene would be at high level, but transient in duration [18]. This compares with plasmid DNA vaccines where plasmid DNA persists for months and expression of the antigen-coding gene is low level.

Another factor in the equation of how efficacious the vaccine will eventually be is whether the subjects to be immunized have pre-existing immunity to the virus vector. However, it may be assumed that humans will not have pre-existing immunity to canarypox virus on which ALVAC is based. Even those people immunized against smallpox with a vaccinia vaccine between the 1950s and 1970s will have little immunity left against vaccinia or its attenuated derivatives, such as MVA. Indeed, it has been shown recently that priming and amplification of immune responses specific for heterologous antigens expressed following administration of recombinant MVA can be achieved. Nevertheless, the primary use of recombinant MVA or ALVAC vaccines will, in most cases, trigger some immunity against their secondary usage, and thus potentially reduce their efficacy when subsequently used in the same subjects. The outlook for recombinant adenoviral vector vaccines is similar; there is often some pre-existing immunity to human adenoviruses in the population, as these are the cause of common respiratory infections. The use of less-prevalent human adenoviruses – for example, Ad6, Ad35, Ad11 or Ad24 [95, 100, 101], or more exotic animal adenoviruses [22] – as bases for the development of recombinant adenoviral vector vaccines may circumvent the potential problem of pre-existing immunity. Again, the primary use of such recombinant adenoviral vector vaccines may preclude any further secondary use with the same vaccine, or another vaccine based upon the same adenovirus, because of immunity to the adenoviral vector "backbone".

In summary, the predictive power of antigen-expression tests *in vitro* remains limited and uncertain. It is evident that it is difficult to mimic the *in-vivo* situation at the site of immunization within a single cultured cell line. As the exact phenotype of cells infected by the viral vector is not easy to establish, the antigen-expression tests in a particular cell line *in vitro* can only establish the level of antigen expression there; expression at the immunization site, while appearing probable, is uncertain both in the level and length of time over which expression may occur. The possibility of pre-existing immunity to the viral vector itself can add to the uncertainty of the efficacy of the vaccination. Therefore, measurements of antigen expression *in vitro* are only able to serve as a guide to potential expression *in vivo*. They do not by themselves reliably indicate the immunogenicity or, ultimately, the efficacy of the vaccine. Other tests are required to correlate with efficacy; the results of these may relate back to antigen expression *in vitro* and the combination of results may be of more practical value than those of the individual tests alone.

6.3 Post-Vaccination Testing

Many preventive vaccines function by stimulating the development of neutralizing antibodies to the immunogen(s), these antibodies being sufficient to protect against a subsequent infection by the pathogen. These vaccines are designed to induce a "humoral" or B-cell response that translates into an enduring production of protective, high-affinity, neutralizing antibodies against the pathogen [23]. The measurement of these specific antibodies in serum can be directly correlated with the efficacy of a vaccine to provide immunity against distinct pathogens [24]. Traditionally, such measurements are made by immunoassays, with no requirements for cell-based assays. However, "B-cell vaccination strategies" do not provide effective immunity against all pathogens, especially against rapidly mutating viruses, bacteria, and parasites. The control and elimination of these pathogens is subject to the effectiveness of the infected host's T-cell responses and cellular immune mechanisms.

For all vaccines, the development of antigen-specific T cells – as helper T cells, CTL and memory T cells – is essential for initiating and maintaining humoral and cellular immunity. The classic view is that peptides derived from the invading pathogen are recognized in combination with major histocompatibility complex (MHC) class II molecules by CD4+ T lymphocytes. These help B-cells in the induction of antibody responses and are crucial for the development of CD8+ T-cell-mediated protective immunity, while cytotoxic CD8+ T lymphocytes (CTL) – which kill infected cells and vanquish infective pathogens, especially viruses – recognize peptides in combination with MHC class I molecules. Indeed, in some infections (e.g., by HIV-1), cellular immunity due to CTL has been considered of greater importance than the presence of neutralizing antibodies (see below). It is evident from many studies of HIV-1-infected individuals that, although they develop neutralizing antibodies to the primary infecting HIV-1 strain, these rapidly become ineffective as the virus mutates [25, 26]. Moreover, the only HIV-1 antigen – recombinant envelope glycoprotein gp120 – to have been fully assessed in clinical trials induced neutralizing antibodies, but completely failed to provide a protective effect [27, 28, 118].

In continuing to use HIV-1 infection for illustration purposes, there is evidence that CD8+ T cells are involved in providing some resistance to the infective process by the virus. For instance, following primary infection with HIV-1, a strong CD8+ CTL response develops during the acute viremia phase, which generally persists into the chronic phase [29–31]. In HIV-1-infected individuals that do not progress to AIDS, strong HIV-1-specific CD8+ CTL activity, as well as CD4+ T-cell proliferative responses, against multiple epitopes have been measured [32, 33]. Furthermore, individuals that are multiply-exposed to HIV-1 (e.g., sex workers in the Gambia and Kenya) but do not contract the infection, have been shown to have HIV-1-specific CD8+ CTL [34, 35]. This observation, in addition to other supporting evidence from HIV-exposed seronegatives [36, 37] and from the SIV/rhesus-macaque model system [38–41], strongly suggest that CTL have a protective role against HIV-1 infection. However, more recent data on the Kenyan sex workers indicate that, while CTL were independently associated with age and recent HIV-1 exposure, they were not prospectively associated with protection [42]. Furthermore, CTL appear to kill HIV-1-infected cells only slowly *in vivo* [43], which has implications for vaccines designed to induce a lytic CTL response. For example, unless the vaccine-induced CTL response is several-fold greater than the natural response to HIV-1 infection, it is unlikely it will prevent infection, or mediate complete viral clearance in the case of a "therapeutic" vaccine (one that is administered to HIV-1 seropositives). However, a CTL response elicited by a "therapeutic" vaccine might reduce viral load and lengthen the asymptomatic period, and thus have some beneficial effect.

Despite the uncertainty over the effectiveness of CTL in conferring protection against HIV-1 infection [118], research is continuing into modern vector technologies with the potential for generating a cellular immune response to HIV-1. The hope is that a vaccine will emerge that induces a strong cellular immune response (especially a CTL response) which, while not being able to provide total protection from HIV infection, should enable "vaccines" to limit viral replication, reduce virus load, and slow the progression towards disease.

Based on the assumption that CTL do play some protective role against HIV-1 infection, and because of the ineffectiveness of neutralizing antibody inducing vaccines, much recent research has focused on developing HIV-1 vaccines that induce cellular immunity in the form of HIV-1-specific CTL [95, 118]. Starting with SIV in the rhesus-macaque model, potent CTL responses and protection against the pathogenic SIV-HIV hybrid virus, SHIV-89.6P, have been achieved with plasmid DNA [44, 45] or recombinant (MVA) poxvirus [46] vaccines expressing SIVmac239 Gag and HIV-1 89.6P Env. In addition, an effective CTL response and pronounced attenuation of subsequent SHIV challenge infection was demonstrated using either a replication-incompetent Ad5 viral vector expressing SIV *gag* [47] or an attenuated vesicular stomatitis virus (VSV) viral vector expressing *env* and *gag* genes and boosting with vectors having viral envelope glycoproteins from different VSV serotypes [48]. Although protection against SHIV 89.6P challenge has been repeatedly shown, it has been found to be more difficult to protect against other immunodeficiency viruses (e.g., SIVmac239) with these types of vaccines [49]. Numerous plasmid DNA and recombinant viral vector vaccines have been developed for immunizing against HIV-1; details of these T-cell vaccines are summarized in Table 6.1. An accurate monitoring of T-cell activation is required in order to evaluate their effectiveness in stimulating T-cell responses, for correlating vaccine-induced T-cell status with overall protection, and for making future improvements. However, a number of specific tests are applicable to this monitoring, and these are described in the following section.

6.3.1
Ex-Vivo Detection of Antigen-Specific T Cells

Originally, the only means of detecting antigen-specific T cells was to separate the T lymphocytes from whole blood and to expose predetermined numbers of cells deposited into wells of microtiter plates to synthetic antigen plus growth stimulatory cytokines, such as interleukin-2 (IL-2) and/or autologous feeder cells [50]. In order to convert this method into an assay for determining the number of antigen-specific cells, sequential dilution of T lymphocytes in 96-well plates, expansion of antigen-specific T cells and subsequent functional assays (e.g., proliferation or cytotoxicity assays) are required to ascertain the number of wells containing antigen-specific clones or lines. Unfortunately, this was found to be a cumbersome procedure which took several days to complete and often produced highly variable results. Furthermore, more recent studies have shown that this so-called "limiting dilution assay" (LDA) often significantly underestimated the "true" *in-vivo* frequencies of antigen-specific T cells [51–53]. This underestimation most likely occurs partly due to apoptosis of immature antigen-specific T cells *in vitro*.

Recently, considerable improvement of the determination of antigen-specific T-cell frequencies has been achieved by the development of assay methods requiring only relatively short *in-vitro* re-stimulation, such as the enzyme-linked immunospot assay (ELISPOT) [54, 55] and intracellular cytokine staining [51]. Details of these and other relevant assays (e.g., CTL cytotoxicity assays) are outlined below.

6.3.1.1 ELISPOT Assay

This assay depends on T cells responding specifically to antigen by the synthesis and secretion of cytokines, particularly IFN-γ. The principle of the ELISPOT assay is that secreted IFN-γ from stimulated antigen-specific T cells is captured by an anti-IFN-γ monoclonal antibody immobilized on a nitrocellulose membrane. The captured IFN-γ is then detected by a second anti-IFN-γ antibody conjugated to an enzyme; "cytokine spots" on the membrane which represent individual IFN-γ-secreting T cells are visualized and enumerated following enzyme–substrate reactions (Fig. 6.1).

The IFN-γ ELISPOT is relatively straightforward to perform, is reasonably sensitive, and has found a place for measuring T-cell response in vaccine trials [56, 57]. However, it is now perceived to have several shortcomings. First, the enumeration of antigen-specific T cells is merely a reflection of T-cell reactivity to recombinant antigens (not as expressed on antigen-presenting cells, APC, *in vivo* during a natural HIV-1 infection), and does not necessarily correspond to CTL cytotoxicity against HIV-infected cells. Second, IFN-γ has little anti-HIV-1 activity *per se* and, therefore, may not be the best cytokine to be tested [58]. Third, although the IFN-γ ELISPOT works well for strong immune responses to acute viral infections – including Epstein-Barr virus (EBV), cytomegalovirus (CMV), influenza virus, and HIV-1 – responses to vaccines are often far weaker, and this

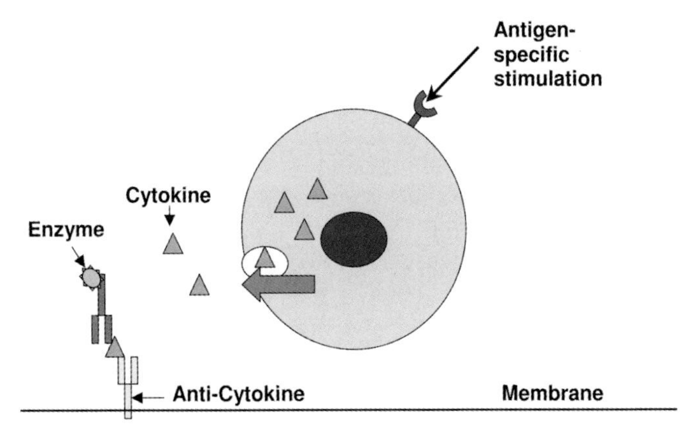

Fig. 6.1 Schematic diagram illustrating the enzyme-linked immunospot (ELISPOT) assay. Antigenically stimulated T lymphocytes are triggered to produce cytokines, for example, IFN-γ. The secreted cytokine is captured by an immobilized anti-cytokine antibody on the surface of a nitrocellulose membrane; the bound cytokine is detected by a second anti-cytokine antibody and cytokine spots are visualized by enzyme/substrate.

results in decreased sensitivity. Finally, variations in cell sampling times following immunization, variations in the reagents, the length of time required to develop the assay plates, and variations in environmental conditions can each significantly affect both the size and number of spots and the level of "background" response [59]. The lack of IFN-γ ELISPOT standardization/harmonization reagents and conditions may, to some degree, account for differences in the results reported for HIV-antigen-stimulated T-cell responses. For example, the magnitude of IFN-γ CD8+ T-cell responses to the gag and p24 HIV-1 proteins in chronically infected patients has been reported to correlate inversely with viral loads and directly to absolute CD4+ T-cell counts [60]. However, in other studies it has been shown that the frequencies of HIV-1-specific IFN-γ-secreting T cells correlated positively with viral loads [32, 61, 62]; also, measures of disease progression did not correlate with the magnitude of HIV-1-specific IFN-γ responses [63–65]. Therefore, *ex-vivo* measurements of IFN-γ secretion may not necessarily correspond to CTL cytotoxicity [80, 84] or to containment of the HIV infective process during the chronic phase [43]. In the cases of preventive vaccines against HIV-1, it is possible to demonstrate for most that, using the IFN-γ ELISPOT assay, they do stimulate the development of HIV-1 antigen-specific CD8+ T cells [56, 57] (see Table 6.1). Such measurements would serve as a measure of vaccine activity or potency *in vivo*, but since the "jury is out" on whether IFN-γ-secreting CD8+ T cells play a significant protective role against infection by HIV-1, they do not in themselves predict the protective efficacy of a vaccine [118].

6.3.1.2 Cytokine Capture Assay and Intracellular Cytokine Staining

In addition to the measurement of secreted cytokines, other methods aim to measure cell-associated cytokines, either in a cytokine capture assay (CCA) where cytokines (e.g., IFN-γ, IL-4) are detected on the cell surface (Fig. 6.2) [66], or more commonly by intracellular cytokine staining (ICS), where cytokine secretion is blocked by drugs such as brefeldin A or monensin, and cytokine accumulates in the cytoplasm (Fig. 6.3) [51, 67, 68].

For both methods, fluorescence-conjugated antibodies are used to bind cytokine and fluorescing cells analyzed by flow cytometry techniques; the cells are fixed

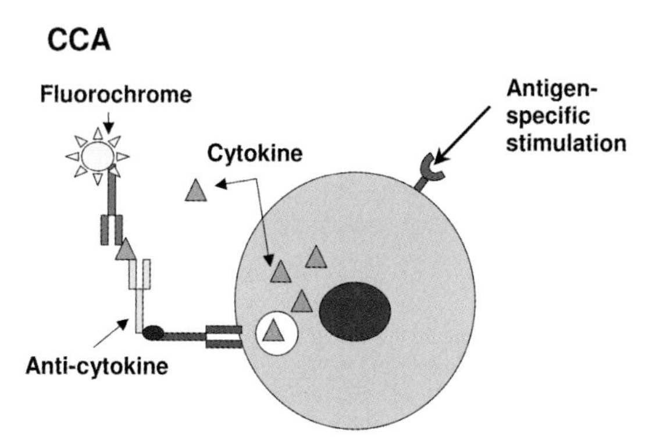

Fig. 6.2 Schematic diagram illustrating the cytokine-capture assay (CCA). These are based on the same principle as the ELISPOT assay, but secreted cytokine is captured on the cell surface and detected with fluorescent chromogen-labeled antibodies; the "stained" T cells remain viable.

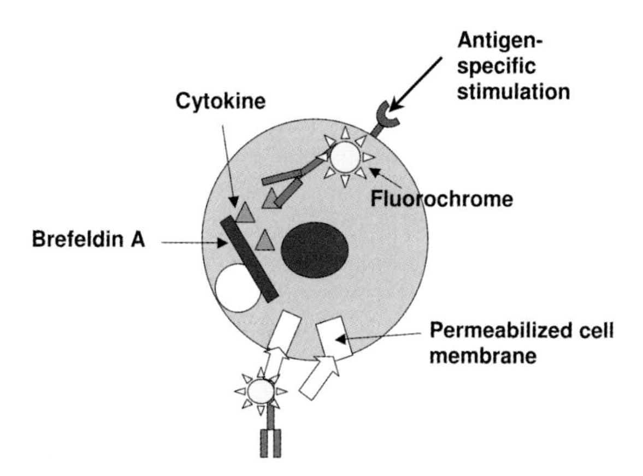

Fig. 6.3 Schematic diagram illustrating the intracellular staining (ICS) method. Cytokine secretion is blocked by either brefeldin A or monensin. The accumulated intracellular cytokine is detected following permeabilization of cells to permit entry of fluorescent chromogen-labeled anti-cytokine antibodies.

and permeabilized prior to cytokine staining in the ICS method (Figs. 6.2 and 6.3). These methods have the advantage that cells can be double-stained to identify the CD antigen phenotypes of cytokine-expressing cells. However, since in the CCA cells remain alive, cell-separation techniques such as FACS or sorting with paramagnetic beads may be applied to facilitate the purification of antigen-specific T cells and further characterization. CCA and ICS are very sensitive, permitting the *ex-vivo* detection of antigen-specific T cells with frequencies as low as 0.02% of the total cells. One disadvantage compared to the ELISPOT assay is that both CCA and ICS require relatively large numbers of cells. Moreover, they are both also subject to variations in cell sampling times, reagents, re-stimulation and other experimental conditions that apply to ELISPOT; fluctuating background responses may also be problematic [69].

6.3.1.3 Measurement of T-Cell Cytotoxicity

Cytotoxicity, which has been shown to be important in clearing many viral infections [70–72], may be a more reliable indicator of the effector function of antigen-specific CD8+ T cells than cytokine synthesis. It could be important, for example, in controlling HIV-1 pathogenesis, though this has recently been questioned as HIV-1 antigen-specific CD8+ T cells appear to kill infected cells at a slow rate *in vivo* [43]. As previously alluded to, measuring IFN-γ-secreting CD8+ T cells does not provide an estimate of cytotoxic potential, as these two activities can be independent of one another [73]. For instance, virus-specific CTL (CD8+ T cells) populations have been shown to vary in their capacity to secrete IFN-γ and/or kill target cells within 5 h *ex vivo* [74], or after re-stimulation *in vitro* [75]. However, as with the IFN-γ ELISPOT data, results from studies in which CTL cytotoxicity has been quantified during HIV infection are conflicting. Some have reported that *ex-vivo* CTL cytotoxicity is not detectable [76, 77], whereas others have found it in HIV-1-specific CD8+ T cells from either acute [31] or chronically infected patients [78–83]. This diverse information may possibly be due to the types of cytotoxicity assay employed in these studies. Most have used the bulk lysis assay approach in which cytotoxicity is measured in whole populations of T cells, such as the chromium release assay [85]. Alternatively, caspase activation has been measured in target cells undergoing lysis [86]. Such assays may provide results that are not correlated with other CD8+ T-cell functions, and thus cause misleading interpretations. The introduction of the Lysispot assay [73] to measure the frequencies of CTL in HIV-infected patients should provide more reliable results, as it can be used simultaneously with IFN-γ ELISPOT assays. For the Lysispot assay, ELISPOT plates are coated with anti-β-galactosidase and autologous B cells loaded with HIV-1 peptide pools, and transfected with herpes simplex virus-Lac amplicons added as target cells. Each HIV-1-specific CTL migrates in the medium to attack and kill one target cell during an incubation time of 4 h, whereupon β-galactosidase is released and binds to the immobilized anti-β-galactosidase. CTL spots are visualized and enumerated following the sequential addition of a biotinylated anti-β-galactosidase, streptavidin-alkaline phosphatase and substrate mix (Fig. 6.4) [73, 84].

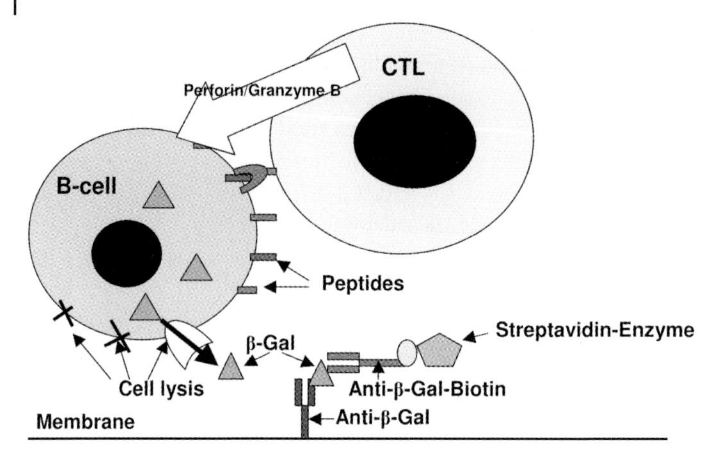

Fig. 6.4 Schematic diagram illustrating the Lysispot assay. Here, ELISPOT plates are coated with anti-β-galactosidase. Autologous B cells are loaded with peptide pools and transfected with herpes simplex virus-Lac amplicons added as target cells. Each peptide-specific cytotoxic T lymphocyte (CTL) migrates in the medium to attack and kill one target cell during an incubation time of 4 h. Following B-cell lysis, β-galactosidase is released and binds to the immobilized anti-β-galactosidase. CTL spots are visualized and enumerated following sequential addition of biotinylated anti-β-galactosidase, streptavidin-alkaline phosphatase and substrate mix.

By using the Lysispot and IFN-γ ELISPOT assays together, Synder-Cappione and colleagues [84] showed that the relative frequencies of IFN-γ-secreting and CTL varied markedly between different HIV peptide pools within the same patient, and some T cells lysed targets without secreting IFN-γ. This finding indicates that the measurement of IFN-γ production alone is probably insufficient to evaluate the breadth of the HIV-specific T-cell response. The further analysis of the frequency of directly cytotoxic HIV-specific T cells – for which the Lysispot assay provides a sensitive tool – may be of enhanced value, compared with IFN-γ production, in assessing the potential efficacy of preventive HIV vaccines.

In addition to the Lysispot assay, novel flow cytometric assays for measuring cell-mediated cytotoxicity are being developed. For example, active human CTL can be detected by an increased cell-surface expression of CD107a due to degranulation and the release of granzyme B, an effector of cytotoxicity in target cells [87, 88]. The parallel increase of annexin V binding to target cells due to target cell apoptosis (cell killing) can also be studied within the same assay. This flow cytometric assay has been designated the "LiveCount Assay", and permits measurement of CTL activation and frequency as well as of target cell death in the same sample [88]. Although aimed primarily at the *ex-vivo* detection of tumor antigen-specific T cells, the LiveCount Assay should also prove useful for detecting/measuring viral antigen-specific CTL.

6.3.2
Current Knowledge on T-Cell Responses in Vaccine Trials

Knowledge gained from analyzing T-cell responses in natural infections, as in HIV-1 infection previously described (above), has provided valuable insights into what is needed for vaccine design and vaccination schedules. Today, novel technological advances are being made, leading in turn to improved ways of measuring CD8+ T-cell responses. Increasingly, these responses are being tested for in animal models and clinical trials of new vaccines. Against HIV infection, much of the recent effort to derive efficacious prophylactic vaccines has focused on the design and development of recombinant poxviral- and adenoviral-vector vaccines. These have been used either singly or together, or in combination with plasmid DNA vaccines in prime-boost regimes [95, 118]. For example, many clinical trials have already been conducted in volunteers (total over 1800) with ALVAC candidate prophylactic vaccines. The latter are usually designed to express multiple HIV antigens, including gp120, gag, and pro (vCP205, vCP1521). An analysis of data from eight volunteer trials showed that no dose response for CTL development could be demonstrated with ALVAC-HIV, but that the number of immunizations and the vaccine dose were associated with the likelihood of developing CTL responses [89]. It was however impossible to distinguish the effect of either the number of immunizations or the vaccine dose on CTL responses, as these two study characteristics were highly associated with each other [89]. In a more recent clinical trial of an ALVAC-HIV vaccine (vCP1452), both the chromium-release assay for T-cell cytotoxicity and the IFN-γ ELISPOT assay gave similar results for HIV-specific CTL responses, and clearly these were shown to be higher (15–16%) in the regular-dose recipients than in the high-dose (about eightfold higher than the regular-dose) recipients (8%) [90]. This outcome suggests that high-dose ALVAC-HIV vaccines do not necessarily provoke higher CTL responses; moreover, recipients of the higher dose had greater local and systemic reactions. The value of the study is that it indicates the need for further studies to develop a highly immunogenic ALVAC-HIV vaccine in which vector and insert design are optimized, as well as ways to increase dosage without increasing reactogenicity.

Since the immune response to ALVAC-HIV vaccines is still relatively low, other vaccination strategies are being evaluated in clinical trials. For example, the immunogenicities of candidate plasmid DNA- and recombinant MVA-vectored HIV vaccines have been evaluated on their own and in a prime-boost regimen in seronegative volunteers [91, 92]. These vaccines contained a common immunogen, HIVA, which consists of HIV-1 clade A Gagp24/p17 proteins fused to a string of clade A epitopes recognized by CTLs. The IFN-γ ELISPOT assay was applied to measure HIV-1 specific T-cell frequencies (see Section 6.3.1.1). Encouragingly, the DNA and MVA vaccines alone and the DNA prime-MVA boost stimulated CTL responses in 14/18, 7/8, and 8/9 volunteers, respectively [91], and have also been shown to be safe [93]. Phase I/II trials with this and other recombinant MVA vaccines, mostly in prime-boost regimens, are continuing [94, 95], but overall in terms of CTL responses, with less than 35% of vaccinees scoring positive in the

IFN-γ ELISPOT test at any one time, the immunogenicity of the current poxvirus-based HIV vaccines has been modest. Better results have recently been achieved with recombinant adenoviral vectored HIV vaccines in nonhuman primates and human volunteers [96–99]. Over 50% of volunteers receiving a monovalent Ad5-*gag* vaccine developed longlasting HIV-1-specific CTL responses to HIV-1 peptides, as determined by cytokine production. A trivalent recombinant Ad5-*gag/pol/nef* has been recently developed and is currently being trialled in volunteers, but results are not expected until 2008 (see Table 6.1) [95].

Although DNA-Ad5 and Ad5-poxvirus prime boost regimes are also being tested [119], pre-existing immunity to Ad5 may prove problematic, and this has led to investigations to develop candidate recombinant Ad-HIV vaccines based on less prevalent Ads, serotypes such as Ad6 and Ad35 [95, 100, 119]. Numerous other

Table 6.1 Prophylactic HIV-1 vaccines in clinical trials

Vaccine[a]	Immunogenic entity	HIV-1 clade	T-cell responses/ Method	Ref.
DNA.HIVA[a]	Gagp24/p17 plus clade A epitopes	A	CTL/IFN-γ ELISPOT	[91–93]
MVA.HIVA[a]	Gagp24/p17 plus clade A epitopes	A	CTL/IFN-γ ELISPOT	[91–93]
ALVAC.vCP1452	Env-Gag-Pol-CTL epitopes	B	CTL/IFN-γ ELISPOT; chromium release	[90]
ALVAC.vCP205	Env-Pro-Gag-CTL epitopes	B	CTL/IFN-γ ELISPOT	[89, 113, 114]
Recombinant Fowlpox virus (rFPV)-HIV-B[a]	Gag-Pol	B	CTL/IFN-γ ELISPOT	[115, 116]
DNA.pHIS-HIV-B[a]	Mutated Gag-Pol-Env-Vpu-Tat-Rev	B	CTL/IFN-γ ELISPOT	[115, 116]
MVA.HIVC	Env-Gag-Tat-Rev-Nef	C	Under trial in India	[95]
MVA.HIVA/E	Env-Gag-Pol	A/E	Under trial in USA/ Thailand	[95]
Ad5.HIV-Gag monovalent	Gag	Codon-biased	CTL/IFN-γ, IL-2 and TNF-α secretion	[95, 99, 119]
Ad5.HIV-trivalent	Gag-Pol-Nef	?	Under trial in several countries	[95, 99]
Ad5.HIV-multivalent	Env (A,B,C); Gag-Pol-Nef (B)	A,B,C	Under trial in USA, Brazil, South Africa	[95, 117]

a) Most of the candidate HIV-1 vaccines have been used in plasmid DNA-viral vector prime-boost regimens, as in the vaccine pairs indicated here. For clinical trials information, the recommended websites are http://chi.ucsf.edu.vaccines and http://www.iavi.org.

TNF = tumor necrosis factor.

nonreplicating viral vector-based HIV vaccines are in development as candidate preventative HIV-1 vaccines [102]. Of these, perhaps the one that shows most promise is that based on an attenuated VSV vector expressing the Gag and Env proteins. This was found to provide complete protection against CD4+ T-cell loss and disease progression in the SHIV/macaque model when, advantageously, the vaccine was administered intranasally [103, 104].

The use of poxvirus-based vectored vaccines is also finding favor for the development of preventive vaccines against other intracellular pathogens, such as malaria [105]. Immunization with plasmid DNA encoding malarial antigens followed by boosting with a recombinant MVA vaccine [106], a novel attenuated fowlpox virus [107], or a recombinant replication-defective adenovirus [108] expressing malarial antigens in animal models, has been shown to enhance CD8+ T-cell induction. Durable, protective memory T-cell responses have also been quantified by ELISPOT assays after such prime-boost vaccination regimes against malaria in human volunteers [109].

6.4 Future Directions

The measurement of vaccine antigen content and development of T-cell responses has proved a useful guide in the evaluation of vaccine immunogenicity and efficacy in animal model systems, where laboratory strains of pathogen can be used as antigen/challenge. However, these parameters have proved much less secure for evaluating candidate preventive vaccines for clinical use in humans [118]. Here, the course of infectious diseases that vaccines are aimed at preventing are more variable and the "wild" strains of pathogens are often rapidly mutating so that the immune system is less effective in counteracting and eliminating them. Vaccines are, therefore, required to boost the immune defenses more than that normally occurring during natural infections. Ideally, they should boost both CD4+ helper T cells and CD8+ CTLs, generate the development of durable memory T cells, and elicit longlasting neutralizing antibody production. Such tests to quantify these functions can, at present, only be carried out post-vaccination, and involve the removal of blood and lymphoid cells from animals or humans. Both animal usage and the variability of such testing would be reduced if vaccine functions/activities could be modeled in artificially created cell-based systems *in vitro*. Today, investigations are ongoing into the creation of synthetic lymphoid tissue-like organoids, which may find uses for this purpose. For example, a tissue-engineered, lymphoid tissue-like organoid, which was constructed by transplanting stromal cells embedded in biocompatible scaffolds into the renal subcapsular space in mice, had an organized tissue structure comparable to that of secondary lymphoid organs [110]. Compartmentalized B-cell and T-cell clusters, high endothelial venule-like vessels, germinal centers and follicular dendritic cell networks were all present in the organoid. Furthermore, IgG antibody formation could be induced by antigenic stimulation. Although this organoid was transplantable into naïve normal or severe immunodeficiency (SCID) mice, maintenance *ex vivo* in a fully functioning form would be the next logical step for establishing a system to test vaccines. Another example is that of a tissue-engineered thymic organoid [111]. In this case, a three-dimensional framework of a tantalum-coated carbon matrix was seeded with murine thymic stroma in an attempt to provide a structure that supports the reconstitution of functioning thymic tissue. When seeded with human bone marrow-derived hematopoietic progenitor cells, the thymic organoid environment reproducibly generated mature functional T cells within 14 days. The thymic organoid may, with further developments, also lead to test systems for determining the immunogenicity of vaccines.

It is clear that much remains to be done to create artificial, synthetic organoids capable of providing valid test systems to quantify the immunogenicity/potency of preventive vaccines. Perhaps, this enterprise could be aided by computer modeling of immune functions. For instance, "model system heuristics" integrate known characteristics and processes of cells, pathogens, antigens, antibodies, and cytokines; the system may then generate qualitative simulations of primary immune responses to bacterial and viral infections [112]. Such a technique could, therefore, serve as a tool for testing assumptions about cellular functions in immune responses *in silica*, which may be informative for creating appropriate organoids for vaccine testing.

References

1 www.abbottdiagnostics.com/About_Us/Hepatitis_Retrovirus/.

2 Ferguson, M., Heath, A. and Minor, P. D. Report of a study for assessing the potency of hepatitis vaccines. *Biologicals* **1990**, *18*, 345–350.

3 Wood, D. J., Heath, A. B. and Sawyer, L. A. A WHO collaborative study on assays of the antigenic content of inactivated poliovirus vaccines. *Biologicals* **1995**, *23*, 83–94.

4 Wood, D. J. and Heath, A. B. A WHO collaborative study of immunogenicity assays of inactivated poliovirus vaccines. *Biologicals* **1995**, *23*, 301–311.

5 Mackett, M. Vaccinia virus vectors. In: Murray, J. A. H. (Ed.), *Transgenesis: applications of gene transfer*. John Wiley & Sons Ltd, Chichester, UK, **1992**, pp. 155–186.

6 Moss, B. Genetically engineered poxviruses for recombinant gene expression, vaccination, and safety. *Proc. Natl. Acad. Sci. USA* **1996**, *93*, 11341–11348.

7 Broeder, C. C. and Earl, P. L. Recombinant vaccinia viruses: Design, generation and isolation. *Mol. Biotechnol.* **1999**, *13*, 223–245.

8 Mastrangelo, M. J., Eisenlohr, L. C., Gomella, L. and Lattime, E. C. Poxvirus vectors: orphaned and underappreciated. *J. Clin. Invest.* **2000**, *105*, 1031–1034.

9 Paoletti, E., Taylor, J., Meignier, C., et al. Highly attenuated poxvirus vectors: NYVAC, ALVAC and TROVAC. *Dev. Biol. Stand.* **1995**, *84*, 159–163.

10 Pincus, S., Tartaglia, J. and Paoletti, E. Poxvirus-based vectors as vaccine candidates. *Biologicals* **1995**, *23*, 159–164.

11 Egan, M. A., Parlat, W. A. and Tartaglia, J. Induction of human immunodeficiency virus type 1 (HIV-1)-specific cytotoxic T-lymphocyte responses in seronegative adults by a non-replicating, host-range-restricted canary pox vector (ALVAC) carrying the HIV-1mn env gene. *J. Dis. Infect.* **1995**, *171*, 1623–1627.

12 Tine, J. A., Lanar, D. E., Smith, D., et al. NYVAC-Pf7: a poxvirus-vectored, multiantigen, multistage vaccine candidate for *Plasmodium falciparum* malaria. *Infect. Immun.* **1996**, *64*, 3833–3844.

13 Sutter, G. and Staib, C. Vaccinia vectors as candidate vaccines: the development of modified vaccinia virus Ankara for antigen delivery. *Curr. Drug Targets Infect. Disord.* **2003**, *3*, 263–271.

14 Marovich, M. A. ALVAC-HIV vaccines: clinical trial experience focusing on progress in vaccine development. *Expert Rev. Vaccines* **2004**, *3* (Suppl. 4), S99–S104.

15 WHO informal consultation on characterization and quality aspect of vaccines based on live viral vectors. Meeting report, **2003**. www.who.int/vaccine_research/documents/vvreport/en/index11.html.

15 Connelly, S. Adenoviral vectors. In: Meager, A. (Ed.), *Gene therapy technologies, applications and regulations: from laboratory to clinic.* John Wiley & Sons Ltd, Chichester, UK, **1999**, pp. 87–107.

16 Yeh, P. and Perricaudet, M. Advances in adenoviral vectors: from genetic engineering to their biology. *FASEB J.* **1997**, *11*, 615–623.

17 Lundstrom, K. Alphavirus-based vaccines. *Curr. Opin. Mol. Ther.* **2002**, *4*, 28–34.

18 Pushko, P., Parker, M., Ludwig, G. V., et al. Replicon-helper systems from attenuated Venezuelan equine encephalitis virus: expression of heterologous genes *in vitro* and immunisation against heterologous pathogens *in vivo*. *Virology* **1997**, *239*, 389–401.

19 Perri, S., Greer, C. E., Thudium, K., et al. An alphavirus replicon particle chimera derived from Venezuelan equine encephalitis and Sindbis viruses is a potent gene-base vaccine delivery system. *J. Virol.* **2003**, *77*, 9710–9715.

20 Davis, N. L., Caley, I. J., Brown, K. W., et al. Vaccination of macaques against a pathogenic simian immunodeficiency virus with Venezuelan equine encephalitis virus replicon particles. *J. Virol.* **2000**, *74*, 371–378.

21 Khatri, A., Xu, Z. Z. and Both, G. W. Gene expression by atypical recombinant ovine adenovirus vectors during abortive infection of human and animal cells *in vitro*. *Virology* **1997**, *239*, 226–237.

22 Slifka, M. K. and Ahmed, R. Long-lived plasma cells: a mechanism for maintaining persistent antibody production. *Curr. Opin. Immunol.* **1998**, *10*, 252–258.

24 Griffen, J. F. A strategic approach to vaccine development: animal models monitoring vaccine efficacy, formulation and delivery. *Adv. Drug Deliv. Rev.* **2002**, *54*, 851–861.

25 Moog, C., Fleury, H. J., Pelligrin, I., et al. Autologous and heterologous neutralizing antibody responses following initial seroconversion in human immunodeficiency virus type-1-infected individuals. *J. Virol.* **1997**, *71*, 3734–3741.

26 Richman, D. D., Wrin, T., Little, S. J. and Petropoulos, C. J. Rapid evolution of the neutralizing antibody response to HIV type infection. *Proc. Natl. Acad. Sci. USA* **2003**, *100*, 4144–4149.

27 Connor, R. I., Korber, B. T., Graham, B. S., et al. Immunological and virological analyses of persons infected by human immunodeficiency virus type 1 while participating in trials of recombinant gp120 subunit vaccines. *J. Virol.* **1998**, *72*, 1552–1576.

28 Pantophlet, R. and Burton, D. R. GP120: target for neutralizing HIV-1 antibodies. *Annu Rev Immunol.* **2006**, *24*, 739–769.

29 Walker, B. D. and Korber, B. T. Immune control of HIV: the obstacles of HLA and viral diversity. *Nat. Immunol.* **2001**, *2*, 473–475.

30 Harrer, T., Harrer, E., Kalams, S. A., et al. Strong cytotoxic T cell and weak neutralizing antibody responses in a subset of persons with stable non-progressing HIV type 1 infection. *AIDS Res. Hum. Retroviruses* **1996**, *12*, 585–592.

31 Ogg, G. S., Xin, J., Bonhoeffer, S., et al. Quantitation of HIV-1-specific cytotoxic T lymphocytes and plasma viral RNA load. *Science* **1998**, *279*, 2103–2106.

32 Betts, M. R., Ambrozak, D. R., Douek, D. C., et al. Analysis of total immunodeficiency virus (HIV)-specific CD4(+) and CD8(+) T-cell responses: relationship to viral load in untreated HIV infection. *J. Virol.* **2001**, *75*, 11983–11991.

33 Rosenberg, E. S., Billingsly, J. M., Caliendo, A. M., et al. Vigorous HIV-1-specific CD4+ T-cell responses associated with control of viraemia. *Science* **1997**, *278*, 1447–1450.

34 Rowland-Jones, S., Sutton, S. J., Ariyoshi, K., et al. HIV-specific cytotoxic T-cells in HIV-exposed but uninfected Gambian women. *Nat. Med.* **1995**, *1*, 59–64.

35 Rowland-Jones, S., Dong, T., Fowke, K. R., et al. Cytotoxic T cell responses to multiple conserved HIV epitopes in HIV-resistant seronegative prostitutes in Nairobi. *J. Clin. Invest.* **1998**, *102*, 1758–1765.

36 Pinto, L. A., Sullivan, J., Berkovsky, J. A., et al. ENV-specific cytotoxic T lymphocyte responses in HIV seronegative health-care workers occupationally exposed to HIV-contaminated body fluids. *J. Clin. Invest.* **1995**, *96*, 867–876.

37 Shearer, G. M. and Clerici, M. Protective immunity against HIV infection: Has nature done the experiment for us? *Immunol. Today* **1996**, *17*, 21–24.

38 Schmitz, J. E., Kuroda, M. J., Santra, S., et al. Control of viremia in simian immunodeficiency virus infection by CD8+ lymphocytes. *Science* **1999**, *283*, 857–860.

39 Kent, S. J., Woodward, A. and Zhao, A. Human immunodeficiency virus type 1 (HIV-1)-specific T-cell responses correlate with the control of acute HIV-1 infection in macaques. *J. Infect. Dis.* **1997**, *176*, 1188–1197.
40 Matano, T., Shibata, R., Siemon, C., et al. Administration of an anti-CD8 monoclonal antibody interferes with the clearance of chimeric simian/human immunodeficiency virus during primary infections of macaques. *J. Virol.* **1998**, *72*, 164–169.
41 Jin, X., Bauer, D. E., Tuttleton, S. E., et al. Dramatic rise in plasma viremia after CD8+ T-cell depletion in simian immunodeficiency virus-infected macaques. *J. Exp. Med.* **1999**, *189*, 991–998.
42 Kaul, R., Rutherford, J., Rowland-Jones, S. L., et al. HIV-1 Env-specific cytotoxic T-lymphocyte responses in exposed, uninfected Kenyan sex workers: a prospective analysis. *AIDS* **2004**, *18*, 2087–2089.
43 Asquith, B., Edwards, C. T. T., Lipsitch, M., et al. Inefficient cytotoxic T lymphocyte-mediated killing of HIV-1 infected cells *in vivo*. *PLoS Biol.* **2006**, *4*, e90.
44 Barouch, D. H., Santra, S., Schmitz, J. E., et al. Control of viremia and prevention of clinical AIDS in rhesus monkeys by cytokine-augmented DNA vaccination. *Science* **2000**, *290*, 486–492.
45 Barouch, D. H., Craiu, A., Santra, S., et al. Elicitation of high frequency cytotoxic T-lymphocyte responses against both dominant and subdominant simian-human immunodeficiency virus epitopes by DNA vaccination of rhesus monkeys. *J. Virol.* **2001**, *75*, 2462–2467.
46 Barouch, D. H., Santra, S., Kuroda, M. J., et al. Reduction of simian-human immunodeficiency virus 89.6P viremia in rhesus monkeys by recombinant modified vaccinia virus Ankara vaccination. *J. Virol.* **2001**, *75*, 5151–5158.
47 Shiver, J. W., Fu, T. M., Chen, L., et al. Replication-incompetent adenoviral vaccine vector elicits effective anti-immunodeficiency-virus immunity. *Nature* **2002**, *415*, 331–335.
48 Rose, N. F., Marx, P. A., Luckay, A., et al. An effective AIDS vaccine based on live attenuated vesicular stomatitis virus. *Cell* **2001**, *106*, 539–549.
49 Horton, H., Vogel, T. U., Carter, D. K., et al. Immunization of rhesus macaques with a DNA prime/modified vaccinia virus Ankara boost regime induces broad simian immunodeficiency virus (SIV)-specific T-cell responses and reduces initial viral replication but does not prevent disease progression following challenge with pathogenic SIV-mac239. *J. Virol.* **2002**, *76*, 7187–7202.
50 Pimsler, M. and Foreman, J. Estimates of the precursor frequency of cytotoxic T lymphocytes against antigens controlled by defined regions of the H-2 gene complex: comparison of the effect of H-2 differences due to intra-H-2 recombination vs. mutation. *J. Immunol.* **1978**, *121*, 1302–1305.
51 Murali-Krishna, M., Altman, J. D., Suresh, D. J. D., et al. Counting antigen-specific CD8+ T-cells: a re-evaluation of bystander activation during viral infection. *Immunity* **1998**, *8*, 177–188.

52 Butz, E. A. and Bevan, M. J. Massive expansion of antigen-specific CD8+ T-cells during an acute virus infection. *Immunity* **1998**, *8*, 167–176.

53 Busch, D. H., Pilip, I. M., Vijh, S. and Pamer, E. G. Coordinate regulation of complex T-cell populations responding to bacterial infection. *Immunity* **1998**, *8*, 353–362.

54 Czerkinsky, C., Nilsson, L-Å., Tarkowski, A., et al. The solid phase enzyme-linked immunospot assay (ELISPOT) for enumerating antibody-secreting cells: Methodology and applications. In: Kemeny, D. M. (Ed.), *ELISA and other solid phase immunoassays*. John Wiley & Sons, Chichester, **1988**, pp. 218–239.

55 Miyahara, Y., Murata, K., Rodriguez, D., et al. Quantification of antigen specific CD8+ T-cells using an ELISPOT assay. *J. Immunol. Methods* **1995**, *181*, 45–54.

56 Lalvani, A., Brookes, R., Hambleton, S., et al. Rapid effector function in CD8+ memory T-cells. *J. Exp. Med.* **1997**, *186*, 859–865.

57 Mwau, M., McMichael, A. J. and Hanke, T. Design and validation of an ELISPOT assay for use in clinical trials of candidate HIV vaccines. *AIDS Res. Human Retroviruses* **2002**, *18*, 611–618.

58 Coccia, E. M., Krust, B. and Hovanessian, A. G. Specific inhibition of viral protein synthesis in HIV-infected cells in response to interferon treatment. *J. Biol. Chem.* **1994**, *269*, 23087–23094.

59 Chichester, J. A., Feitelson, M. A. and Calkins, C. E. Different response requirements for IFNgamma production in ELISPOT assay by CD4+ T-cells from mice early and late after immunization. *J. Immunol. Methods* **2006**, *309*, 99–107.

60 Edwards, B. H., Bansal, A., Sabbaj, S., et al. Magnitude of functional CD8+ T-cell responses to the gag protein of human immunodeficiency virus type 1 correlates inversely with viral load in plasma. *J. Virol.* **2002**, *76*, 2298–2305.

61 Buseyne, F., Scott-Algara, D., Porrot, F., et al. Frequencies of *ex vivo*-activated human immunodeficiency virus type 1-specific γ-interferon-producing CD8+ T-cells in infected children correlate positively with plasma viral load. *J. Virol.* **2002**, *76*, 12414–12422.

62 Trabattoni, D., Piconi, S., Biasin, M., et al. Granule-dependent mechanisms of lysis are defective in CD8 T-cells of HIV-infected, antiretroviral therapy-treated individuals. *AIDS* **2004**, *18*, 859–869.

63 Dalod, M., Dupuis, M., Deschemin, J. C., et al. Broad, intense anti-human immunodeficiency virus (HIV) *ex vivo* CD8+ responses in HIV type 1-infected patients: comparison with anti-Epstein-Barr virus responses and changes during antiretroviral therapy. *J. Virol.* **1999**, *73*, 7108–7116.

64 Addo, M. M., Yu, X. G., Rathod, A., et al. Comprehensive epitope analysis of human immunodeficiency virus type 1 (HIV-1)-specific T-cell responses directed against the entire HIV-1 genome demonstrate broadly directed responses, but no correlation to viral load. *J. Virol.* **2003**, *77*, 2081–2092.

65 Gea-Banacloche, J. C., Migueles, S. A., Martino, L., et al. Maintenance of large numbers of virus-specific CD8+ T-cells in HIV-infected progressors and long-term nonprogressors. *J. Immunol.* **2000**, *165*, 1082–1092.

66 Manz, R., Assenmacher, M., Pfluger, E., et al. Analysis and sorting of live cells to secreted molecules, relocated to a cell surface affinity matrix. *Proc. Natl. Acad. Sci. USA* **1995**, *92*, 1921–1925.

67 Picker, L. J., Singh, M. K., Zdraveski, Z., et al. Direct demonstration of cytokine synthesis heterogeneity among human memory/effector T-cells by flow cytometry. *Blood* **1995**, *86*, 1408–1419.

68 Sun, Y., Iglesias, E., Samri, A., et al. A systematic comparison of methods to measure HIV-1 specific CD8 T-cells. *J. Immunol. Methods* **2003**, *272*, 23–34.

69 Dunn, H. S., Haney, D. J., Ghanekar, S. A., et al. Dynamics of CD4 and CD8 T-cell responses to cytomegalovirus in healthy human donors. *J. Infect. Dis.* **2002**, *186*, 15–22.

70 Kagi, D., Lederman, B., Burki, K., et al. Cytotoxicity mediated by T-cells and natural killer cells is greatly impaired in perforin-deficient mice. *Nature* **1994**, *369*, 31–37.

71 Kagi, D., Seiler, P., Pavlovic, J., et al. The roles of perforin- and Fas-dependent cytotoxicity in protection against cytopathic and noncytopathic viruses. *Eur. J. Immunol.* **1995**, *25*, 3256–3262.

72 Topham, D. J., Tripp, R. A. and Doherty, P. C. CD8+ T-cells clear influenza virus by perforin- or Fas-dependent processes. *J. Immunol.* **1997**, *159*, 5197–5200.

73 Synder, J. E., Bowers, W. J., Livingstone, A. M., et al. Measuring the frequency of mouse and human cytotoxic T-cells by the Lysispot assay: independent regulation of cytokine secretion and short term killing. *Nat. Med.* **2003**, *9*, 231–235.

74 Masopust, D., Vezys, V., Marzo, A. L. and Lefrancois, L. Preferential localization of effector memory cells in lymphoid tissue. *Science* **2001**, *291*, 2413–2417.

75 Regner, M., Lobigs, M., Blanden, R. V. and Mullbacher, A. Effector cytolytic function but not IFN-γ production in cytotoxic T-cells triggered by virus-infected target cell *in vitro*. *Scand. J. Immunol.* **2001**, *54*, 366–374.

76 Shankar, P., Russo, M., Harnisch, B., et al. Impaired function of circulating HIV-specific CD8+ T-cells in chronic human immunodeficiency virus infection. *Blood* **2000**, *96*, 3094–3101.

77 Trimble, L. A. and Lieberman, J. Circulating CD8 T lymphocytes in human immunodeficiency virus-infected individuals have impaired function and downmodulate CD3 ζ, the signaling chain of the T-cell receptor complex. *Blood* **1998**, *91*, 585–594.

78 Mueller, Y. M., Bojczuk, M., Halstead, E. S., et al. IL-15 enhances survival and function of HIV-specific CD8+ T-cells. *Blood* **2003**, *101*, 1024–1029.

79 Mueller, Y. M., DeRosa, S. C., Hutton, J. A., et al. Increased CD95/Fas-induced apoptosis of HIV-specific CD8+ T-cells. *Immunity* **2001**, *15*, 871–882.

80 Lichterfield, M., Yu, X. G., Waring, M. T., et al. HIV-1-specific cytotoxicity is preferentially mediated by a subset of CD8+ T-cells producing both interferon-γ and tumour necrosis factor-α. *Blood* **2004**, *104*, 487–494.

81 Keoshkerian, E., Ashton, L. J., Smith, D. G., et al. Effective HIV-specific cytotoxic T-lymphocyte activity in long-term nonprogressors: associations with viral replication and progression. *J. Med. Virol.* **2003**, *71*, 483–491.

82 Appay, V., Nixon, D. F., Donahoe, S. M., et al. HIV-specific CD8+ T-cells produce antiviral cytokines but are impaired in cytolytic function. *J. Exp. Med.* **2000**, *192*, 63–75.

83 Gray, C. M., Lawrence, J., Schapiro, J. M., et al. Frequency of class I HLA-restricted antiHIV CD8+ T-cells in individuals receiving highly active antiretroviral therapy (HAART). *J. Immunol.* **1999**, *162*, 1780–1786.

84 Synder-Cappione, J. E., Divekar, A. A., Maupin, G. M., et al. HIV-specific cytotoxic cell frequencies measured directly *ex vivo* by the Lysispot assay can be higher or lower than the frequencies of IFN-γ-secreting cells: anti-HIV cytotoxicity is not generally impaired relative to other chronic virus responses. *J. Immunol.* **2006**, *176*, 2662–2668.

85 Brunner, K. T., Mauel, J., Cerotti, J. C. and Chapuis, B. Quantitative assay of the lytic action of immune lymphoid cells on 51-Cr-labelled allogeneic target cells *in vitro*: inhibition by iso antibody and by drugs. *Immunology* **1968**, *14*, 181–186.

86 Lui, L., Chadroudi, A., Silvestri, G., et al. Visualization and quantification of T-cell mediated cytotoxicity using cell-permeable fluorogenic caspase substrates. *Nat. Med.* **2002**, *8*, 185–189.

87 Burkett, M. W., Shafer-Weaver, K. A., Strobl, S., et al. A novel flow cytometric assay for evaluating cell-mediated cytotoxicity. *J. Immunother.* **2005**, *28*, 396–402.

88 Devevre, E., Romero, P. and Mahnke, Y. D. LiveCount Assay: concomitant measurement of cytolytic and phenotypic characterisation of CD8(+) T-cells by flow cytometry. *J. Immunol. Methods* **2006**,*311*, 31–46.

89 Edupuganti, S., Weber, D. and Poole, C. Cytotoxic T-lymphocyte responses to canarypox vector-based HIV vaccines in HIV-seronegative individuals: a meta-analysis of published studies. *HIV Clin. Trials* **2004**, *5*, 259–268.

90 Goepfert, P. A., Horton, H., McErath, M. J., et al. High dose Canarypox vaccine expressing HIV-1 protein, in seronegative human subjects. *J. Dis. Infect.* **2005**, *192*, 1249–1259.

91 Mwau, M., Cebere, I., Sutton, J., et al. A human immunodeficiency virus type 1 (HIV-1) clade A vaccine in clinical trials: stimulation of HIV-specific T-cell responses by DNA and recombinant modified vaccinia virus Ankara (MVA) vaccines in humans. *J. Gen. Virol.* **2004**, *85*, 911–919.

92 Estcourt, M. J., McMichael, A. J. and Hanke, T. DNA vaccines against human immunodeficiency virus type 1. *Immunol. Rev.* **2004**, *199*, 144–155.

93 Cebere, I., Dorell, L., McShane, H., et al. Phase I clinical trial safety of DNA- and modified virus Ankara-vectored human immunodeficiency virus type I (HIV-1) vaccines administered alone and in prime-boost regime to healthy HIV-1 uninfected volunteers. *Vaccine* **2006**, *24*, 417–425.

94 Im, E. J. and Hanke, T. MVA as a vector for vaccines against HIV-1. *Expert Rev. Vaccines* **2004**, *3* (Suppl 4), S89–S97.

95 Girard, M. P., Osmanov, S. K. and Kieny, M. P. A review of vaccine research and development: the human immunodeficiency virus (HIV). *Vaccine* **2006**, *24*, 4692–4700.

96 Santra, S., Seaman, M. S., Xu, L., et al. Replication-defective adenovirus serotype 5 vectors elicit durable cellular and humoral responses in nonhuman primates. *J. Virol.* **2005**, *79*, 6516–6522.

97 Shiver, J. W., Fu, T. M., Chen, L., et al. Replication-incompetent adenoviral vaccine vector elicits effective anti-immunodeficiency-virus immunity. *Nature* **2002**, *415*, 331–335.

98 Casimiro, D. R., Chen, L., Fu, T. M., et al. Comparative immunogenicity in rhesus monkeys of DNA plasmid, recombinant vaccinia virus, and replication-defective adenovirus vectors expressing a human immunodeficiency virus type 1 gag gene. *J. Virol.* **2003**, *77*, 6305–6313.

99 Shiver, J. W. and Emini, E. A. Recent advances in the development of HIV-1 vaccines using replication-incompetent adenovirus vectors. *Annu. Rev. Med.* **2004**, *55*, 355–372.

100 Kotense, S., Koudstaal, W., Sprangers, M., et al. Adenovirus types 5 and 35 seroprevalence in AIDS risk groups supports type 35 as a vaccine vector. *AIDS* **2004**, *18*, 1213–1216.

101 Barouch, D. H., Pau, M. G., Custers, J. H., et al. Immunogenicity of recombinant adenovirus serotype 35 vaccine in the presence of pre-existing anti-Ad5 immunity. *J. Immunol.* **2004**, *172*, 6290–6297.

102 Sauter, S. L., Rahman, A. and Muralidhar, G. Non-replicating viral vector-based AIDS vaccines: interplay between viral vectors and the immune system. *Curr. HIV Res.* **2005**, *3*, 157–181.

103 Rose, N. F., Marx, P. A., Luckay, A., et al. An effective AIDS vaccine based on live attenuated vesicular stomatitis virus recombinants. *Cell* **2001**, *106*, 539–549.

104 Publicover, J., Ramsburg, E. and Rose, J. K. A single-cycle vaccine vector based on vesicular stomatitis virus can induce immune responses comparable to those generated by a replication competent vector. *J. Virol.* **2005**, *79*, 13231–13238.

105 Schneider, J., Gilbert, S. C., Hannan, C. M., et al. Induction of CD8+ T-cells using heterologous prime-boost immunisation strategies. *Immunol. Rev.* **1999**, *170*, 29–38.

106 Schneider, J., Gilbert, S. C., Blanchard, T. J., et al. Enhanced immunogenicity for CD8+ T-cell induction and complete protective efficacy of malaria DNA vaccination by boosting with modified vaccinia virus Ankara. *Nat. Med.* **1998**, *4*, 397–402.

107 Anderson, R. J., Hannan, C. M., Gilbert, S. C., et al. Enhanced CD8+ T-cell immune responses and protection elicited against *Plasmodium berghei* malaria by prime boost immunization regimens using a novel attenuated fowlpox virus. *J. Immunol.* **2004**, *172*, 3094–3099.

108 Gilbet, S. C., Schneider, J., Hannan, C. M., et al. Enhanced CD8 T-cell immunogenicity and protective efficacy in a mouse malaria model using a recombinant adenoviral vaccine in heterologous prime-boost immunization regimes. *Vaccine* **2002**, *20*, 1039–1045.

109 Keating, S. M., Bejon, P., Berthoud, T., et al. Durable human memory T-cells quantifiable by cultured enzyme-linked immunospot assays are induced by heterologous prime boost immunization and correlate with protection against malaria. *J. Immunol.* **2005**, *175*, 5675–5680.

110 Suematsu, S. and Watanabe, T. Generation of a synthetic lymphoid tissue-like organoid in mice. *Nat. Biotechnol.* **2004**, *22*, 1539–1545.

111 Poznansky, M. C., Evans, R. H., Foxall, R. B., et al. Efficient generation of human T-cells from a tissue-engineered thymic organoid. *Nat. Biotechnol.* **2000**, *18*, 714–715.

112 Trelease, R. B. and Park, J. Qualitative process modeling of cell-cell-pathogen interactions in the immune system. *Comput. Methods Programs Biomed.* **1996**, *51*, 171–181.

113 Ferrari, G., Humphrey, W., McElrath, M. J., et al. Clade B-based HIV-1 vaccines elicit cross-clade cytotoxic T lymphocyte reactivities in uninfected volunteers. *Proc. Natl. Acad. Sci. USA* **1997**, *94*, 1396–1401.

114 Cao, H., Kaleebu, P., Hom, D., et al. Immunogenicity of a recombinant human immunodeficiency virus (HIV)-canarypox vaccine in HIV-sero-negative Ugandan volunteers: results of the HIV Network for Prevention Trials 007 Vaccine Study. *J. Dis. Infect.* **2003**, *187*, 887–895.

115 Coupar, B. E., Purcell, D. F., Thomson, S. A., et al. Fowlpox virus vaccines for HIV and SHIV clinical and preclinical trials. *Vaccine* **2006**, *24*, 1378–1388.

116 Kelleher, A. D., Puls, R. L., Bebbington, M., et al. A randomized, placebo-controlled phase I trial of DNA prime, recombinant fowlpox boost prophylactic vaccine for HIV-1. *AIDS* **2006**, *20*, 294–297.

117 Mascola, J. R., Sambor, A., Beaudry, K., et al. Neutralizing antibodies elicited by immunization of monkeys with DNA plasmids and recombinant adenoviral vectors expressing human immunodeficiency virus type 1 proteins. *J. Virol.* **2005**, *79*, 771–779.

118 Spearman, P. Current progress in the development of HIV vaccines. *Curr. Pharmaceut. Design* **2006**, *12*, 1147–1167.

119 Casimiro, D. R., Bett, A. J., Fu, T. M., et al. Heterologous human immunodeficiency virus type 1 priming-boosting immunization strategies involving replication-defective adenovirus and poxvirus vaccine vectors. *J. Virol.* **2004**, *78*, 11434–11438.

7
Designer Cells Derived from Primary Tissue and Designed Cell Lines as a Sustainable Cell Source for Drug Discovery and Safety Assessment

Volker Sandig and Ingo Jordan

7.1
Introduction

Assays based on cultured cells are moving to the center of the drug discovery process, and in recent years have gained a substantial share in safety testing. Primary cells and cell lines may yield complex and often more meaningful data compared to biochemical assays, provided that the cell system is chosen or designed to fit the purpose of the assay. These requirements vary substantially however, and often a compromise must be made between practicability on one hand and a close match with the natural response on the other hand. Permanent cell lines with their capability of continuous self-renewal and amplification provide a sustainable source of standardized material, much in contrast to primary tissue. However, proteome or transcriptome wide screens for pleiotropic effects suffer from substantial differences between immortalized cells and primary cells, or cultivated and primary tissue samples. The target of first, maintaining the physiological properties of a given tissue, and second, immortalizing the cells that regenerate that tissue *in vitro* to provide a continuous supply, represents a formidable challenge that requires complex and careful manipulations of several interlinked cellular pathways.

Once an immortalized cell line is obtained, additional layers of complexity to the screen can be added by genetic manipulation not possible with primary cells. These cell lines are designed specifically to express a single target protein or proteins assembling into a functional pathway. In addition, they may be designed to exhibit novel functional properties not found in Nature, for example to improve detection and quantification in drug screens. The convenience of set-up and handling, robustness, and the stability of assays with designed cell lines is an important advantage over primary material. Increased complexity provides the opportunity for very elaborate and unique assays, but this usually comes at the cost of further remoteness from the natural situation, such as loss of certain differentiation states or tissue markers. The focal point of this chapter is to discuss the approaches for immortalization, the application of designer cell lines, and their limitations.

Drug Testing In Vitro: Breakthroughs and Trends in Cell Culture Technology
Edited by Uwe Marx and Volker Sandig

ISBN: 978-3-527-31488-1

7.2 Suitability and Limitations of Primary Cells as Physiologic Models

Growing numbers of hits from primary high-throughput discovery programs generate the need for early, fast and robust lead candidate identification. In order to avoid creating a bottleneck at this stage, decisions on compound selection/rejection must be made before animal testing, with the required throughput and minimal substance use. For the accurate prediction of compound specificity and related toxicity, these assays must rely on physiological models that clearly reflect the situation in the target tissue. Miniaturized culture systems of primary cells or tissue samples – preferably of human origin – should best suit this purpose. Indeed, the human primary hepatocyte is one of the most extensively used *in-vitro* models in toxicology because of its central role in drug metabolism and detoxification and as a sensitive detector for generalized cytotoxicities relevant to other organs [1]. Since human hepatocytes are obtained from donor livers not used in transplantation, the availability of fresh cells is unpredictable and limited (http://www.unos.org./). Frozen hepatocytes have become increasingly available as a more reliable source, and have also improved in quality due to sophisticated cryopreservation protocols [2]. However, they still suffer from highly variable attachment to the culture dish and changes in gene activity affecting apoptosis, differentiation, and metabolic enzymes [3]. The expression pattern is not just different between the liver and the cultured hepatocyte; rather, it changes progressively, with the time between isolation and analysis further complicating validation [4]. Many of these changes result from interactions with immune cells in response to injury during the isolation procedure itself. Liver slices which *a priori* would be expected to reflect *in-vivo* gene expression better, as they maintain much of the tissue architecture, strongly induce proinflammatory genes and do not reflect compound-induced changes better than primary hepatocytes [5]. Assay variation is also affected by gender and age differences between donors, as well as polymorphisms determining the activity of metabolic enzymes such as cytochrome P450s. Up to 25 variants have been found for a single enzyme (CYP2D6) [6]. These challenges might in time direct the focus to rodent hepatocytes, where a homogenous supply can be assured.

However, even when the changes in gene expression are taken into account, human primary cells often reproduce the patient's response better than rodent cells or rodent *in-vivo* models. For instance, compounds that act as agonists for PPARalpha (peroxisome proliferator- activated receptor) which regulate the expression of lipid and xenobiotic metabolism genes induce hepatic and pancreatic tumors in rats and mice. However, tumorigencity is absent in man and monkey because of low receptor density and reduced response of regulated genes (for a review, see [7]).

7.3 Tumor Cell Lines: Sometimes an Alternative

Compared to primary cells, permanent cell lines would represent continuous, stable and rather homogeneous test systems. Nevertheless, cell lines are – with few exceptions – less popular metabolic models. Of the available liver cell lines, only in HepG2 (the dominant cell line in drug metabolism) and in the more recently identified hepatoma BC2, can phase I and II enzymes be induced, while basal activity remains very low (reviewed in [8]).

The human Caco-2 cell line, developed from an intestinal carcinoma, is a valuable model for the study of metabolic and transport parameters in the human intestine with induction of major intestinal P450 CY3A4 [9–11]. Cell lines typically originate from naturally occurring tumors, are often dedifferentiated, and show only a fraction of the physiologic response. A very few cell lines can be redifferentiated under appropriate culture conditions. As an example, Ntera2 [12] can generate postmitotic neurons expressing multiple specific markers [13] and even forming neuronal networks [14]. The differentiation of liver cell lines remains challenging, and has yielded only limited success; the search for highly differentiated hepatic tumors continues [15].

The differentiation of embryonic stem cells (ESC) *in vitro* could provide an unlimited supply for cell types of interest using the replicative potential of the starter cell. Ethical concerns against the use of ESC have prevented widespread routine generation of cellular models in drug testing from stem cells. Recent findings that adult mouse spermatogonial stem cells (SSC) can obtain ESC-like features and differentiate into derivatives of the three different germ layers may provide an ethical alternative to ESC if transferable to human SSC [16].

Designing cell lines by the immortalization of primary cells using known mechanisms is another viable alternative. While this approach also represents substantial challenges, it offers not only well-defined, unlimited test material but also the potential for generating sets of similar cell lines on variable genetic background for studies in a pharmacogenomic context.

7.4 Immortalization by Design: Infinite Proliferation and a Differentiated Phenotype?

7.4.1 Telomerase: the Primary Target in Human Cells

It has long been known that human cells are far more difficult to transform *in vitro* than rodent cells. Immortalization (the acquisition of an infinite lifespan) is the first step in transformation, and molecular events required for immortalization are more tightly regulated in some species compared to others.

In human cells the available life span is controlled by the progressive shortening of telomeres; these are repetitive DNA elements that protect the chromosome ends from degradation and fusion with other chromosomes. When telomeres are eroded below a certain threshold, they fail to protect chromosome ends, and this causes senescence and crisis. In contrast, rodent somatic cells possess stable and substantially longer telomeres. The control of telomere length offers an additional level of protection against tumor formation in animals with a higher life expectancy (for a review, see [17]). Dogs are valuable animal models in drug testing, as they are less expensive than nonhuman primates and the results can be better extrapolated to human physiology. Cellular systems derived from dog primary cells are desirable to improve interpretation and study design. However, similar to human – and in contrast to rodent primary cells – dog telomerase activity also is tightly regulated [18].

Human ESC have an infinite lifespan because the telomeres are continuously maintained: in contrast to somatic cells, stem cells stably express functional telomerase, a ribonucleoprotein with template-specific reverse trancriptase activity. Human telomerase regulation is based on expression of the protein component (the TERT catalytic subunit) by control of promoter activity and alternative splicing [19]. The RNA component of the enzyme and other proteins involved in telomere maintenance (POT1, TRF1, and TRF2) are ubiquitously expressed. The discovery that human telomerase activity is regulated via a single protein subunit suggested that an unlimited supply of differentiated cells lines from any tissue could be obtained by ectoptic expression of the *tert* gene. Indeed, several groups have shown that this approach allows some human cell types, including fibroblasts, retinal pigment epithelial cells and mesothelia (mesoderm-derived epithelial cells) to bypass senescence and become immortalized in a single step [20, 21]. However, for other cell types such as endothelial cells, immortalization by *tert* alone remains controversial [22, 23]; for example, mammary epithelial cells and keratinocytes require additional genetic changes for immortalization [21].

Although adult tissue stem cells have a high reproductive capacity, telomerase levels are mostly insufficient to maintain telomere length [24]. Neuronal stem cells and mesenchymal stem cells have a silent *tert* gene and require exogenous gene transfer or endogenous gene activation [25, 26]. Only hematopoietic stem cells (CD34-positive) retain full telomerase activity [27].

The function of telomerase in immortalization seems to extend beyond the maintenance of telomere length. It has been shown that telomerase can cooperate with other oncogenes for transformation in cells that maintain telomeres by an alternative mechanism [28]. The pleiotropic effects of TERT are tissue-specific: it has been found that telomerase activates cell cycle-promoting genes and down-regulates apoptosis-promoting genes in fibroblasts, whereas it fails to do so in endothelial cells [29]. Furthermore, very high-level ectopic expression of telomerase alone may induce, rather than prevent, senescence in fibroblasts [30].

7.4.2
Inactivation of Rb and p53 Pathways

Telomere erosion is an important – but not the only – block against immortalization in human cells. In the culture dish most cell types (with the exception of fibroblasts) cease division within few population doublings, irrespective of telomere length. This arrest is termed M_0 [31], and is observed in human and mouse cells, although the underlying mechanism of the two species is different.

Multiple stress factors associated with the cultivation of primary cells *in vitro* result in activation of the CDKN2A locus in both systems. The locus codes for independent genes via alternative spicing, P16 INK4a and ARF. These two gene products link the main pathways in cell cycle control and tumor protection: P16 (INK4a) inhibits cyclin-dependent kinases CDK4 and CDK6 to maintain the Rb protein in a hypophosphorylated state.

As phosphorylation of Rb is required for cell cycle progression, the expression of P16 (INK4a) therefore locks cells into G_1 of the cell cycle. The ARF protein prevents p53 degradation and thus is responsible for p53-dependent halt of the cell cycle and induction of apoptosis.

Although p16 and ARF genes are induced simultaneously in mouse cells, it is only the activation of the ARF/p53 axis that prevents immortalization in mice. In contrast, mainly p16 is activated in human cells. However, even in the presence of active telomerase it is not sufficient only to inhibit p16: deactivation of p16 causes cell cycle progression as Rb is allowed to be phosphorylated. This in turn triggers release of the pleiotropic E2F transcription factor which is normally sequestered by Rb. The release of E2F induces cell cycle progression and activation of ARF, and this leads to elevated levels of p53. As p53 accumulates, cell cycle progression is halted, and if p53 activity increases further then apoptosis is induced. In summary, the immortalization of human cells requires not only telomerase expression but also inactivation of two additional pathways, Rb and p53.

As a logical consequence, viral oncogenes from small DNA tumor viruses known to inactivate both Rb and p53 pathways are frequently employed in immortalization protocols, the most popular of which is SV40 large T antigen. Most human cells can be immortalized with SV40 large T antigen in the presence of telomerase, including keratinocytes, endothelial cells, neuronal cells, and even hepatocytes. It is important to use the spliced T antigen gene instead of the complete SV40 early region, which also encodes small T antigen because its coexpression induces full transformation via interaction with protein phosphatase 2A [32].

Instead of SV40 large T antigen, E6 and E7 genes from human high-risk papillomaviruses (HPV16 and 18) may be used, which interact separately with either p53 or Rb, respectively. The genes of the E1 region from human nontumorigenic C group adenoviruses represent yet another alternative. In addition to E1A, which binds to and inactivates Rb and E1B 55k which converts p53 from a transcriptional activator to a repressor, a third protein is coexpressed: the E1B 19k protein is an analogue of Bcl2 and blocks cytochrome c release from the mitochondrial membrane, thus preventing apoptosis.

The oncoproteins from SV40, HPV and adenovirus fulfill complex functions not limited to the binding of Rb and p53. In fact, they act as master regulators of the viral life cycle and prepare the cell metabolism for virus replication. It is therefore not surprising that the interaction with additional targets will modulate the protein's capacity to immortalize or transform. For instance, the effects of SV40 T antigens in transgenic mice range from hyperplasia to invasive carcinoma accompanied by metastasis, depending on the tissue in which they are expressed [33]. HPV E6 and E7 genes efficiently immortalize keratinocyte epithelial and endothelial cells, while adenovirus genes transform neuronal cells best. The generation of HEK293 (a cell line used extensively in biotechnology applications) illustrates this phenomenon: starting from embryonic kidney a vast number of transfections resulted in a single cell line (HEK293) with clear neuronal characteristics such as expression of p75NGFR ([34] and own data).

In this regard, we generated cell lines from fetal brain, liver and mesenchymal material using adenovirus type 5 E1 genes. The expression pattern of the resulting cell lines deviate from that of their respective ancestors (see Fig. 7.1). Depending on the immortalizing gene, these changes may either eliminate or preserve a tissue-specific pattern. The E1 genes were found to direct an immortal cell towards epithelial characteristics: vimentin, a mesenchymal marker strongly expressed in the primary source, is gradually lost during immortalization while EpCAM, an epithelial marker, is only present in the E1 cell line. In contrast, endothelial differentiation is well supported by E1 [35], making it a suitable tool to derive functional endothelial cell lines.

The pleiotropic effects of the viral oncoproteins may even overcome the need to co-introduce teleomerase into human cells. Myc, a downstream target of Rb, strongly activates the telomerase promoter [36]. Although E1A does not bind DNA directly, it participates in transcription complexes and also activates *tert*. However, transactivation depends on the accessibility of the *tert*-locus. This explains why embryonic tissue enriched with stem cells is better suited in immortalization approaches that do not require exogenous *tert*: although already depressed, the *tert* promoter is not yet extensively methylated and can therefore be more easily reactivated. Furthermore, E1A, T antigen and HPV E6 also interact with p300, a histone acetylase which is part of the polymerase II complex and induces or fixates global changes in gene expression.

In order to preserve features of the differentiated cell, one should limit the interference with cell metabolism to required pathways. While such attempts have been made, for example by dissecting functions in SV40 T antigen [37], a specific block of RB or p53 pathways by small interfering RNA has not yet found broad application, perhaps because the role of additional pathways is underestimated.

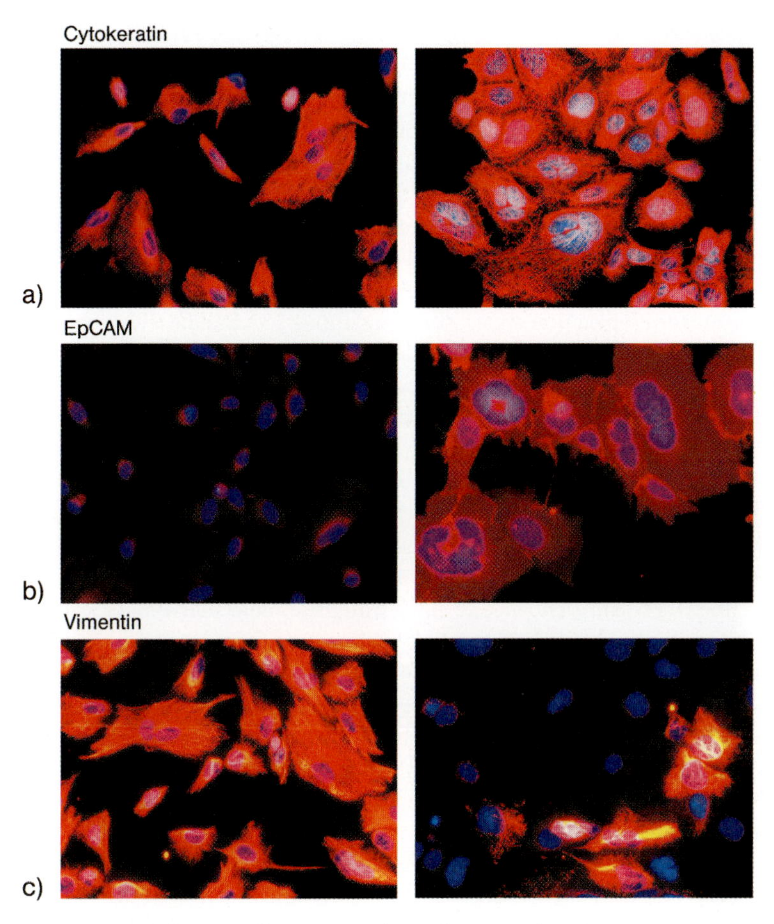

Fig. 7.1 Immortalization with E1A/E1B induces changes in the expression pattern of tissue-specific markers.
(a) Maintained expression of cytokeratin in primary human mesenchymal cells after immortalization with adenovirus E1.
(b) Gain of EpCAM marker expression.
(c) Loss of vimentin expression.
The primary cells are on the left side of each panel, and immortalized cells on the right side.

7.4.3
Conditional Immortalization

The precise inactivation of a minimal number of genes may only in some cases improve the applicability of designer cell lines in drug testing. In general, re-entry of a resting, terminally differentiated cell into proliferation will inevitably modify expression patterns. This raises the question of whether immortalization can be reversed after the required cell mass has been generated.

Temperature-dependent inactivation of mutant (tsA58) large T antigen represents the classical approach. Cells are grown at the permissive temperature of 32 °C and shifted to 39 °C or even 40 °C for redifferentiation. This approach has been applied to multiple tissues, ranging from thyroid epithelium [38] to osteocytes [39], glomerular epithelial cells [40] and (rat) hepatocytes [41].

The high restrictive temperature limits the value of the approach to generate "normal" cells under physiological conditions. Therefore, tight transcription regulation, for example by the Tet system [42] or excision by site-specific recombinases Cre/loxP and the Flp/FRT, yield better results [43–45]. Care must be taken to limit the number of replication cycles, since mutations accumulate with extended passaging due to inactivation of p53-dependent DNA repair control [41]. Moreover, when applied to E7/E1A, chromosomal changes are likely to accumulate because both genes interact with the Ran GTPase uncoupling the centrosome cycle from the cell cycle [46].

Immortal cell lines in some way resemble stem cells from the source tissue. Removal of the immortalizing gene function is just one part on the path to redifferentiation; just as for stem cells, medium supplements or scaffolds that allow a three-dimensional, tissue-like structure may be required to revert an immortal line. Increasing knowledge concerning the molecular events leading to immortalization has helped substantially to fine tune these approaches and to provide a large number of cell lines that may serve as models in the drug screening process. For one highly relevant model of drug-related toxicity, the kidney (proximal tubular epithelium), a human cell line (HK-2) immortalized using HPVE6/E7 [47], has found widespread application (over 60 publications). However, for the most relevant target for drug toxicity – the human liver – no suitable model has yet become available.

As fascinating as these novel substrates are in theory, a reverted previously immortal cell is an artificial product that needs to be compared to the natural tissue in preliminary assays before it can be applied in large screens. While some physiological responses are faithfully reproduced, others may be disturbed, and it is important to keep these limitations in mind.

7.5 Designed Cells in Complex Drug Tests

7.5.1 Cell Properties Required for Complex Screening Systems

The generation of cellular systems for complex high-throughput screens face very different hurdles, far downstream of substrate generation. The starting cell line often is well established and suitable for automated screening procedures; it is robust enough for reproducible handling independent of the operator, and grows fast enough to be cloned for genetic homogeneity and to provide assay material in reasonable time (for a review, see Chapter 5).

The gene of interest (a receptor, metabolic enzyme or virus protein) can easily be inserted using conventional vectors, transfection methods, and selection systems. Reasonable screening effort yields a cell line of appropriate characteristics with respect to expression level and stability. Challenges arise, however, when downstream effectors, cofactors and detection systems for automated readout have to be stably coexpressed at adjusted levels. Often, analogous and comparable systems must be generated for multiple members of the same protein family, or isoforms of the same protein. Complexity further increases if the screen is directed against interactions among cells or between host cells and parasites in the search for therapeutics against infectious diseases.

7.5.2
Complex Designer Cells in Screens

Today, some 40% of current pharmaceuticals target G protein-coupled receptors (GPCRs), and this huge family of proteins remains central to ongoing and intense pharmaceutical research [48]. As yet, fewer than one hundred of the receptor genes out of 800–1000 members of the superfamily have been well characterized, so that additional promising targets should emerge in time.

Cell lines suitable for screens against drugs that target GPCRs require the receptors and – depending on the cell type and the subfamily of the receptor – either the corresponding or a promiscuous G protein alpha subunit [49]. Furthermore, for high-throughput acceptable machine readability an additional gene may be introduced, usually a reporter gene functionally linked to the investigated event. Convenient reporter systems include calcium-sensitive photoprotein (aequorin) or a β-arrestin2-linked GFP or artificially split reporter enzymes where subunits need to interact to produce a signal. To further complicate the situation, the cell line cannot be chosen based entirely on convenience because the endogenous expression of multiple receptors can interfere with the assay [50].

The generation of a cell clone with desired expression levels for a single foreign gene out of a pool of stably transfected cells requires a nontrivial effort. The task becomes formidable with the required co-expression of several foreign genes that need to assemble into a target of several subunits that triggers or inhibits reporter expression in response to drugs. Developments in promoter architecture and the prevention of de-activation of foreign sequences by cellular mechanisms, as well as improved methods for the introduction of expression cassettes by directed recombination, facilitate the generation of designer cells for very complex screens.

For historic reasons, viral promoters such as the SV40 early promoter, the Rous sarcoma virus (RSV) long terminal repeat (LTR) and, most frequently, the immediate early promoter from human cytomegalovirus (hCMV), are used to drive the expression of foreign genes. However, despite the 10- to 50-fold different promoter activity in transient assays (expression measured at 2–3 days post introduction of recombinant DNA), stable producer clones containing the strongest promoter (hCMV) have no clear advantage over clones derived with other viral promoters. Moreover, expression levels vary greatly between individual clones containing the same vector, and in many clones the expression declines with prolonged propagation. One explanation for this observation is that viral promoters integrated into the host genome preferentially become inactivated by DNA methylation or progressive deacetylation of histones H3 and H4 [51]. Both processes are linked: DNA methylation induces deacetylation of histones, making the region inaccessible to transcription factors, while extensive acetylation is able to prevent methylation at promoter sites [52]. In particular, the hCMV promoter is affected so strongly that only very few clones maintain expression at a medium or higher level. The search for stable and highly expressing clones after random integration of the vector, which makes cell line generation so tedious and time-consuming, simply identifies rare genomic sites that are protected from this effect. Accordingly, sequences from such sites such as the chicken HS4 insulator [53] or matrix attachment regions [54] can be inserted into the vector flanking the expression unit to achieve stable expression, even from the hCMV promoter.

Protective sequences are not necessarily separated elements, but in some cases represent an integral part of a promoter region. Such protected promoters have been found among ubiquitously expressed, housekeeping genes, for example the TATA box binding protein (TBP) promoter [55], elongation factor 1alpha promoter [56, 57], and the phosphoglycerate kinase or the elongation factor2 promoter (our data). Similar to viral promoters, they can be used across species and tissues. Specifically chosen combinations such as PGK, EF1 and EF2, allow the co-introduction of several genes from a single vector, thus avoiding competition between neighboring promoters [58]. While some of these promoters lag behind the hCMV promoter (this may actually advantageous when physiologic expression levels are important), others confer similar or higher activity (Fig. 7.2).

To reduce the effort for generation of multiple test lines even further, a system that does not require screening for individual producer clones would be of significant advantage. The influence of the genomic locus on expression level and stability makes homologous recombination and integration by site-specific recombinases attractive in cell line design. Typically, a reporter gene (e.g., β-galactosidase) linked to a site for a recombinase such as Flp or Cre is used to identify a preferable locus for integration. When this site is targeted by co-transfection of a shuttle vector containing the gene of interest and a recombinase site with a second vector expressing the recombinase, the reporter gene is separated from its promoter and becomes inactivated. Insertion of the gene in all surviving clones is assured by a promoter trap. The system and a number of cell lines with active loci labeled for insertion are commercially available from Invitrogen for Flp recombinase (Flp-In™).

The generation of new cell lines using this system is complicated by the integration of multiple vector copies at separate positions and, more frequently, as concatamers as a result of high cellular ligase activity Although copy numbers can be influenced by the amount of DNA transfected and the transfection method used, single copy integrations are difficult to achieve. Moreover, not all loci expressing the reporter gene are easily accessible by the recombinase. The first problem can be addressed by vector insertion by retroviral gene transfer [59].

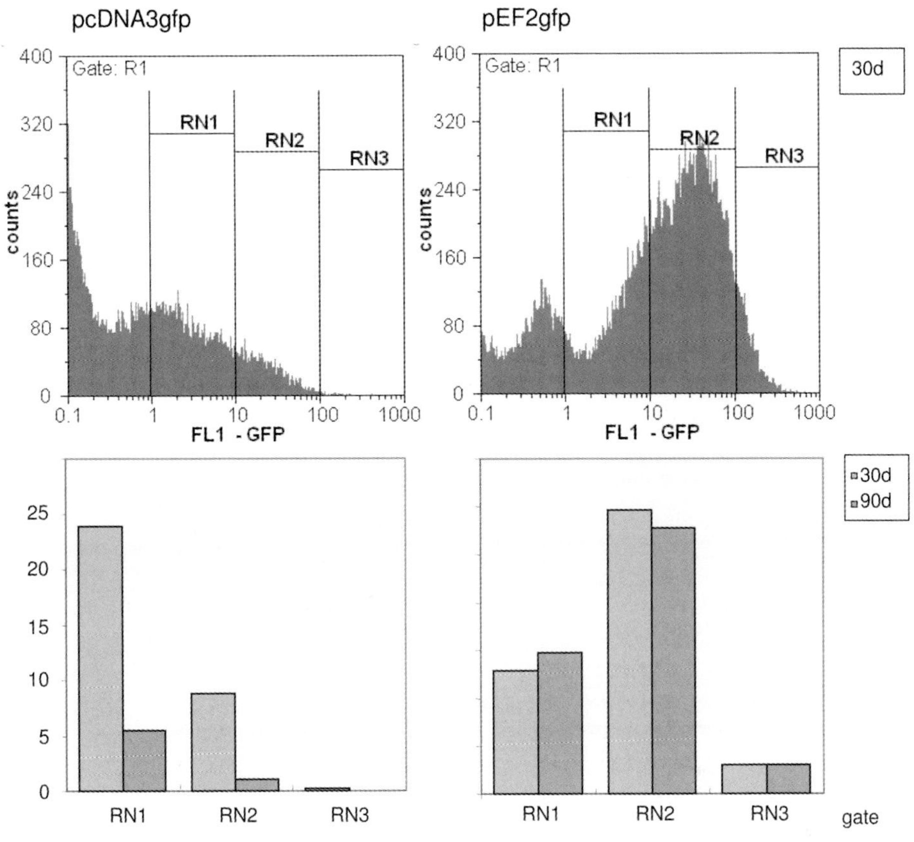

Fig. 7.2 The human EF2 promoter as an example of a cellular promoter that drives strong and stable expression in the majority of clones. pEF2 gfp was constructed replacing the hCMV promoter in pcDNA3 by a 3700-bp human elongation factor 2 promoter region and compared to the commonly used CMV promoter in the parental plasmid. Gfp expression was analyzed by FACS in pools of stably transfected Chinese hamster ovary (CHO) cells after low stringency G418 selection at 30 and 90 days after transfection. Clones were grouped according to expression levels. Whereas the rate of moderate- and high-producers drops substantially between 30 and 90 days for pcDNA3, it remains stable for pEF2gfp.

Alternatively, we are using a set of two consecutive recombination procedures to first identify clones that either already contain a single copy of the replacement cassette, or to reduce concatamers to a single copy (see Fig. 7.3). This procedure of consecutive recombination also confirms that the locus is accessible to targeting as the second step screens for expression levels. To favor gene exchange over excision, we use mutated frt sites that differ in the core sequence from the wild-type site [60]: these sites efficiently recombine with identical frt sites, but fail to interact with each other. We positioned a promoterless ATG-deficient *neo* gene outside of the replacement cassette that is activated by a minimal promoter, and an in-frame start codon was introduced with the recombination cassette.

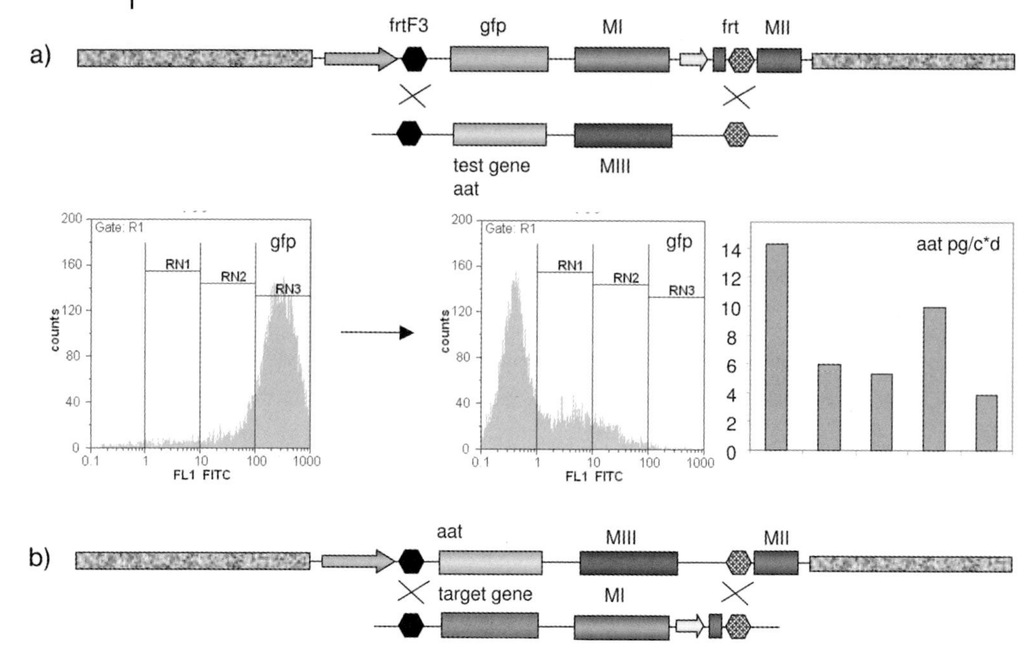

Fig. 7.3 Minimizing the effort towards multiple recombinant cell lines with predefined features. Derivation of starter clones for efficient gene replacement by consecutive recombinations. As an example, BHK cells are shown. (a) Pools of clones expressing high gfp levels are generated using stringent double selection (FACS analysis, left). In a first recombination, gfp is replaced for alpha-1-antitrypsin (aat) inactivating selection markers MI and MII and activating marker MIII. After recombination, most clones lack gfp expression, while in a fraction (containing multiple vector copies) gfp is still active (FACS analysis, center). Only clones that have undergone successful recombination and are devoid of gfp are ranked by aat expression (right). (b) The chosen clone is susceptible to replacement of aat by multiple target genes. Efficiency is ensured by reactivation of markers MI and MII.

In addition, the cassette contained a second selection marker which needs to be activated by the promoter residing at the integration locus, thus allowing for multiple exchange procedures.

For routine gene exchange the combined use of two selection markers assures a high efficiency of recombination. Variations in the expression level that still occur among clones originating from a single individual recipient cell can be attributed to perturbance of the architecture of a previously stable locus, caused by epigenomic phenomena. The phenomenon is less prominent when the cellular promoters are used instead of the CMV promoter. We have successfully incorporated the system into mouse 3T3, Chinese hamster ovary and baby hamster kidney cells, as well as designer cell lines of human and avian origin.

7.5.3
Viruses and Host Cells in Drug Tests

Screens for drugs against targets that consist of several subunits, or that assemble into linked pathways, require the generation of designer cells with multiple expression cassettes. Although complicated in the generation, the advantage of the established assay is isolation of the target from overlapping biochemical pathways in the native context and the opportunity for robust and fast quantification of drug effects via coupled reporter genes.

Desired screen complexity that can be solved with designed cells is also illustrated by assays directed against interactions among cells, or between host cells and parasites, in the search for therapeutics against infectious diseases. The range and variability of examples from the area of screening for antivirals against HIV illustrate the versatility of designed cells. Cell-based drug screens help in medium-throughput screens of libraries with hundreds of compounds, in small-scale applications for detailed characterization of selected promising drug candidates, and in the evaluation of the results from clinical trials with virus isolates from patients treated with new compounds. Drug screening against HIV is discussed as an example because of the complexity of HIV pathogenesis and the large amount of knowledge already available. Lessons learned here may be applicable to screens for drugs against other complex infectious diseases or tumor processes involving specific transactivation events.

Among the very first drugs to be described against HIV was the nucleoside analogue AZT (azidothymidine or zidovudine), and this is still in use today [61]. This compound was synthesized [62] well before discovery of HIV, and its suitability against HIV was demonstrated in what can be considered the first stage towards establishment of antiviral drug screens with designed cells. For this, T cells were cloned and immortalized with HTLV-I virus (an oncogenic human retrovirus in a different genus than HIV) that was released from co-cultivated lethally irradiated natural tumor cells. An immortalized clone with high susceptibility to HIV (at that time still called HTLV-III/LAV) was selected for studies where HIV infection was allowed in the presence or absence of AZT. The effect of AZT was quantified by counting the number of viable cells and of cells positive for viral antigens after 8–10 days of culture. Such experiments are time-consuming, and provide good confirmation of observed effects, but they are not suitable for medium-throughput screens of large libraries. Furthermore, promising but unstable drugs may not be identified in screens that require days to develop. Finally, the virus-induced cytopathic effect and potential toxicity of a drug may interfere and overlap, further complicating the interpretation of results. All of these caveats can be addressed in screens with designed cells.

7.5.4
Viruses and Designed Host Cells

Improved assays were feasible with increased understanding of molecular events in the infectious cycle: upon entry into the cell, the viral genomic RNA molecule is reverse-transcribed into DNA, and this copy is integrated into the host chromosome as a provirus. Integration of the proviral DNA is the main mechanism of retroviruses to establish a persistent infection, as cell division maintains a pool of infected cells even in the absence of actively replicating virus. Induction of the HIV provirus towards production and release of virus particles is initiated by the viral Tat (transactivator of transcription) protein. Because Tat is unique to HIV and pivotal in productive infectious cycles, it has been appreciated very early as a means towards providing robust assays in the search for antivirals [63, 64]. The LTR, an important regulatory and Tat-responsive stretch of proviral DNA, was fused to the gene for a reporter (lacZ or CAT), and this cassette was stably inserted into cells susceptible to HIV infection. Upon productive infection with HIV, the viral Tat protein activates reporter gene expression, and this event can be quantified in the presence or absence of proposed inhibitor molecules, for example the venerable AZT.

7.5.5
Defined Viral and Cellular Pathways and Designed Host Cells

In a related system with the stably integrated Tat-responsive reporter cassette, the activity of AZT was further characterized by transfection of an expression plasmid for Tat in the presence or absence of AZT [65]. Such a system where isolated viral events are probed in cells allows drug screens without the hazards of viable virus, and this may be advantageous for more dangerous pathogens than HIV. However, as AZT interferes with reverse transcription the reporter signal strength was not affected. Only drugs that target the transactivation step of HIV can be detected in this particular set-up of the LTR-reporter system.

This apparent limitation provides a very important opportunity: the screen for drugs specifically targeting a defined node or phase in the infectious cycle. This is especially important for complex pathogens that frequently change appearance (due to high mutation rates or complicated sequence of generations, as in the eukaryotic parasite *Plasmodium*) or alternate between dormant states and active replication. HIV has a high mutation rate and spawns escape mutants in single-drug approaches. Furthermore, drugs that target early phases of the infectious cycle (from attachment to insertion of proviral DNA) are without effect on cells that already carry proviral inserts and continuously shed virus. Finally, although HIV is chronically produced in massive numbers (even in clinically latent phases of disease), the virus also remains in a truly latent state as inactive provirus in a subset of infected cells. Current therapeutic approaches therefore aim at a combination of drugs that specifically target different pathways of complex pathogens, and highly active anti-retroviral therapy (HAART) against AIDS is a very good example of the success of such an approach.

Whilst HAART has significantly increased the quality of life and extended the life expectancy of patients infected with HIV, it cannot yet provide a cure for AIDS. Hence, pharmaceutical research is still faced with the challenge to refine screens in order to identify complementary drugs that specifically target only defined aspects in the life cycle of HIV. Patients infected with hepatitis B virus (HBV), hepatitis C virus (HCV) or certain herpesviruses may also benefit from complementary chemotherapy; other pathogens that may be tackled by such approaches include parasites such *Plasmodium*, the cause of malaria. The principle behind drug screening against HIV therefore is instructive also for other diseases.

Rational drug design, molecular modeling, and enzymatic, cell-free systems each represent extremely powerful methods of developing new therapies. Moreover, designed cell-based systems provide a means of testing these drug candidates in a setting with a complexity which is just one step removed from animal experiments or actual clinical studies.

Indeed, a drug which was rationally designed to interfere with viral trans-activation of transcription (ALX 40-4C) was shown, in a cellular fusion assay, not to target the intended intracellular event but rather to interfere with viral entry [66, 67].

As discussed briefly above, the Tat protein is also a preferred target in complementary chemotherapy. Here, the inhibition of Tat prevents activation of the provirus and thus forces the virus in a dormant state to control further viremia; the T cells already infected may be rescued with Tat inhibitors. It is perhaps surprising today that a great variety of assays and a significant number of drug leads still has not produced a released compound against Tat. Nevertheless, the lessons learned from these experiments may be applicable to screens for inhibitors of other viral or cellular trans-activators of transcription.

When the Tat inhibitor Ro 5-3335 [68] was assayed in animal studies, some toxicity was encountered. Consequently, 400 analogues of Ro 5-3335 were synthesized and probed for interference with Tat activity by transient co-transfection of LTR-driven SEAP reporter plasmid with TAT expression plasmid. Among the promising analogues was Ro 24-7429. Interference with trans-activation was demonstrated using cells chronically infected with HIV and which had down-regulated levels of CD4 receptor so that reinfection with a released virus was inhibited. Thus, a decrease in released particles correlated with the interference of provirus activation. Virion production was quantified via release of p24 viral antigen and cell-associated RNA.

Tat is known to interact with several cellular proteins, and this interaction is essential for trans-activation. This constrains the number of available escape mutations. Indeed, a cell-based screen for the emergence of escape mutants was performed by passaging virus on acutely infected cells in the presence of inhibitor; the inhibitor dose was adjusted to 50% to 95% of depression of p24 antigen expression. Again, Ro 24-7429 showed much promise as it did not allow escape mutants to emerge even after 100 passages in this *in-vitro* assay.

Ro 24-7429 failed in clinical trials, however, as it was found to be toxic before antiviral concentrations could be reached. One explanation is that the availability of the drug was reduced by binding to human serum; in fact, cellular assays where cultivation was performed in the presence of human serum instead of fetal bovine serum suggested a severely reduced intracellular concentration of Ro 24-7429 [69, 70].

The latter result highlights an additional advantage of the complexity in cell-based systems as opposed to theoretical or enzymatic approaches: physiological inhibitors or modifications towards activation can be considered in the set-up and thus observed at a level well below those in actual animal or clinical studies. Choosing the adequate host cell for the assays will reveal whether this activation can be expected in the patient. For example, nucleoside analogue reverse transcriptase inhibitors (e.g., ddC and AZT) must be activated in the host cell by phosphorylation, but not all HIV-susceptible cells perform this modification [71].

The early cellular systems for medium- to high-throughput screens for interference with defined viral pathways have been further refined [72–75]. Among others, the CAT and lacZ reporter genes have been replaced with EGFP, SEAP or luciferase to provide fast readouts with minimal handling and development times. Another complication of HIV pathogenesis is a shift in the requirement for co-receptors on target cells: early in seroconversion, the predominant HIV-1 population in the patient is specific for the CCR5 co-receptor, whereas increasing numbers of CXCR4-specific viruses emerge in later stages [76, 77]. The patterns of co-receptor expression varies among T-cell populations [78], including the generally accepted immortalized Jurkat T cell used in some drug screens. The shift in co-receptor specificity is important for pathogenesis and epidemiology. Consequently, the indicator cells are designed or selected to express either or both co-receptors at high levels to assess specific or both types of viruses.

7.5.6
Virus Field Isolates and Designed Host Cells

Sometimes it is necessary not to use laboratory viruses but to follow the evolution of strains within a swarm of viruses from actual patients during therapy or disease progression. Using these viral isolates as "black boxes", without the need for sequencing or detailed characterization, is an advantage of cell-based assays. Enzymatic assays with recombinant proteins or molecular modeling require material and knowledge about each viral strain that significantly complicate the test system, or delay its implementation.

For example, a designed cell line provided important insight into the properties of virus isolates from patients under monotherapy with a reverse transcriptase drug: HeLa was made susceptible to HIV infection by stable expression of CD4 receptor. Monolayers of the CD4-positive HeLa readily form plaques upon HIV infection. Quantification of the plaques with crystal violet-stained cells is a measurement of infectious input, and reduction of plaque formation a measurement of drug efficiency [79]. Although this readout is cumbersome and not amenable to high-throughput systems with current technologies (it may be in the future with improved automated image interpretation), there is a decisive advantage already: most laboratory strains of HIV are selected for fast replication and high cytotoxicity, as these properties facilitate research. Field isolates however – as devastating as they are in human patients – can be relatively slow and innocuous in cell cultures [79, 80]. For such viruses (as opposed to laboratory strains), high-throughput-amenable quantification of the cytopathic effect via reduction of metabolic conversion of dye (e.g., the tetrazolium salt, MTT) by viable cells produces an inadequate signal. However, plaque reduction after the addition of drug allowed the quantification of increased resistance to zidovudine of HIV in human patients. This assay further allowed the susceptibility of various strains to various reverse-transcriptase inhibitors to be compared. The results suggested that resistance to one compound does not imply resistance to related reverse-transcriptase inhibitors [81]. Narrow opportunities for the development of resistance across different compounds therefore allows combination therapy with several reverse-transcriptase inhibitors (e.g., the nucleoside analogues AZT, 3TC, and abacavir are combined into single formulations).

7.5.7 Designed Viruses and Designed Host Cells

T-20 or enfuvirtide was rationally designed to interfere with fusion of the viral envelope and the plasma membrane by inhibiting a required conformational change in the transmembrane protein gp41. In clinical trials, the emergence of escape mutants was observed within 14 days of monotherapy with T-20. The method to assay field isolates for the evolution of resistance to this drug required a more complex approach: HeLa cells were stably transfected with LTR-driven reporter and with CD4 and CCR5 receptors, and the reporter signal was shown to decrease in a T-20 dose-dependent fashion in cells challenged with HIV. To demonstrate that indeed viral entry is the target of T-20, the viral envelope proteins from laboratory HIV particles were replaced with envelope proteins from the escape mutants, and these pseudotyped viruses were used in the screen in the presence of escalating doses of T-20. As expected, virus replication – and thus reporter gene activity – was stronger for a given T-20 concentration in pseudotyped viruses when compared to laboratory strain reference [75], demonstrating that T-20 drives HIV into selection of envelope-protein escape mutants. T-20 therefore should be a good candidate for combination therapy with drugs that target other aspects of the viral life cycle, such as reverse transcription or morphogenesis via the viral protease.

Another approach using recombinant viruses on designed cells was performed to search for inhibitors of transactivation and to characterize the mechanism of inhibition. This approach is instructive, as this principle should be applicable to other screens where a specific trans-activator and promoter interaction is to be targeted by a drug: a replication-deficient adenovirus (a DNA virus vector completely unrelated to retroviruses) containing the LTR-reporter cassette was used as tester on HeLa cells that had been engineered to constitutively express the Tat protein. Following infection with the adenoviral vector, the inhibitor was added [82] and the reduction of reporter expression was measured. Using an adenoviral vector to transduce the reporter allows better efficiency and consistency compared to transfection.

7.5.8
Designed Host Cells Combined

Yet further increased complexity is realized in assays that rely on mixtures of different designed cells or cell types [83]. Although these assays are complicated, they yield insights into molecular mechanisms at the cell membrane, such as fusion events that are required for enveloped viruses to enter the target cell. Interactions at lipid bilayers with trans-membrane or integral proteins are difficult to simulate without cellular *in-vitro* systems. If the fusion event can be quantified, a screen for drugs that specifically target this very first step of infection is possible.

As mentioned above, the compound ALX 40-4C was rationally designed to interfere with transcription, but subsequently shown to inhibit fusion of the viral envelope with the cell membrane. The interference of ALX 40-4C with membrane fusion was performed by first generating two populations of cells: quail fibrosarcoma QT6-C5 target cells were transiently co-transfected with cellular receptors (CD4 and co-receptors) and a luciferase expression cassette under control of the bacteriophage T7 promoter. HeLa effector cells were infected with vaccinia virus vectors expressing HIV envelope protein and T7 polymerase. The two cell populations were mixed in the presence or absence of inhibitor, and the luciferase signal from cell lysates was assayed after 8–10 h [66]. In the absence of inhibitor, the viral envelope protein in the plasma membrane of effector cells and the transiently expressed receptor on the surface of target cells interacted, leading to activation of the viral fusion domain and thus formation of multinucleated syncitia. The cytoplasm of the syncitia then contained both T7-dependent reporter construct and cognate T7 polymerase, thereby allowing expression of the reporter and thus quantification of fusion activity.

A set-up that facilitates the fusion assay and, at the same time, combines assays for inhibitors of viral entry and transactivation in a single screen [84, 85] has been developed using a mixture of designed cells with cells chronically infected with HIV (a chronic infection, as opposed to latent infection, is characterized by a continuous release of virus). The indicator cells are adherent HeLa cells stably equipped with CD4 receptor and LTR-driven reporter, in this case lacZ. The chronically infected cells express viral envelope protein in the plasma membrane, but have down-regulated levels of CD4. If the indicator and the effector cells are co-cultivated, the interaction of CD4 on the HeLa derivatives and envelope protein on the producers cause fusion of the cell membranes. This leads to formation of the syncitia that can be assayed by microscopic examination. The transfer of Tat protein from the producer cells furthermore activates the reporter cassette in the indicator cells, and this can be visualized histologically or quantified by using spectroscopic galactosidase dyes. In the presence of the test substance, three possible readouts can be distinguished (Fig. 7.4): blue syncitia if the drug is without

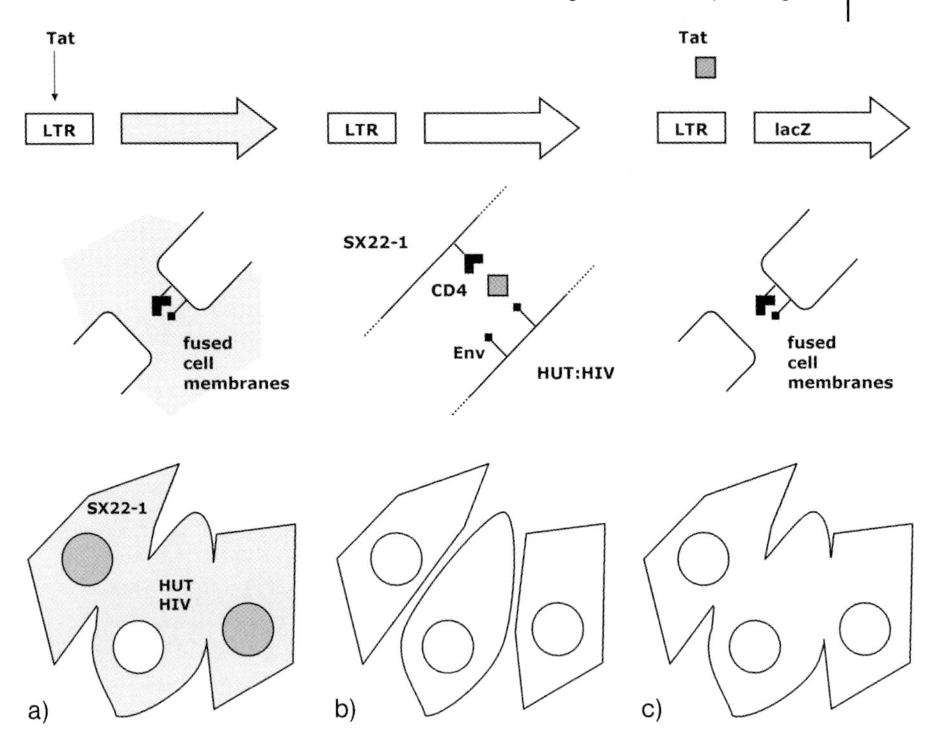

Fig. 7.4 Fusion-induced gene stimulation (FIGS) assay. The SX22-1 indicator cells stably carry a TAT transactivator-responsive lacZ-cassette and express recombinant CD4 receptor molecule for HIV. The viral envelope protein (Env) is displayed in the membrane of the chronically infected HUT effector cells. (a) When the two cell types are mixed, interaction of CD4 and Env causes fusion of indicator and effector cells. This leads to formation of syncitia and transfer of Tat from the effector to the indicator cell, thus activating lacZ expression that can be visualized or quantified. (b) If a drug interferes with cell fusion (or virus entry), no syncitia are formed and, as Tat is not transferred, the cells also do not express lacZ. (c) If a drug interferes with Tat activity, fusion (and therefore syncitia formation) is not suppressed but reporter gene expression is not transactivated, leading to formation of unstained syncitia.

effects; white (lacZ-negative) syncitia if the drug interferes with transactivation; and conventional unstained monolayer of cells if the drug prevents membrane fusion and virus entry.

Lessons learned from HIV screens can be transferred to other fields of infectious diseases. For example, the recent development of replicon systems for HCV has led to *in-vitro* drug screens [86] against this serious chronic disease that affects 170 million people worldwide. The HCV replicon is a subviral RNA molecule maintained within the host cell without extracellular states. The variations for cell-based screens include systems with specially selected cells that support robust replicon expression, and replicons that carry and express reporter genes to facilitate quantification of interfering drugs in high-throughput systems. Highly illustrative with respect to this chapter, HCV replicons have been created that release HIV Tat to activate a stably inserted LTR-reporter cassette in HCV-permissive Huh-7 cells [87]. The *in-vitro* screens have already yielded a number of promising compounds directed against viral proteases and polymerases. In addition, the *in-vitro* study of the complete life cycle of HCV has finally become accessible, with recent developments suggesting that cell culture-based screens against viral entry and uncoating or morphogenesis also should be possible in the near future [88]. Such studies will, hopefully, be guided and accelerated by the experience with HIV that dates back to the 1980s.

References

1. Searfoss, G. H., T. P. Ryan, and R. A. Jolly. The role of transcriptome analysis in pre-clinical toxicology. *Curr. Mol. Med.* **2005**, *5*, 53–64.
2. Terry, C., A. Dhawan, R. R. Mitry, S. C. Lehec, and R. D. Hughes. Preincubation of rat and human hepatocytes with cytoprotectants prior to cryopreservation can improve viability and function upon thawing. *Liver Transpl.* **2006**, *12*, 165–177.
3. Kafert-Kasting, S., K. Alexandrova, M. Barthold, B. Laube, G. Friedrich, L. Arseniev, and J. G. Hengstler. Enzyme induction in cryopreserved human hepatocyte cultures. *Toxicology* **2006**, *220*, 117–1125.
4. Boess, F., M. Kamber, S. Romer, R. Gasser, D. Muller, S. Albertini, and L. Suter. Gene expression in two hepatic cell lines, cultured primary hepatocytes, and liver slices compared to the in vivo liver gene expression in rats: possible implications for toxicogenomics use of in vitro systems. *Toxicol. Sci.* **2003**, *73*, 386–402.
5. Jessen, B. A., J. S. Mullins, A. De Peyster, and G. J. Stevens. Assessment of hepatocytes and liver slices as in vitro test systems to predict in vivo gene expression. *Toxicol. Sci.* **2003**, *75*, 208–222.
6. MacGregor, J. T., J. M. Collins, Y. Sugiyama, C. A. Tyson, J. Dean, L. Smith, M. Andersen, R. D. Curren, J. B. Houston, F. F. Kadlubar, G. L. Kedderis, K. Krishnan, A. P. Li, R. E. Parchment, K. Thummel, J. E. Tomaszewski, R. Ulrich, A. E. Vickers, and S. A. Wrighton. In vitro human tissue models in risk assessment: report of a consensus-building workshop. *Toxicol. Sci.* **2001**, *59*, 17–36.
7. Klaunig, J. E., M. A. Babich, K. P. Baetcke, J. C. Cook, J. C. Corton, R. M. David, J. G. DeLuca, D. Y. Lai, R. H. McKee, J. M. Peters, R. A. Roberts, and P. A. Fenner-Crisp. PPARalpha agonist-induced rodent tumors: modes of action and human relevance. *Crit. Rev. Toxicol.* **2003**, *33*, 655–780.
8. Brandon, E. F., C. D. Raap, I. Meijerman, J. H. Beijnen, and J. H. Schellens. An update on in vitro test methods in human hepatic drug biotransformation research: pros and cons. *Toxicol. Appl. Pharmacol.* **2003**, *189*, 233–246.

9 Meunier, V., M. Bourrie, Y. Berger, and G. Fabre. The human intestinal epithelial cell line Caco-2; pharmacological and pharmacokinetic applications. *Cell Biol. Toxicol.* **1995**, *11*, 187–194.

10 Bock, K. W., and C. Kohle. UDP-glucuronosyltransferase 1A6: structural, functional, and regulatory aspects. *Methods Enzymol.* **2005**, *400*, 57–75.

11 Berger, V., A. F. Gabriel, T. Sergent, A. Trouet, Y. Larondelle, and Y. J. Schneider. Interaction of ochratoxin A with human intestinal Caco-2 cells: possible implication of a multidrug resistance-associated protein (MRP2). *Toxicol. Lett.* **2003**, *140–141*, 465–476.

12 Pleasure, S. J., C. Page, and V. M. Lee. Pure, postmitotic, polarized human neurons derived from NTera 2 cells provide a system for expressing exogenous proteins in terminally differentiated neurons. *J. Neurosci.* **1992**, *12*, 1802–1815.

13 Megiorni, F., B. Mora, P. Indovina, and M. C. Mazzilli. Expression of neuronal markers during NTera2/cloneD1 differentiation by cell aggregation method. *Neurosci. Lett.* **2005**, *373*, 105–109.

14 Gortz, P., W. Fleischer, C. Rosenbaum, F. Otto, and M. Siebler. Neuronal network properties of human teratocarcinoma cell line-derived neurons. *Brain Res.* **2004**, *1018*, 18–25.

15 O'Connor, J. E., A. Martinez, J. V. Castell, and M. J. Gomez-Lechon. Multiparametric characterization by flow cytometry of flow-sorted subpopulations of a human hepatoma cell line useful for drug research. *Cytometry A* **2005**, *63*, 48–58.

16 Guan, K., K. Nayernia, L. S. Maier, S. Wagner, R. Dressel, J. H. Lee, J. Nolte, F. Wolf, M. Li, W. Engel, and G. Hasenfuss. Pluripotency of spermatogonial stem cells from adult mouse testis. *Nature* **2006**, *440*, 1199–1203.

17 Hahn, W. C., and R. A. Weinberg. Modelling the molecular circuitry of cancer. *Nat. Rev. Cancer* **2002**, *2*, 331–341.

18 Nasir, L., P. Devlin, T. McKevitt, G. Rutteman, and D. J. Argyle. Telomere lengths and telomerase activity in dog tissues: a potential model system to study human telomere and telomerase biology. *Neoplasia* **2001**, *3*, 351–359.

19 Cerezo, A., H. Kalthoff, M. Schuermann, B. Schafer, and P. Boukamp. Dual regulation of telomerase activity through c-Myc-dependent inhibition and alternative splicing of hTERT. *J. Cell Sci.* **2002**, *115*, 1305–1312.

20 Bodnar, A. G., M. Ouellette, M. Frolkis, S. E. Holt, C. P. Chiu, G. B. Morin, C. B. Harley, J. W. Shay, S. Lichtsteiner, and W. E. Wright. Extension of life-span by introduction of telomerase into normal human cells. *Science* **1998**, *279*, 349–352.

21 Dickson, M. A., W. C. Hahn, Y. Ino, V. Ronfard, J. Y. Wu, R. A. Weinberg, D. N. Louis, F. P. Li, and J. G. Rheinwald. Human keratinocytes that express hTERT and also bypass a p16(INK4a)-enforced mechanism that limits life span become immortal yet retain normal growth and differentiation characteristics. *Mol. Cell. Biol.* **2000**, *20*, 1436–1447.

22 O'Hare, M. J., J. Bond, C. Clarke, Y. Takeuchi, A. J. Atherton, C. Berry, J. Moody, A. R. Silver, D. C. Davies, A. E. Alsop, A. M. Neville, and P. S. Jat. Conditional immortalization of freshly isolated human mammary fibroblasts and endothelial cells. *Proc. Natl. Acad. Sci. USA* **2001**, *98*, 646–651.

23 Yang, J., E. Chang, A. M. Cherry, C. D. Bangs, Y. Oei, A. Bodnar, A. Bronstein, C. P. Chiu, and G. S. Herron. Human endothelial cell life extension by telomerase expression. *J. Biol. Chem.* **1999**, *274*, 26141–26148.

24 Flores, I., R. Benetti, and M. A. Blasco. Telomerase regulation and stem cell behaviour. *Curr. Opin. Cell Biol.* **2006**, *18*, 254–260.

25 Parsch, D., J. Fellenberg, T. H. Brummendorf, A. M. Eschlbeck, and W. Richter. Telomere length and telomerase activity during expansion and differentiation of human mesenchymal stem cells and chondrocytes. *J. Mol. Med.* **2004**, *82*, 49–55.

26 Villa, A., B. Navarro-Galve, C. Bueno, S. Franco, M. A. Blasco, and A. Martinez-Serrano. Long-term molecular and cellular stability of human neural stem cell lines. *Exp. Cell Res.* **2004**, *294*, 559–570.

27 Mulloy, J. C., J. Cammenga, F. J. Berguido, K. Wu, P. Zhou, R. L. Comenzo, S. Jhanwar, M. A. Moore, and S. D. Nimer. Maintaining the self-renewal and differentiation potential of human CD34+ hematopoietic cells using a single genetic element. *Blood* **2003**, *102*, 4369–4376.

28 Stewart, S. A., W. C. Hahn, B. F. O'Connor, E. N. Banner, A. S. Lundberg, P. Modha, H. Mizuno, M. W. Brooks, M. Fleming, D. B. Zimonjic, N. C. Popescu, and R. A. Weinberg. Telomerase contributes to tumorigenesis by a telomere length-independent mechanism. *Proc. Natl. Acad. Sci. USA* **2002**, *99*, 12606–12611.

29 Kumazaki, T., K. Hiyama, T. Takahashi, H. Omatsu, K. Tanimoto, T. Noguchi, E. Hiyama, Y. Mitsui, and M. Nishiyama. Differential gene expressions during immortalization of normal human fibroblasts and endothelial cells transfected with human telomerase reverse transcriptase gene. *Int. J. Oncol.* **2004**, *24*, 1435–1442.

30 Gorbunova, V., A. Seluanov, and O. M. Pereira-Smith. Evidence that high telomerase activity may induce a senescent-like growth arrest in human fibroblasts. *J. Biol. Chem.* **2003**, *278*, 7692–7698.

31 Romanov, S. R., B. K. Kozakiewicz, C. R. Holst, M. R. Stampfer, L. M. Haupt, and T. D. Tlsty. Normal human mammary epithelial cells spontaneously escape senescence and acquire genomic changes. *Nature* **2001**, *409*, 633–637.

32 Hahn, W. C., S. K. Dessain, M. W. Brooks, J. E. King, B. Elenbaas, D. M. Sabatini, J. A. DeCaprio, and R. A. Weinberg. Enumeration of the simian virus 40 early region elements necessary for human cell transformation. *Mol. Cell. Biol.* **2002**, *22*, 2111–2123.

33 Ahuja, D., M. T. Saenz-Robles, and J. M. Pipas. SV40 large T antigen targets multiple cellular pathways to elicit cellular transformation. *Oncogene* **2005**, *24*, 7729–7745.

34 Shaw, G., S. Morse, M. Ararat, and F. L. Graham. Preferential transformation of human neuronal cells by human adenoviruses and the origin of HEK 293 cells. *FASEB J.* **2002**, *16*, 869–871.

35 Giampietri, C., M. Levrero, A. Felici, A. D'Alessio, M. C. Capogrossi, and C. Gaetano. E1A stimulates FGF-2 release promoting differentiation of primary endothelial cells. *Cell Death Differ.* **2000**, *7*, 292–301.

36 Batsche, E., M. Lipp, and C. Cremisi. Transcriptional repression and activation in the same cell type of the human c-MYC promoter by the retinoblastoma gene protein: antagonisation of both effects by SV40 T antigen. *Oncogene* **1994**, *9*, 2235–2243.

37 Freed, W. J., P. Zhang, J. F. Sanchez, O. Dillon-Carter, M. Coggiano, S. L. Errico, B. D. Lewis, and M. E. Truckenmiller. Truncated N-terminal mutants of SV40 large T antigen as minimal immortalizing agents for CNS cells. *Exp. Neurol.* **2005**, *191* (Suppl 1), S45–S59.

38 Wynford-Thomas, D., J. A. Bond, F. S. Wyllie, J. S. Burns, E. D. Williams, T. Jones, D. Sheer, and N. R. Lemoine. Conditional immortalization of human thyroid epithelial cells: a tool for analysis of oncogene action. *Mol. Cell. Biol.* **1990**, *10*, 5365–5377.

39 Bodine, P. V., S. K. Vernon, and B. S. Komm. Establishment and hormonal regulation of a conditionally transformed preosteocytic cell line from adult human bone. *Endocrinology* **1996**, *137*, 4592–4604.

40 Satchell, S. C., C. H. Tasman, A. Singh, L. Ni, J. Geelen, C. J. von Ruhland, J. O'Hare, M. A. Saleem, L. P. van den Heuvel, and P. W. Mathieson. Conditionally immortalized human glomerular endothelial cells expressing fenestrations in response to VEGF. *Kidney Int.* **2006**, *69*, 1633–1640.

41 Kim, B. H., S. R. Sung, E. H. Choi, Y. I. Kim, K. J. Kim, S. H. Dong, H. J. Kim, Y. W. Chang, J. I. Lee, and R. Chang. Dedifferentiation of conditionally immortalized hepatocytes with long-term in vitro passage. *Exp. Mol. Med.* **2000**, *32*, 29–37.

42 May, T., H. Hauser, and D. Wirth. Transcriptional control of SV40 T-antigen expression allows a complete reversion of immortalization. *Nucleic Acids Res* **2004**, *32*, 5529–5538.

43 Paillard, F. Reversible cell immortalization with the Cre-lox system. *Hum. Gene Ther.* **1999**, *10*, 1597–1598.

44 Weber, A. Immortalization of hepatic progenitor cells. *Pathol. Biol. (Paris)* **2004**, *52*, 93–96.

45 Noguchi, H., N. Kobayashi, K. A. Westerman, M. Sakaguchi, T. Okitsu, T. Totsugawa, T. Watanabe, T. Matsumura, T. Fujiwara, T. Ueda, M. Miyazaki, N. Tanaka, and P. Leboulch. Controlled expansion of human endothelial cell populations by Cre-loxP-based reversible immortalization. *Hum. Gene Ther.* **2002**, *13*, 321–334.

46 Lavia, P., A. M. Mileo, A. Giordano, and M. G. Paggi. Emerging roles of DNA tumor viruses in cell proliferation: new insights into genomic instability. *Oncogene* **2003**, *22*, 6508–6516.

47 Ryan, M. J., G. Johnson, J. Kirk, S. M. Fuerstenberg, R. A. Zager, and B. Torok-Storb. HK-2: an immortalized proximal tubule epithelial cell line from normal adult human kidney. *Kidney Int.* **1994**, *45*, 48–57.

48 Reetz, M. T. An overview of high-throughput screening systems for enantioselective enzymatic transformations. *Methods Mol. Biol.* **2003**, *230*, 259–282.

49 Kostenis, E. Potentiation of GPCR-signaling via membrane targeting of G protein alpha subunits. *J. Recept. Signal Transduct. Res.* **2002**, *22*, 267–281.

50 Ames, R., P. Nuthulaganti, J. Fornwald, U. Shabon, H. van-der-Keyl, and N. Elshourbagy. Heterologous expression of G protein-coupled receptors in U-2 OS osteosarcoma cells. *Receptors Channels* **2004**, *10*, 117–124.

51 Hodges, B. L., K. M. Taylor, M. F. Joseph, S. A. Bourgeois, and R. K. Scheule. Long-term transgene expression from plasmid DNA gene therapy vectors is negatively affected by CpG dinucleotides. *Mol. Ther.* **2004**, *10*, 269–278.

52 Cervoni, N., and M. Szyf. Demethylase activity is directed by histone acetylation. *J. Biol. Chem.* **2001**, *276*, 40778–40787.

53 Mutskov, V. J., C. M. Farrell, P. A. Wade, A. P. Wolffe, and G. Felsenfeld. The barrier function of an insulator couples high histone acetylation levels with specific protection of promoter DNA from methylation. *Genes Dev.* **2002**, *16*, 1540–1554.

54 Zahn-Zabal, M., M. Kobr, P. A. Girod, M. Imhof, P. Chatellard, M. de Jesus, F. Wurm, and N. Mermod. Development of stable cell lines for production or regulated expression using matrix attachment regions. *J. Biotechnol.* **2001**, *87*, 29–42.

55 Antoniou, M., L. Harland, T. Mustoe, S. Williams, J. Holdstock, E. Yague, T. Mulcahy, M. Griffiths, S. Edwards, P. A. Ioannou, A. Mountain, and R. Crombie. Transgenes encompassing dual-promoter CpG islands from the human TBP and HNRPA2B1 loci are resistant to heterochromatin-mediated silencing. *Genomics* **2003**, *82*, 269–279.

56 Wakabayashi-Ito, N., and S. Nagata. Characterization of the regulatory elements in the promoter of the human elongation factor-1 alpha gene. *J. Biol. Chem.* **1994**, *269*, 29831–29837.

57 Running Deer, J., and D. S. Allison. High-level expression of proteins in mammalian cells using transcription regulatory sequences from the Chinese hamster EF-1alpha gene. *Biotechnol. Prog.* **2004**, *20*, 880–889.

58 Cai, H. N., Z. Zhang, J. R. Adams, and P. Shen. Genomic context modulates insulator activity through promoter competition. *Development* **2001**, *128*, 4339–4347.

59 Wirth, D., and H. Hauser. Flp-mediated integration of expression cassettes into FRT-tagged chromosomal loci in mammalian cells. *Methods Mol. Biol.* **2004**, *267*, 467–476.

60 Schlake, T., and J. Bode. Use of mutated FLP recognition target (FRT) sites for the exchange of expression cassettes at defined chromosomal loci. *Biochemistry* **1994**, *33*, 12746–12751.

61 Mitsuya, H., K. J. Weinhold, P. A. Furman, M. H. St Clair, S. N. Lehrman, R. C. Gallo, D. Bolognesi, D. W. Barry, and S. Broder. 3′-Azido-3′-deoxythymidine (BW A509U): an antiviral agent that inhibits the infectivity and cytopathic effect of human T-lymphotropic virus type III/lymphadenopathy-associated virus in vitro. *Proc. Natl. Acad. Sci. USA* **1985**, *82*, 7096–7100.

62 Lin, T. S., and W. H. Prusoff. Synthesis and biological activity of several amino analogues of thymidine. *J. Med. Chem.* **1978**, *21*, 109–112.

63 Felber, B. K., and G. N. Pavlakis. A quantitative bioassay for HIV-1 based on trans-activation. *Science* **1988**, *239*, 184–187.

64 Bonnerot, C., N. Savatier, and J. F. Nicolas. [Toward an unpublished method of detecting human retroviruses: activation of HIV-1 LacZ recombinant provirus by the tat gene product]. *C. R. Acad. Sci. III* **1988**, *307*, 311–316.

65 Chandra, A., I. Demirhan, S. K. Arya, and P. Chandra. D-penicillamine inhibits transactivation of human immunodeficiency virus type-1 (HIV-1) LTR by transactivator protein. *FEBS Lett.* **1988**, *236*, 282–286.

66 Doranz, B. J., K. Grovit-Ferbas, M. P. Sharron, S. H. Mao, M. B. Goetz, E. S. Daar, R. W. Doms, and W. A. O'Brien. A small-molecule inhibitor directed against the chemokine receptor CXCR4 prevents its use as an HIV-1 coreceptor. *J. Exp. Med.* **1997**, *186*, 1395–1400.

67 O'Brien, W. A., M. Sumner-Smith, S. H. Mao, S. Sadeghi, J. Q. Zhao, and I. S. Chen. Anti-human immunodeficiency virus type 1 activity of an oligocationic compound mediated via gp120 V3 interactions. *J. Virol.* **1996**, *70*, 2825–2831.

68 Hsu, M. C., A. D. Schutt, M. Holly, L. W. Slice, M. I. Sherman, D. D. Richman, M. J. Potash, and D. J. Volsky. Inhibition of HIV replication in acute and chronic infections in vitro by a Tat antagonist. *Science* **1991**, *254*, 1799–1802.

69 Haubrich, R. H., C. Flexner, M. M. Lederman, M. Hirsch, C. P. Pettinelli, R. Ginsberg, P. Lietman, F. M. Hamzeh, S. A. Spector, and D. D. Richman. A randomized trial of the activity and safety of Ro 24-7429 (Tat antagonist) versus nucleoside for human immunodeficiency virus infection. The AIDS Clinical Trials Group 213 Team. *J. Infect. Dis.* **1995**, *172*, 1246–1252.

70 Cupelli, L. A., and M. C. Hsu. The human immunodeficiency virus type 1 Tat antagonist, Ro 5-3335, predominantly inhibits transcription initiation from the viral promoter. *J. Virol.* **1995**, *69*, 2640–2643.

71 Richman, D. D., R. S. Kornbluth, and D. A. Carson. Failure of dideoxynucleosides to inhibit human immunodeficiency virus replication in cultured human macrophages. *J. Exp. Med.* **1987**, *166*, 1144–1149.

72 Ochsenbauer-Jambor, C., J. Jones, M. Heil, K. P. Zammit, and O. Kutsch. T-cell line for HIV drug screening using EGFP as a quantitative marker of HIV-1 replication. *Biotechniques* **2006**, *40*, 91–100.

73 Miyake, H., Y. Iizawa, and M. Baba. Novel reporter T-cell line highly susceptible to both CCR5- and CXCR4-using human immunodeficiency virus type 1 and its application to drug susceptibility tests. *J. Clin. Microbiol.* **2003**, *41*, 2515–2521.

74 Spenlehauer, C., C. A. Gordon, A. Trkola, and J. P. Moore. A luciferase-reporter gene-expressing T-cell line facilitates neutralization and drug-sensitivity assays that use either R5 or X4 strains of human immunodeficiency virus type 1 *Virology* **2001**, *280*, 292–300.

75 Wei, X., J. M. Decker, H. Liu, Z. Zhang, R. B. Arani, J. M. Kilby, M. S. Saag, X. Wu, G. M. Shaw, and J. C. Kappes. Emergence of resistant human immunodeficiency virus type 1 in patients receiving fusion inhibitor (T-20) monotherapy. *Antimicrob. Agents Chemother.* **2002**, *46*, 1896–1905.

76 Xiao, H., C. Neuveut, H. L. Tiffany, M. Benkirane, E. A. Rich, P. M. Murphy, and K. T. Jeang. Selective CXCR4 antagonism by Tat: implications for in vivo expansion of coreceptor use by HIV-1. *Proc. Natl. Acad. Sci. USA* **2000**, *97*, 11466–11471.

77 Berger, E. A., R. W. Doms, E. M. Fenyo, B. T. Korber, D. R. Littman, J. P. Moore, Q. J. Sattentau, H. Schuitemaker, J. Sodroski, and R. A. Weiss. A new classification for HIV-1. *Nature* **1998**, *391*, 240.

78 Bleul, C. C., L. Wu, J. A. Hoxie, T. A. Springer, and C. R. Mackay. The HIV coreceptors CXCR4 and CCR5 are differentially expressed and regulated on human T lymphocytes. *Proc. Natl. Acad. Sci. USA* **1997**, *94*, 1925–1930.

79 Chesebro, B., and K. Wehrly. Development of a sensitive quantitative focal assay for human immunodeficiency virus infectivity. *J. Virol.* **1988**, *62*, 3779–3788.

80 Anand, R., F. Siegal, C. Reed, T. Cheung, S. Forlenza, and J. Moore. Non-cytocidal natural variants of human immunodeficiency virus isolated from AIDS patients with neurological disorders. *Lancet* **1987**, *2*, 234–238.

81 Larder, B. A., B. Chesebro, and D. D. Richman. Susceptibilities of zidovudine-susceptible and -resistant human immunodeficiency virus isolates to antiviral agents determined by using a quantitative plaque reduction assay. *Antimicrob. Agents Chemother.* **1990**, *34*, 436–441.

82 Hsu, M. C., U. Dhingra, J. V. Earley, M. Holly, D. Keith, C. M. Nalin, A. R. Richou, A. D. Schutt, S. Y. Tam, M. J. Potash, et al. Inhibition of type 1 human immunodeficiency virus replication by a tat antagonist to which the virus remains sensitive after prolonged exposure in vitro. *Proc. Natl. Acad. Sci. USA* **1993**, *90*, 6395–6399.

83 Nussbaum, O., C. C. Broder, and E. A. Berger. Fusogenic mechanisms of enveloped-virus glycoproteins analyzed by a novel recombinant vaccinia virus-based assay quantitating cell fusion-dependent reporter gene activation. *J. Virol.* **1994**, *68*, 5411–5422.

84 Wyatt, J. R., T. A. Vickers, J. L. Roberson, R. W. Buckheit, Jr., T. Klimkait, E. DeBaets, P. W. Davis, B. Rayner, J. L. Imbach, and D. J. Ecker. Combinatorially selected guanosine-quartet structure is a potent inhibitor of human immunodeficiency virus envelope-mediated cell fusion. *Proc. Natl. Acad. Sci. USA* **1994**, *91*, 1356–1360.

85 Hamy, F., E. R. Felder, G. Heizmann, J. Lazdins, F. Aboul-ela, G. Varani, J. Karn, and T. Klimkait. An inhibitor of the Tat/TAR RNA interaction that effectively suppresses HIV-1 replication. *Proc. Natl. Acad. Sci. USA* **1997**, *94*, 3548–3553.

86 Horscroft, N., V. C. Lai, W. Cheney, N. Yao, J. Z. Wu, Z. Hong, and W. Zhong. Replicon cell culture system as a valuable tool in antiviral drug discovery against hepatitis C virus. *Antivir. Chem. Chemother.* **2005**, *16*, 1–12.

87 Yi, M., F. Bodola, and S. M. Lemon. Subgenomic hepatitis C virus replicons inducing expression of a secreted enzymatic reporter protein. *Virology* **2002**, *304*, 197–210.

88 Lindenbach, B. D., P. Meuleman, A. Ploss, T. Vanwolleghem, A. J. Syder, J. A. McKeating, R. E. Lanford, S. M. Feinstone, M. E. Major, G. Leroux-Roels, and C. M. Rice. Cell culture-grown hepatitis C virus is infectious in vivo and can be recultured in vitro. *Proc. Natl. Acad. Sci. USA* **2006**, *103*, 3805–3809.

8
How Human Embryonic Stem Cell Research Can Impact *In-Vitro* Drug Screening Technologies of the Future

André Schrattenholz and Martina Klemm

8.1
Introduction

New challenges for hazard and risk assessment in pharmaceutical and chemical industries and problematic animal testing require the development of novel *in-vitro* models for the molecular characterization of drug- or chemical-related effects. On the side of toxicity testing, in Europe REACH (Registration, Evaluation and Authorization of Chemicals) has passed the first hurdle to become legislation in November 2005 [1], and estimates of related costs and time lines for testing the approximately 30 000 eligible substances produced at quantities above 1 t in the European Union (EU), range from 2.4 to 8 billion Euros and 11 to 40 years, and are supposed to consume between 4 and 40 million test animals [2]. There are related projects in the US (voluntary EPA-High Production Volume Challenge program = HPV) [3] and Japan (TTREC = Toxicity Testing Reports for Environmental Chemicals) [4], and thus validated alternative test methods are urgently needed for this upcoming dramatic surge of safety tests. Whereas a number of animal tests for topical toxicity have been successfully replaced by alternative methods, systemic and in particular reproductive toxicity require essentially novel test strategies in order to achieve adequate safety levels [5].

On the other hand, next to toxicity, efficacy must be tested before compounds can be applied to patients in clinical phases. Related animal models for some of the most important areas of human disease, for example, cancer and neurodegeneration, require very laborious, sophisticated and often extremely disruptive, irksome and even cruel procedures. Complex functional, molecular and behavioral read-outs are often an absolute necessity for preclinical therapeutic trials. Some examples include xenograft models for various human cancers [6–8]; MCAO-models for stroke by induction of severe cerebral ischemia by permanent *o*cclusion of the *m*iddle *c*erebral *a*rtery in a variety of species [9–11]; various transgenic rodent models for conditions such as Alzheimer's disease (AD) [12–15], or amyotrophic lateral sclerosis (ALS) [16], or extremely tough *e*xperimental models for *a*utoimmune *e*ncephalomyelitis induced by vaccination of test animals with the

Drug Testing In Vitro: Breakthroughs and Trends in Cell Culture Technology
Edited by Uwe Marx and Volker Sandig

ISBN: 978-3-527-31488-1

Table 8.1 A compilation of current published protocols for differentiated endpoints from murine and human embryonic stem cells (ESC).[a)]

Germ layer	Tissue	Cell type	Ref.
Murine embryonic stem cells			
Ectoderm	Skin	Melanocytes	[96]
		Keratinocytes	[97]
	Neural	Neurons (dopaminergic)	[98]
		Neurons (dopaminergic)	[99]
		Neurons (dopaminergic)	[100]
		Neurons (dopaminergic)	[101]
		Neurons	[102]
		Neurons	[103]
		Neurons/glial cells	[104]
		Neurons/glial cells	[105]
		Neurons/glial cells	[106]
		Oligodendrocytes	[107]
		Glial cells	[108]
Mesoderm	Blood	Hematopoietic cells	[109]
		Hematopoietic cells	[110]
		Mast cells	[111]
		Dendritic cells	[112]
		Macrophages	[113]
		Lymphoid precursors	[114]
	Adipose tissue	Adipocytes	[115]
	Endothelial	Vascular progenitors	[116]
		Vascular progenitors	[117]
	Cardiac muscle	Cardiomyocytes	[118]
		Cardiomyocytes	[119]
		Cardiomyocytes	[120]
		Cardiomyocytes	[121]
		Cardiomyocytes	[122]
	Skeletal/smooth muscle	Muscle cells	[123]
		Muscle cells	[124]
		Smooth muscle cells	[125]
	Cartilage	Chondrocytes	[126]
	Bone	Osteoblasts	[127]
Endoderm	Pancreas	Pancreatic cells	[128]
		Pancreatic cells	[129]
		Pancreatic cells	[130]
	Liver	Hepatocytes	[131]

Table 8.1 (continued)

Germ layer	Tissue	Cell type	Ref.
Human embryonic stem cells			
Ectoderm	Skin	Keratinocytes	[132]
		Melanocytes, keratinocytes	[133]
	Neural	Motor neurons	[134]
		Motor neurons	[135]
		Neurons	[136]
		Neurons	[137]
		Neural precursors	[138]
Mesoderm	Blood	Lymphohematopoietic cells	[139]
		Hematopoietic cells	[140]
		Hematopoietic cells	[141]
		Hematopoietic cells	[136]
	Cardiac muscle	Cardiomyocytes	[142]
		Cardiomyocytes	[143]
		Cardiomyocytes	[144]
	Skeletal/smooth muscle	Muscle cells	[145]
	Cartilage	Chondrocytes	[145]
	Bone	Osteoblasts	[145]
	Endothelium	Vascular progenitors	[146]
Endoderm	Pancreas	Pancreatic cells	[147]
		Pancreatic cells	[148]
		Pancreatic cells	[149]
		Pancreatic cells	[133]
	Liver	Hepatocytes	[150]
		Hepatocytes	[133]

a) The list may not be comprehensive, and it should be noted that, in most cases, the characterization of tissue-specific cell types mentioned has been performed on the level of markers. Only more recently have more studies additionally included functional read-outs. In this list, ESC derived from other organisms (e.g., Rhesus monkeys) are not included, essentially because rodents are the most widely used preclinical animals in various models, and human *in-vitro* systems would be much more desirable than any other in overcoming potential inter-species variations.

autoantigen *m*yelin *o*ligodendrocyte *g*lycoprotein (MOG-EAE) [17–19] as preclinical models of multiple sclerosis (MS). However sophisticated these models are, there is a general sense of caution towards them, because often their read-outs are insufficient or misleading [20]. Taken together, in both areas of toxicity and efficacy testing, there is a need for the development of novel *in-vitro* methods, which can balance disadvantages in terms of organ-specific barriers and metabolism with advantages regarding the higher relevance of humanized systems, more precise functional and molecular read-outs, and the potential of higher throughput. Given emerging regulatory challenges, the purely diagnostic use of human embryonic stem cell (ESC) models appears to develop into a highly attractive alternative. ESC can be differentiated to organotypic cell cultures and serve as flexible substrate for a variety of functional molecular endpoints. Moreover, they are ideal for validation purposes using modern silencing technologies in a framework of genetically homogeneous differentiations to various "tissue"-like cell culture substrates such as neural cells, cardiomyocytes, various types of muscle cells, and adipocytes. An overview of currently available protocols for differentiations of ESC is provided in Table 8.1.

Indeed, in combination with cutting-edge analytical technologies, and in particular proteomics technologies, there is hope of obtaining surrogate biomarker information on the level of protein signatures, which by taking into account post-translational modifications, can serve as content for next generation, high-throughput screening (HTS) methods [21, 22]. These protein signatures also have the intrinsic potential of representing a comprehensive view of underlying mechanisms, including and relating key protein surrogates for mechanistic aspects (pathways and modes of action), toxicity-related events and/or others of purely descriptive diagnostic value [22, 23].

8.2
First Excursion: Protein Surrogate Biomarker Signatures

Innovative alternative methods such as diagnostic human ESC (hESC) models are embedded in a context of subsequent analytical procedures. Here, we would like to focus as an example on current developments concerning developmental toxicity. The precise functional read-outs of hESC- and murine ESC (mESC)-based tests will be explained in greater detail below. On the molecular level, a quantitative differential analysis of protein expression is a mandatory requirement. This is by no means a trivial task, and this section is meant to provide a basic understanding of underlying problems and adequate methods and technologies for addressing them. Related efforts are directed by international and national regulatory agencies such as ECVAM (European Centre for the Validation of Alternative Methods) and BfR (Federal Institute for Risk Assessment, Germany), sponsored by the European Commission with the participation of major pharmaceutical companies and numerous academic partners [24]. ESC models play a crucial role in these projects [5].

Pharmaco- and toxico- and diagnostic genomics use surrogate biomarkers on the level of nucleic acids. Such methods have been vigorously pursued and developed with huge amounts of investments over the past decade or more, for all types of purposes and disease areas with certain achievements on the diagnostic level, but otherwise limited directly applicable success [22, 23, 25]. However, the knowledge – and in particular the methodology – developed during the course of the Human Genome Project and further whole-genome-sequencing projects are prerequisite for the subsequent field of proteomics. It is this availability of annotated genomic information of important organisms in well-organized and accessible databases, the statistical and bioinformatics tools developed for array-based nucleic acid quantification and cluster analysis, which now help us to understand the unfolding wealth of information essentially emerging from mass spectrometry-based protein identification.

The reason for the relative disappointments of genomics is that information is incredibly condensed on the level of genes, and unfolds to an enormous complexity of protein expression by post-transcriptional and post-translational modifications.

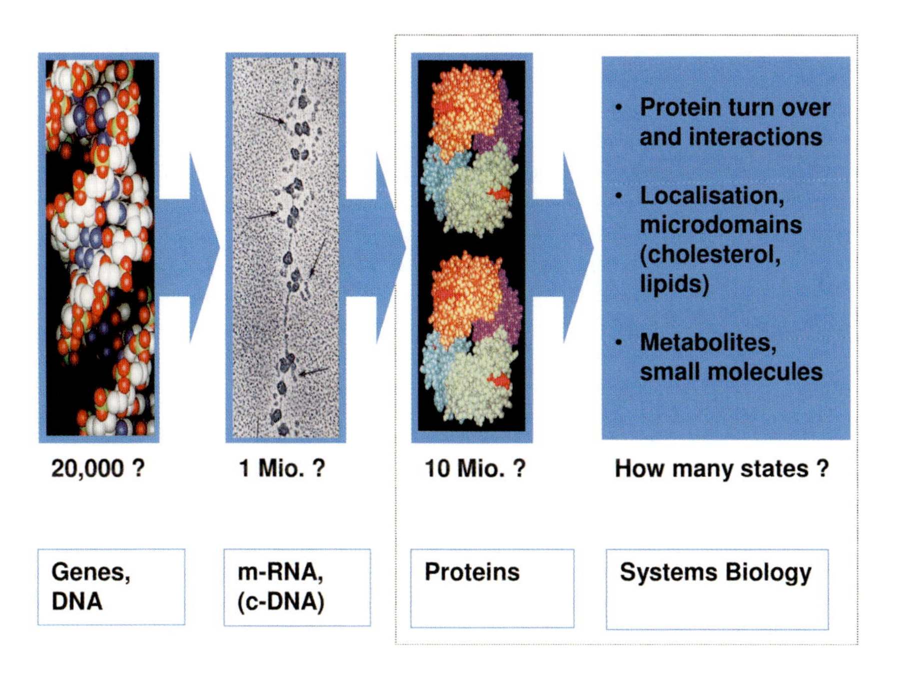

Fig. 8.1 Number relationship of molecular species on the levels of DNA, RNA and proteins: the enormous complexity at the level of proteins provides the basis for the equally enormous flexibility of biological systems. Single genes often lead to hundreds, or even thousands, of functionally modified protein molecules. Thus, the main task of a comprehensive analysis of proteins ("proteomics") is the establishment of a reliable methodology for complexity reduction. Dynamic, cellular processes with compensation and crosstalk require consequentially differential and quantitative identification of correlated protein signatures, rather than single all-or-none targets.

For humans, these mechanisms help to translate roughly 20 000 genes [23, 27] into millions of molecular species on the level of proteins. This functional inflation of complexity is largely due to chemical modifications of amino acid residues of proteins, for example, by phosphorylation, glycosylation, methylation, acetylation, oxidation, or by proteolytic processing. These mechanisms generate very distinct, organ- and process-specific protein embodiments from the very same reading frames provided by uniform genomic information in each nucleus. One of the major difficulties of related proteomic research next to the complexity of sheer numbers, is the dynamic range of possible concentrations (8 to 15 orders of magnitude); another is the time frame of molecular changes (ranging from fractions of seconds to years) [22, 23]. In addition, proteins have a maximum of chemical diversity (acidic, alkaline, hydrophobic, hydrophilic, etc.) and no amplifying method (e.g., PCR for nucleic acids) exists or is even perceivable. The relationships of numbers of molecular species on the levels of DNA, RNA, and proteins are illustrated in Figure 8.1.

The enormous complexity at the level of proteins, provides the basis for the equally enormous flexibility of biological systems.

Consequently, robust and reliable quantification and differential display of protein expression in complex samples is at the core of biomarker definition by proteomics technologies. Unfortunately, at all levels, related methods are laborious and cumbersome. Protein quantification requires the labeling of protein mixtures by radioactive or stable isotopes or fluorescent dyes. Separation techniques are defined by the complexity of samples; highly complex mixtures require two-dimensional polyacrylamide gel electrophoresis (2D-PAGE), multidimensional liquid chromatography (LC) or capillary electrophoresis (CE). Complexity reduction of samples by appropriate fractionation (e.g., albumin depletion, phosphoproteins) has always to be considered before embarking upon a search for protein biomarkers. However, regardless of fractionation strategies, the best way to reduce complexity is a reliable differential display, which depends directly on the linear dynamic range of concentrations expected in a given sample. As shown in Figure 8.2, for an example of rather complex protein mixtures from neural derivatives of mESC, compared for the sake of a molecular signature of ischemic events in neurons, high-resolution 2D-PAGE and radioactive labeling are the best choice [28]. In cases of less-complex samples, for example, after extensive fractionation or from bacterial samples (only few post-translational modifications!), LC- or CE-based methods and stable isotope labeling represent alternatives [29, 30].

A number of recent reviews have described the state of the art in proteomics technologies [22, 23, 25]. Analytical strategies for the various levels of protein biomarker discovery must be considered carefully at every stage, and should be directly correlated to functional parameters and embedded in an iterative bioinformatics processing, which relates incoming results from mass spectrometry directly back to functional biological models used to generate respective samples. Against this background *in-vitro* models have real advantages, because the approach enables an integrated (at least partially) functional validation of the molecular data. The application of genomics tools, such as cluster analysis for the reliable quantitation of protein expression data (from still rather low-throughput proteomic technologies), justifies expectations that here the novel content for future high-throughput devices can be generated.

Given the enormously complex and dynamic nature of protein modifications, and the redundant and pleiotropic organization of almost all major signal transduction pathways, we envisage the emerging importance of biomarker signatures consisting of functionally related sets of post-translational protein isoforms, rather than single targets or surrogate biomarkers. So, one thing is certain: future devices will have to provide information beyond the amino acid backbone of proteins, for instance, post-translational isoforms which are functionally defined and precisely characterized on the molecular level.

8.3
Second Excursion: Validation

The validation of a target or a biomarker before proceeding to further development usually requires the unambiguous demonstration that the candidate really plays the proposed role. One way to achieve this is by the genetic manipulation of biological systems, which turns on or off the expression of candidate proteins, related pathways or associated proteins by site-directed mutagenesis, knock-out or transfection experiments. Recently, one of the most attractive approaches is that of RNA silencing (siRNA), which avoids the obfuscating effects of pleiotropic and redundant signaling pathways which are affected in more global ways by mutational approaches [31–33]. In particular, with regard to highly popular transgenic animal models, siRNA technologies provide a much more targeted and focused approach to validation by aiming at specific m-RNAs at specific stages of development, differentiation, or under well-defined functional conditions (Fig. 8.3).

Embryonic stem cells fulfill these requirements; they can be differentiated under very well controlled conditions to various organotypic cell types, which then represent different embodiments of a single genome, thus providing a perfect substrate for initial validation. Quite a number of recent studies have applied siRNA protocols to hESC and mESC [34–36]. A short description of siRNA procedures is provided in Figure 8.3. Together with the cutting-edge analytical technologies mentioned above, the corresponding use of hESC and mESC could provide a most powerful ensemble for integrated discovery, validation and screening, developing and applying protein biomarker signatures in novel *in-vitro* methods for toxicology and efficacy testing of chemicals and pharmaceuticals. ESC-based *in-vitro* systems can, moreover, be applied directly for the generation of corresponding *in-vivo* models [34–36].

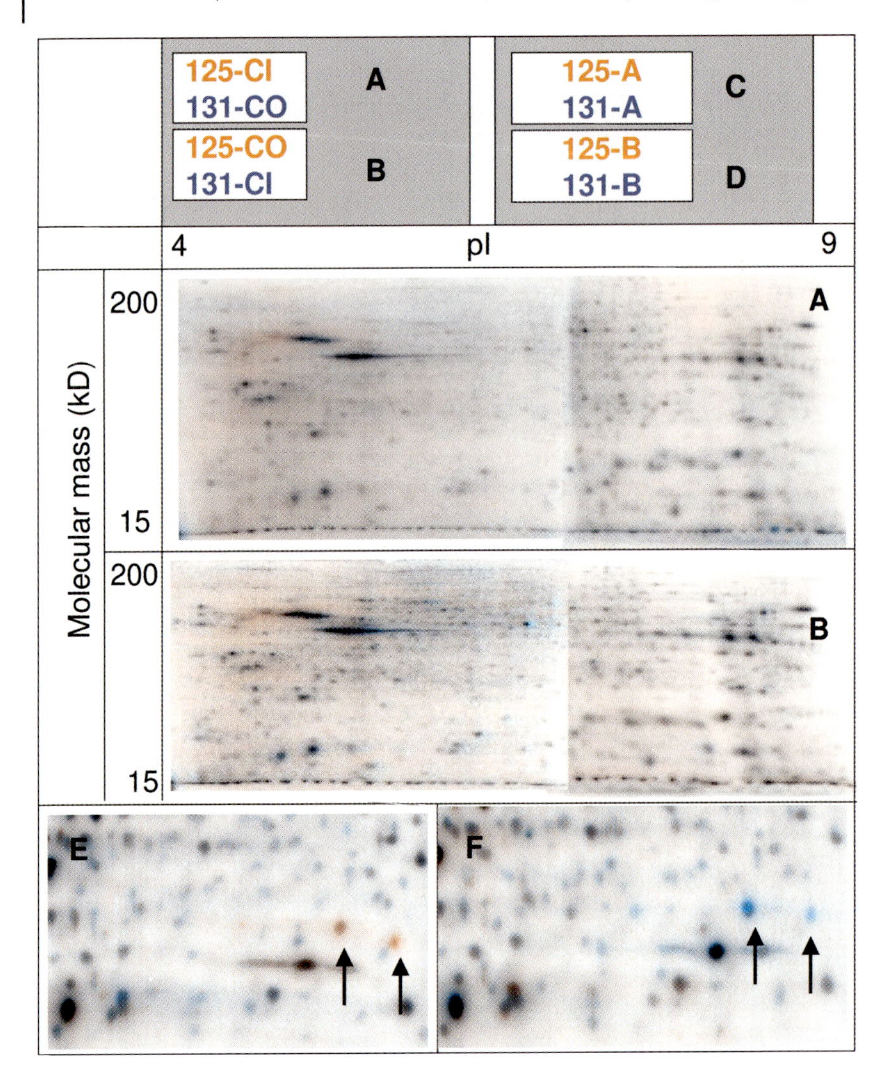

Fig. 8.2 Pattern control of ischemic effects in a neural murine embryonic stem cell (ESC) culture model by quantitative and differential analysis of protein expression: high-resolution 2D gels [151] were used to compare the protein expression pattern of neurons exposed to chemical ischemia (CI) with control neurons (CO). The two single gels in (A) and (B) were obtained without prior fractionation; the labeling pattern is explained in the upper part of the figure. Labeling controls (C) and (D) are not shown. In the lower part of the figure a more detailed view of surrogate two-color representation of radioactively cross-labeled duplicate gel frames comparing phosphoproteomes of control neurons versus neurons exposed to chemical ischemia is shown. The left panel shows control in blue versus ischemia in orange; the right panel shows control in orange versus ischemia in blue. The two phosphoproteins appearing after the induction of chemical ischemia shown here, have the same quantities and positions in both series of the labeling procedure (^{125}I and ^{131}I), and thus represent novel post-translational surrogate markers, in an ESC model related to human neurodegeneration [28].

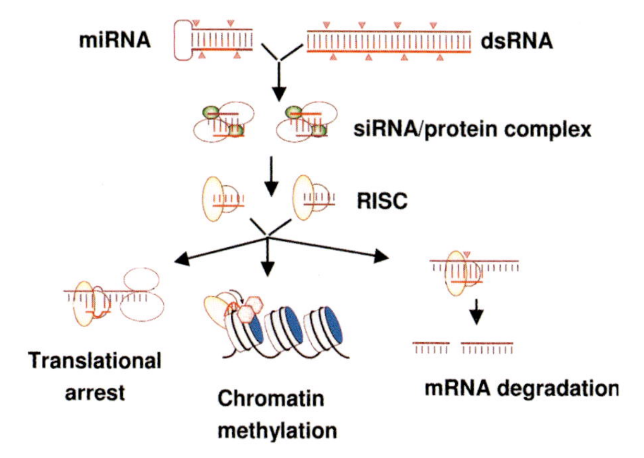

Fig. 8.3 An overview of the silencing mechanisms by small interfering RNA (siRNA). The key event is the generation of double-stranded RNA (dsRNA); this can occur endogenously in the cell, by genomic transcription of antisense RNA to microRNA (miRNA) or long dsRNA or by direct experimental introduction into target cells. The dsRNA is cleaved by intrinsic enzymes to siRNAs at cleavage sites indicated by orange triangles. These siRNA pieces form different types of RNA-induced silencing complexes (RISC, with single-stranded siRNA), which subsequently cause mRNA degradation, chromatin methylation, or translational inhibition [152, 153]; siRNA approaches are versatile because they can be applied *in vitro*, and also used for corresponding systemic or local *in-vivo* mammalian gene silencing.

8.4 Reproductive Toxicology and *In-Vitro* Tests

Despite a broad range of alternative methods, substantial numbers of animals are still required for the *in-vivo* testing of toxicity of drugs, chemicals, and cosmetics [37, 38]. In particular, efforts for the development of alternative tests for reproductive toxicity hazard [5] are unfolding, including the use of mESC and hESC [39–41]. For toxicity testing in general – but reproductive toxicity in particular – current regulatory data requirements vary considerably, being most stringent for pharmaceuticals, food additives and pesticides/biocides. For a number of reproductive key phases, which for these regulatory purposes to date are covered in standardized animal tests, no validated alternatives exist. These *in-vivo* tests have a disadvantage in that molecular information on functional deficits in terms of biomarkers appears difficult [39, 42]. As outlined above, alternative *in-vitro* tests have the potential to provide this urgently needed, more specific molecular and functional information about toxicological mechanisms on the respective target tissues and, moreover, have the chance to translate subsequently to devices for fast and efficient HTS [21].

The complexity of events involved in reproductive and systemic toxicology, organ or tissue interactions, barriers and metabolic activation, is generally inadequately addressed *in vitro*. This limitation can be compensated, for instance, by the addition of metabolizing systems or co-culture of various cell types. On the other hand, the use of hESC provides the unique opportunity to approach the human target species by ruling out variances between species which are one of the major disadvantages of *in-vivo* tests in animals. The current contribution is set in a framework of developing concepts for the validation of innovative alternative systems, namely mESC and hESC or reporter gene-based tests using genetically engineered cells [5].

Due to the current policy for testing chemicals in the European Community (REACH) [1] and elsewhere (EPA-HPV program in the US, TTREC in Japan) [3, 4], animal consumption for toxicity safety testing is expected to increase substantially, and the time lines foreseen are narrow. Reproductive toxicity is one of the crucial endpoints, and so there is an urgent need for the rapid development of alternative methods for at least partial replacement. In this regard, the subtle – but nonetheless potentially severe – impairment of endocrine homeostasis by endocrine disruptors has prompted the development of corresponding *in-vitro* methods. Their respective reliability and relevance needs to be explored, in particular in combination with further *in-vitro* methods and novel sensor technologies, computer-based *q*uantitative *s*tructure–*a*ctivity *r*elationship (QSAR), and methods based on patterns or signatures of surrogate biomarkers. Beyond development and prevalidation, the integration into a conceptual framework for safety toxicology (tiered testing strategy with decision points and various branches) is the ambitious goal of related projects such as ReProTect [5, 24].

8.5 Reproductive Toxicology and hESC

There is a worldwide dedicated research effort towards the continuous improvement of risk assessment and health and safety protection. In the European Commission's regulatory framework of drug approval for treatments of patients, certain sets of standardized animal tests are performed according to the European Council directives [43–45] and the guidelines of the International Conference on Harmonisation [46]. The hierarchical decision-tree approaches defined by the European Commission [47] and the Organisation for Economic Cooperation and Development [48, 49] for international standards of pharmacological and toxicological tests primarily address the adult organism. Currently, although comprehensive multigenerational studies are undertaken to provide information about all aspects of reproductive and developmental toxicity, developmental studies to address the specific risks of the developing embryo are not regulatory requirements and, therefore, appropriate methods for the screening of toxic substance effects during embryonic development (embryo- and fetal toxicity) are essentially absent. For example, there were cases of important toxic side effects with pathological consequences during development, which were only discovered after approval for therapeutic applications in patients (the dire case of

Contergan). Likewise, more than 30 compounds with highly toxic potential had to be withdrawn from the market by the FDA (Food and Drug Administration) during the years 1998 to 2001, in each case without being detected by the required set of prior animal tests (one of the most recent examples was Lipobay). Among these 30 drugs, eight caused substantial side effects, especially in women. Current animal tests are simply not sufficient to detect gender-specific predispositions in humans. Moreover, during the time after Contergan, about 85% of those cases of developed drugs with highly toxic potential could not be predicted safely by animal tests [50, 51].

For human embryotoxicity, and especially in terms of neurodevelopmental processes, the question is whether the use of human embryonic or adult stem cells is better to establish *in-vitro* screening methods. Basically, stem cells have two characteristic properties: (1) they can be maintained undifferentiated in culture for long periods of time; and (2) they can be induced into specific organotypic differentiations. As shown in Figure 8.4, there exist both embryonic and adult tissue-specific stem cells. Embryonic stem cells are established from the inner cell mass (ICM) of 5- to 7-day-old blastocysts [52, 53]; the tissues of origin for adult stem cells are marked in red.

For the long-term cultivation of undifferentiated cells, ESC cultures are maintained in co-culture with so-called "feeder" cells from mouse fibroblasts [54], or cell-free in conditioned media [55]. The major inhibitory factor for differentiation in the murine cell culture system is the cytokine leukemia inhibitory factor (LIF) [56], while the adequate factor for long-term hESC culture is still under investigation. ESC display far-reaching pluripotent properties; for example, they can differentiate into virtually every organ-specific human cell type when cultured as embryoid bodies. Differentiation into diverse cell types of endodermal (e.g., pancreatic and hepatic cells), mesodermal (e.g., bones, muscle and blood cells, cardiomyocytes) or ectodermal origin (e.g., neurons and glial cells) is regulated by LIF deprivation for mESC, and the proper combination of growth factors. The published reports on the differentiation potential of mESC and hESC are summarized in Table 8.1.

Adult stem cells are derived from mature organs of adult individuals such as the epidermis, hair, intestine, liver, hematopoietic system, brain or bone marrow, and have limited possibilities of proliferation and differentiation ("developmental restriction"). Due to their multipotent properties, these stem cells are mostly restricted to one tissue only. The plasticity of the adult stem cells – which means the ability to contribute into cell types characteristic of another organ – was only demonstrated for hematopoietic stem cells, stromal cells of the bone marrow and multipotent adult precursors, which can be generated *in vitro* from certain cells of the bone marrow and certain neural cells [57] (see also Fig. 8.3). The question of whether adult stem cells can be artificially redifferentiated is currently under intense investigation, but remains open for the time being. A further disadvantage of adult stem cells is their limited ability to differentiate and proliferate, which goes together with decreased lifetime *in vitro*. Moreover, only very small numbers of these cells are available (they decrease even further with age), they are difficult

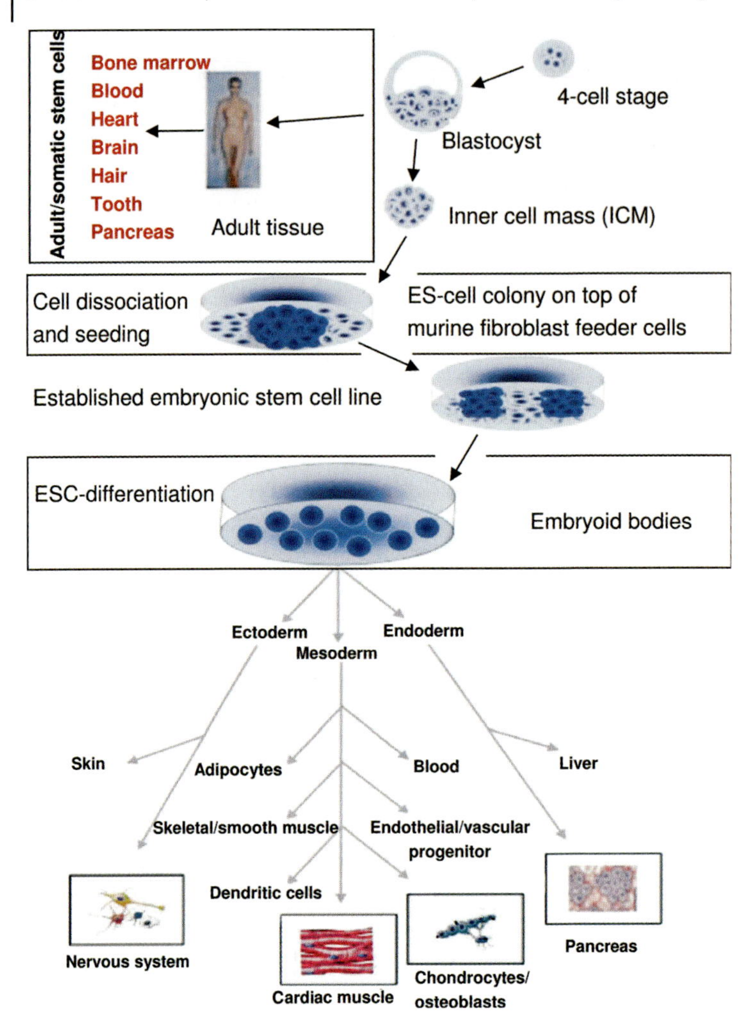

Fig. 8.4 An overview of adult and embryonic stem cells. Starting from the blastocyst stage, the inner cell mass (ICM) is used for the generation of embryonic stem cell (ESC) lines. These ESC lines can be maintained and expanded easily, are pluripotent, and can be differentiated to all types of tissue-specific cell lines of the three germ layers under appropriate conditions (see Table 8.1). In contrast, adult/somatic stem cells derived from tissues of adult organisms are more likely multipotent and, depending on their origin, they can only differentiate to a limited set of endpoints. Moreover, they are difficult to obtain and cannot be maintained and expanded in large-scale quantities.

to distinguish from surrounding cells, and their derivation is either difficult or impossible, depending upon the original organ. To date, it remains unclear as to how adult stem cells function and how their potential might be optimally exploited [58–60].

Consequently, results with direct relevance for human developmental toxicity can only be obtained using hESC approaches. Only these models can provide information about the effects of compounds on all aspects of embryonic neuronal development.

Now, using hESC, there exists for the first time the chance to develop innovative *in-vitro* analytical methods for the early detection of human-specific embryotoxic risks of compounds and chemicals. The combination of hESC-based *in-vitro* test models with cutting-edge molecular analysis holds the promise to deliver – with extreme sensitivity – highly relevant functional and molecular marker endpoints for the early detection of embryotoxic effects of drugs or chemicals. The outstanding potential of hESC-based *in-vitro* models has thus to be viewed in context with results from corresponding mESC-screening systems. In combination with quantitative differential proteomic display techniques, biomarkers for neurotoxicity have already been explored. A variety of ESC-based tests have already been validated and are in use for regulatory purposes [39–42, 61]. In terms of relevance, proteomics is in principle far superior to RNA/DNA-based array technologies, because the relatively static and small human genome of approximately 20 000 genes is translated into a highly dynamic and complex proteome of several million protein molecules of post-translationally modified isoforms. It is on this latter level where dynamic functional aspects of cellular events take place [22, 62–68]. Both neurotoxic and embryotoxic substances bind to proteins, changing interactions in complex functional networks, sometimes on very rapid time scales (e.g., phosphorylation), and the changes induced by this type of intervention are ideally analyzed by proteomics technologies. The hope is that emerging molecular signatures from mESC can in the future also be derived from functionally controlled differentiated hESC, with key biomarker proteins identified and quantified, thus enabling a molecular understanding of side effects and innovative strategies for acceleration in drug development, validation, and toxicity testing [69].

A clear understanding of the neurotoxic effects of medical drugs for human embryos during pregnancy would constitute a major contribution towards improved drug safety and preventive healthcare. The effects of neurotoxic substance application can be defined on the level of distinct molecular consequences in terms of immediate protein expression changes. The advanced proteomics technologies mentioned above, in correlation with synchronized functional/physiological measurements, can provide a comprehensive, precise and quantitative molecular pattern analysis of the underlying mechanisms. Moreover, the availability of human biomarkers for neuronal embryotoxicity could drastically improve early screening procedures for embryotoxic and neurotoxic substances.

A number of projects are ongoing, with hESC being cultivated *in vitro* and differentiated to functional neurons according to published procedures. Undifferentiated hESC are characterized by surface antigens such as SSEA-3, SSEA-4 [70], TRA-1-60, TRA-1-81, Oct-4 [71] and GCTM-2; or by enzyme activities (alkaline phosphatase). Neurons differentiated from precursors are characterized by morphological criteria and immunohistochemistry (antibodies for neuron-specific proteins MAP-2 and synaptophysin). Subsequently – and more importantly – a functional and physiological control of neurons by their response to various neuron-specific stimuli (neurotransmitter, depolarization, etc.) is performed and measured by, for example, single cell Ca-imaging.

At present, the main objective of related projects is the identification of molecular mechanisms underlying the human neurotoxicity of 20 selected substances [5, 37]. The corresponding events take place in early embryonic and specifically neural developmental stages and maturation processes. Thus, an *in-vitro* test system based on hESC is ideally suited for corresponding functional and molecular analysis. Cells will be exposed to various concentrations of toxic compounds and analyzed on the molecular level at various endpoints, representing precursors, early and mature stages of neurons [5]. Dose–response relationships of toxic effects induced by substances are essentially quantified by the influence of these compounds on normal functional signals measured by calcium imaging. Even subtle influences of toxic compounds upon intracellular transient calcium concentration changes due to normal reactions caused by neurotransmitters or other physiological parameters will become apparent on a statistically significant level.

Based on the functional analysis, protein pellets will be generated from treated and untreated neuronal differentiated human stem cells at appropriate endpoint conditions, and subsequently be submitted to a quantitative and differential protein pattern analysis (see Fig. 8.2).

Under the given circumstances of ethical controversy concerning hESC and strict legal guidelines for stem cell importation and use in the European Community (in particular in Germany), it is very sensible to set standards with certain registered and controlled hESC lines (http://stem.nih.gov/research/registry). Arguably, this would not lead to a proliferate generation of ever new and individualized cell lines with associated scarification of embryos, as in the case of therapeutic stem cell approaches, but on the contrary, it would lead to the development of one or a few existing lines for very broad and standardized application [21]. In this way, it is conceivable for the first time to exploit the outstanding features of hESC and establish related innovative screening methods for embryotoxicity and neurotoxicity, including the identification of molecular toxicity biomarkers without using animal-based *in-vitro* or *in-vivo* systems. Moreover, it would avoid some of the burning ethical issues haunting the therapeutic branch of hESC research.

8.6
Efficacy and Mode of Action Studies: Systems Biology Using Embryonic Stem Cell-Based Screening Systems

In terms of efficacy, a variety of *in-vitro* studies employing mESC for testing neuroprotective compounds and conditions have been published. Essentially, the protocols have been adapted from corresponding experiments using primary cultures [72]. Specific agonists and antagonists allow fairly precise pharmacological characterization of the major ionotropic and metabotropic receptors (for glutamate,

GABA, acetylcholine, etc.) and voltage-dependent ion channels in neural cell culture models [72]. The precise control of these physiological responses is prerequisite for defining appropriate experimental windows for the subsequent molecular analysis or pharmacological profiling. For a replacement of animal models for neuroprotection, essentially a variety of cellular insults are performed (such as excitotoxic, ischemic or by oxidative or β-amyloid-related stress). Neuroprotective conditions are then monitored on the levels of molecular markers from differential protein analysis and functional read-outs. The simplest read-out is quantification of neuronal survival by functional vitality controls, fluorescent vital staining, or live cell staining. As shown in Figure 8.5, it follows correlational analysis of functional and molecular data, and the definition of physiological and/or pharmacological/toxic endpoints.

The relevant cell status for effective neurotoxic conditions is defined by the outcome of dose–response curves in calcium-signaling experiments; rescue by neuroprotective conditions quantified in terms of surviving neurons, etc. ESC are thus embedded in a seamless and iterative process, where emerging molecular surrogates can be fed back directly to functional *in-vitro* systems (Validation I), and fed forward to *in-vivo* models generated by using the very same ESC (Validation II), under comparable paradigms of intervention (for example, siRNA). Moreover, conditions tested in neural derivatives can be checked in genetically homogeneous alternative differentiation protocols (Validation III; for example, does a neuroprotective substance have adverse effects in hepatocytes?). ESC models offer the huge advantage of avoiding genetic variation, one of the major factors complicating biomarker validation.

By summarizing the examples in Figure 8.5, the authors would like to emphasize two important points:

Next to a characterization of tissue-specific differentiated endpoints from ESC by molecular markers, a thorough functional characterization and control of corresponding cell cultures is absolutely mandatory. In the case of neurons, efficacy studies need to be performed in a context of pharmacological and physiological consistency.

In contrast to strategies employed in the therapeutic field, "pure" endpoints (e.g., solitary neurons for transplantation) are often far too artificial and short-lived and, moreover, ignore epigenetic events and reprogramming [73].

Thus, the use of hESC *in-vitro* models on the contrary aims at generating stable cell culture conditions, including mixed cultures with multiple cell types which, in some cases, provide a better match of organotypic "physiological" conditions. Mixed neural cell cultures with about 15% neurons and a variety of supporting cells, for example, much more closely represent the situation in a brain. In the case of neurons, the basic requirement would be the presence of major neuronal ion channels (e.g., NMDA-, $GABA_A$- and cholinergic receptors) in appropriate proportions and with adequate pharmacology. These neural cultures could only then be used as models for important human diseases of the nervous system. The corresponding molecular analysis of functional mESC models for excitotoxicity and ischemia, consequently revealed surrogate biomarkers such as superoxide

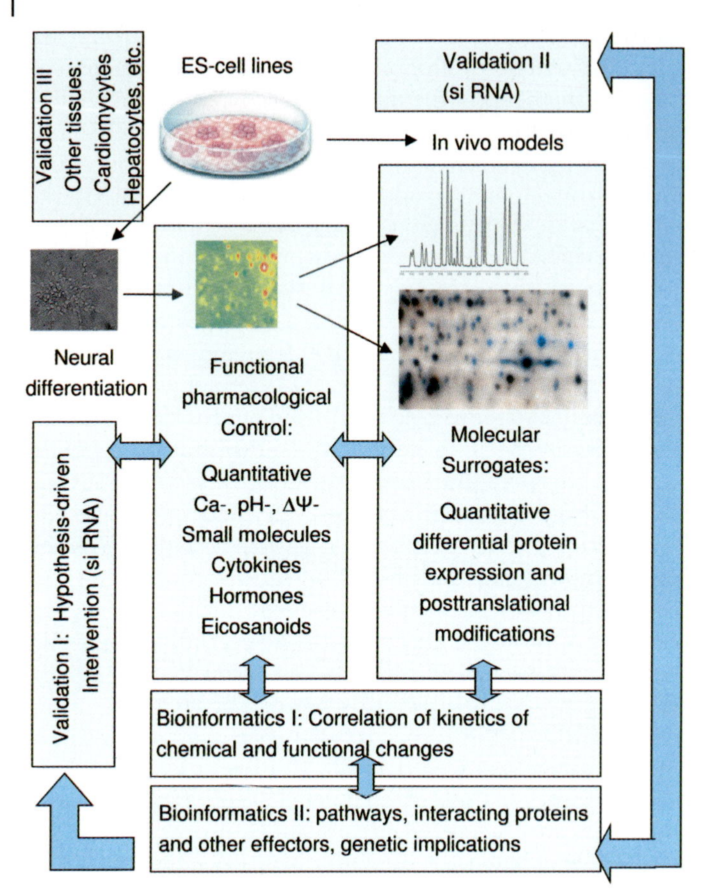

Fig. 8.5 Embryonic stem cell (ESC)-based *in-vitro* systems for neurotoxicity and neuroprotection screening. In a context of cutting-edge analytical technologies, these offer the opportunity to integrate functional read-outs (toxicology-, mode of action- or efficacy-related) with precisely controlled biomarker signatures (proteins, metabolites, or other effectors). The corresponding data mining (mostly automated) generates contextual information from chemical, genetic or literature databases (Bioinformatics II), which in an essentially iterative process can instantly be fed back into experimental procedures or data layers. On a first level (Bioinformatics I), mass spectrometry-based investigation of chemical details underlying differential molecular information produces a refined mechanistic understanding of the functional read-outs. Emerging key surrogate markers can be directly fed back into *in-vitro* and *in-vivo* models (Validations I and II) by, for example, hypothesis-driven silencing technologies (siRNA). ESC-based *in-vitro* models (either murine or human) have one huge advantage, namely that they can be used directly to generate corresponding *in-vivo* models. A further huge advantage is the possibility to test, for example, neuroprotective substances in genetically homogeneous different organ-specific cell types, such as hepatocytes or cardiomyocytes (Validation III). The mechanistic information form (e.g., differential proteomic profiling) helps to refine the control and read-outs from related *in-vivo* models (Western blots, chip-based markers, pharmacology). The combined functional and molecular information of this type can help to refine chemical concepts of intervention (quantitative structure–activity relationship (QSAR), molecular modeling).

dismutase (SOD), neuregulins or dihydropyrimidinase-related protein 2 (DRP-2) [28, 72, 74], which are consistently implicated independently by genomic analysis in human neurodegenerative diseases such as amyotrophic lateral sclerosis (ALS), Alzheimer's disease, and others [75].

8.7 Conclusions and Outlook

Taken together, the current state of research into stem cell models for the replacement of animal tests in toxicology and efficacy studies justifies far-reaching hopes. The validated mESC-assay for testing embryotoxicity appears to represent only a first example en route to further, and eventually humanized, *in-vitro* models [76–78]. The development of these novel models must be embedded in and accompanied by cutting-edge analytical technologies, exploiting their advantages to the fullest extent. The realization of the huge potential to address mechanistic questions (protein surrogate biomarkers, differential pharmacology and silencing technologies, and various tissue-specific but genetically homogeneous cell types) would also require the incorporation of novel concepts concerning *in-vitro* pharmacokinetics and bioavailability into related projects [79–81]. Given the biological complexity in developmental toxicity or neuroprotection, and also taking into account upcoming regulatory requirements, it will likely be a battery of related *in-vitro* tests necessary for the timely, safe, and cost-efficient generation of related information [82]. Remaining problems, such as systemic barriers, endocrine disruptors and metabolism, will certainly be addressed in the foreseeable future by *in-vivo* models, but even there, some of the expected tools from *in-vitro* ESC research could be extremely valuable in the sense of the 'three Rs' (refinement, replacement and reduction [83]). The post-translational protein information emerging from ESC models, for instance, could serve as the content for chip-based HTS methods in assessing *in-vivo* effects.

Technologies and infrastructural support for the maintenance, characterization and control of hESC lines are rapidly improving [84–87], and have the potential to provide a worldwide set of standardized tools for validated alternative methods in toxicity and efficacy testing [5, 41, 69, 78, 88].

The final consideration belongs to ethical questions concerning the generation of hESC, which includes issues ranging from the donation of oocytes, *in-vitro* fertilization, and the need for the continuous sacrifice of human embryos to medical therapies at very early stages and with new scientific challenges of risk assessment [89–94]. This controversy in the therapeutic field [95] is fuelled by immense hopes, promises and commercial interest, and related efforts have suffered a number of recent high-profile set-backs. Yet, all of these topics are essentially absent in the case of hESC-based *in-vitro* screening systems for the replacement of animal tests.

In contrast, the purely diagnostic use of hESC – with its immense experimental advantages (as discussed above), and in particular with the major benefit of being "human" – has a tendency rather to establish a few selected cell lines and validated protocols as worldwide standards, thus avoiding the ethical implications surrounding the generation of new cell lines and therapeutic risks. Today, this is acknowledged even in European countries, where there is very restrictive legislation with regard to hESC (e.g., Germany and Italy). Nevertheless, in these countries substantial public funding has been approved by central ethics committees for related research projects involving regulatory agencies, academic institutions, major pharmaceutical companies, and small biotechnology units. More recently, the media coverage on these projects has been relatively low profile – but essentially positive – and the projects are well under way [5].

References

1 REACH: Commission welcomes Council's agreement on new EU chemical legislation, 17th November 2005; http://europa.eu.int/rapid/pressReleasesAction.do?reference=IP/05/1583&format=HTML&aged=0&language=EN&guiLanguage=en.

2 Hofer, T., Gerner, I., Gundert-Remy, U., Liebsch, M., Schulte, A., Spielmann, H., Vogel, R., and Wettig, K. *Arch. Toxicol.* **2004**, 78, 549–564.

3 EPA-High Production Volume (HPV) Challenge Program, **2006**; http://www.epa.gov/chemrtk/volchall.htm.

4 TTREC: Background to the Publication of Toxicity Testing Reports for Environmental Chemicals and the Framework of the Reports, **2006**; http://wwwdb.mhlw.go.jp/ginc/html/db1.html.

5 Hareng, L., Pellizzer, C., Bremer, S., Schwarz, M., and Hartung, T. *Reprod. Toxicol.* **2005**, *20*, 441–452.

6 O'Connor, O. A., Toner, L. E., Vrhovac, R., Budak-Alpdogan, T., Smith, E. A., and Bergman, P. *Leuk. Lymphoma* **2005**, *46*, 973–992.

7 McCormack, E., Bruserud, O., and Gjertsen, B. T. *Leukemia* **2005**, 19, 687–706.

8 Suggitt, M. and Bibby, C. *Clin. Cancer Res.* **2005**, *11*, 971–981.

9 DeVries, A. C., Nelson, R. J. Traystman, R. J., and Hurn, P. D. *Neurosci. Biobehav. Rev.* **2001**, *25*, 325–342.

10 Yanamoto, H., Nagata, I., Niitsu, Y., Xue, J. H., Zhang, Z., and Kikuchi, H. *Exp. Neurol.* **2003**, *182*, 261–274.

11 Cechetto, D. F. *Stroke* **1993**, *24*, I6–I9.

12 Higgins, G. A. and Jacobsen, H. *Behav. Pharmacol.* **2003**, *14*, 419–438.

13 van Dooren, T., Dewachter, I., Borghgraef, P., and van Leuven, F., *Subcell. Biochem.* **2005**, *38*, 45–63.

14 Van Dam, D., Vloeberghs, E., Abramowski, D., Staufenbiel, M., and De Deyn, P. P. *CNS. Spectr.* **2005**, *10*, 207–222.

15 Jankowsky, J. L., Savonenko, A., Schilling, G., Wang, J., Xu, G., and Borchelt, D. R. *Curr. Neurol. Neurosci. Rep.* **2002**, *2*, 457–464.

16 Grieb, P. *Folia Neuropathol.* **2004**, *42*, 239–248.

17 Sriram, S. and Steiner, I. *Ann. Neurol.* **2005**, *58*, 939–945.
18 Skundric, D. S. *Curr. Neurovasc. Res.* **2005**, *2*, 349–362.
19 Ziemssen, T. and Ziemssen, F. *Autoimmun. Rev.* **2005**, *4*, 460–467.
20 Gruber, F. P. and Hartung, T. *ALTEX* **2004**, *21* (Suppl 1), 3–31.
21 Schrattenholz, A., Klemm, M., and Cahill, M. A. *Altern. Lab. Anim.* **2004**, *32* (1), 123–131.
22 Schrattenholz, A. *Drug Discovery Today – Technologies* **2004**, *1*, 1–8.
23 Zolg, J. W. and Langen, H. *Mol. Cell Proteomics* **2004**, *3*, 345–354.
24 ReProTect Integrated Project Summary, **2004**; http://ecvam.jrc.it/index.htm.
25 Kopec, K. K., Bozyczko-Coyne, D., and Williams, M. *Biochem. Pharmacol.* **2005**, *69*, 1133–1139.
26 Shastry, B. S. *J. Hum. Genet.* **2005**, *50*, 321–328.
27 Little, P. F. *Genome Res.* **2005**, *15*, 1759–1766.
28 Schrattenholz, A., Wozny, W., Klemm, M., Schroer, K., Stegmann, W., and Cahill, M. A. *J. Neurol. Sci.* **2005**, *229–230*, 261–267.
29 Vogt, J. A., Hunzinger, C., Schroer, K., Holzer, K., Bauer, A., Schrattenholz, A., Cahill, M. A., Schillo, S., Schwall, G., Stegmann, W., and Albuszies, G. *Anal. Chem.* **2005**, *77*, 2034–2042.
30 Hunzinger, C., Schrattenholz, A., Poznanovic, S., Schwall, G., and Stegmann, W. *J. Chromatogr. A* **2006**, *1123*, 170–181.
31 Baulcombe, D. *Trends Biochem. Sci.* **2005**, *30*, 290–293.
32 Cejka, D., Losert, D., and Wacheck, V. *Clin. Sci. (Lond)* **2006**, *110*, 47–58.
33 Huppi, K., Martin, S. E., and Caplen, N. J. *Mol. Cell* **2005**, *17*, 1–10.
34 Liu, Y. P., Dambaeva, S. V., Dovzhenko, O. V., Garthwaite, M. A., and Golos, T. G. *Stem Cells Dev.* **2005**, *14*, 487–492.
35 Yang, S., Tutton, S., Pierce, E., and Yoon, K. Mol. Cell. Biol. **2001**, *21*, 7807–7816.
36 Zaehres, H., Lensch, M. W., Daheron, L., Stewart, S. A., Itskovitz-Eldor, J., and Daley, G. Q. *Stem Cells* **2005**, *23*, 299–305.
37 Worth, A. P. and Balls, M. Alternative (non-animal) methods for chemicals testing: current status and future prospects. *Altern. Lab. Anim.* **2002**, *30*, 15–21.
38 Eskes, C. and Zuang, V. Alternative (non-animal) methods for cosmetics testing: current status and future prospects. *Altern. Lab. Anim.* **2005**, *33*, 47–81.
39 Bremer, S. and Hartung, T. *Curr. Pharm. Des.* **2004**, *10*, 2733–2747.
40 Buesen, R., Visan, A., Genschow, E., Slawik, B., Spielmann, H., and Seiler, A. *ALTEX* **2004**, *21*, 15–22.
41 Genschow, E., Spielmann, H., Scholz, G., Pohl, I., Seiler, A., Clemann, N., Bremer, S., and Becker, K. *Altern. Lab. Anim.* **2004**, *32*, 209–244.
42 Pellizzer, C., Bremer, S., and Hartung, T. *ALTEX* **2005**, *22*, 47–57.
43 EC Commission Directive 1999/83/EC of 8th September, 1999, amending the Annex to Council Directive 75/318/EEC on the approximation of the laws of the Member States relating to analytical, pharmacotoxicological and clinical standards and protocols in respect of the testing of medicinal products, **1999**.

44 EC Directive 2001/20/EC and of the Council of 4th April, 2001, on the approximation of the laws, regulations and administrative provisions of the Member States relating to the implementation of good clinical practice in the conduct of clinical trials on medicinal products for human use, **2001**.
45 EC Directive 2001/83/EC of the European parliament and of the council of 6th November, 2001, on the Community code relating to medicinal products for human use, **2001**.
46 ICH International Conference on Harmonisation of Technical Requirements for Registration of Pharmaceuticals for Human Use. ICH Harmonised Tripartite Guideline: Detection of Toxicity to Human Reproduction for Medicinal Products. Recommended for Adoption at Step 4 of the ICH Process on 24 July, 1993, by the ICH Steering Committee. ICH Secretariat, Geneva, **1993**.
47 EC Council Directive 89/341/EEC of 3rd May 1989 amending Directives 65/65/EEC, 75/318/EEC and 75/319/EEC on the approximation of provisions laid down by law, regulation or administrative action relating to proprietary medicinal products, **1989**.
48 OECD (1996). Final Report of the OECD Workshop on Harmonization of Validation and Acceptance Criteria for Alternative Toxicological Test Methods. OECD Paris, July 1996; 60 pp, **1996**.
49 OECD (1982). OECD Guidelines for the testing of chemicals. OECD Paris, **1982**.
50 Heywood, R. *Hum. Reprod.* **1986**, *1*, 397–399.
51 Parke, D. V. *Altern. Lab. Anim.* **2006**, *22*, 207–209.
52 Evans, M. J. and Kaufman, M. H. *Nature* **1981**, *292*, 154–156.
53 Martin, G. R. *Proc. Natl. Acad. Sci. USA* **1981**, *78*, 7634–7638.
54 Robertson, E. In: Robertson, E. (Ed.), *Teratocarcinomas and Embryonic Stem Cells: A Practical Approach*. IRL Press, Oxford, **1987**, pp. 71–112.
55 Smith, A. G. and Hooper, M. L. *Dev. Biol.* **1987**, *121*, 1–9.
56 Nichols, J., Evans, E. P., and Smith, A. G. *Development* **1990**, *110*, 1341–1348.
57 NIH stem cell report (**2001**) www.nih.gov/news/stemcell/scireport.htm.
58 Romano, G. *Drug News Perspect.* **2004**, *17*, 637–645.
59 Vats, A., Bielby, R. C., Tolley, N. S., Nerem, R., and Polak, J. M. *Lancet* **2005**, *366*, 592–602.
60 Young, H. E., Duplaa, C., Katz, R., Thompson, T., Hawkins, K. C., Boev, A. N., Henson, N. L., Heaton, M., Sood, R., Ashley, D., Stout, C., Morgan, III, J. H., Uchakin, P. N., Rimando, M., Long, G. F., Thomas, C., Yoon, J. I., Park, J. E., Hunt, D. J., Walsh, N. M., Davis, J. C., Lightner, J. E., Hutchings, A. M., Murphy, M. L., Boswell, E., McAbee, J. A., Gray, B. M., Piskurich, J., Blake, L., Collins, J. A., Moreau, C., Hixson, D., Bowyer, III, F. P., and Black, Jr., A. C. *J. Cell Mol. Med.* **2005**, *9*, 753–769.
61 Bremer, S., Cortvrindt, R., Daston, G., Eletti, B., Mantovani, A., Maranghi, F., Pelkonen, O., Ruhdel, I., and Spielmann, H. *Altern. Lab. Anim.* **2005**, *33* (Suppl 1), 183–209.

62 Kitano, H. *Foundations of Systems Biology*. MIT Press, Boston, **2001**.
63 Butcher, E. C. *Nat. Rev. Drug Discov.* **2005**, *4*, 461–467.
64 Chen, R. E. and Thorner, *J. Genome Biol.* **2005**, *6*, 235.
65 Fattore, M. and Arrigo, P. *In Silico Biol.* **2005**, *5*, 199–208.
66 Heijne, W. H., Kienhuis, A. S., van Ommen, B., Stierum, R. H. and Groten, J. P. *Expert Rev. Proteomics* **2005**, *2*, 767–780.
67 Strange, K. *Am. J. Physiol. Cell Physiol.* **2005**, *288*, C968–C974.
68 Yengi, L. G. *Pharmacogenomics* **2005**, *6*, 185–192.
69 Bremer, S. and Hartung, T. *Curr. Pharm. Des.* **2004**, *10*, 2733–2747.
70 Solter, D. and Damjanov, I. *Methods Cancer Res.* **1979**, *18*, 277–332.
71 Schöler, H. R., Balling, R., Hatzopoulos, A. K., Suzuki, N., and Gruss, P. *EMBO J.* **1989**, *8*, 2543–2550.
72 Schillo, S., Pejovic, V., Hunzinger, C., Hansen, T., Poznanovic, S., Kriegsmann, J., Schmidt, W. J., and Schrattenholz, A. *J. Proteome Res.* **2005**, *4*, 900–908.
73 Morgan, H. D., Santos, F., Green, K., Dean, W., and Reik, W. *Hum. Mol. Genet.* **2005**, *14* (Spec No. 1), R47–R58.
74 Sommer, S., Hunzinger, C., Schillo, S., Klemm, M., Biefang-Arndt, K., Schwall, G., Putter, S., Hoelzer, K., Schroer, K., Stegmann, W., and Schrattenholz, A. *J. Proteome Res.* **2004**, *3*, 572–581.
75 Schrattenholz, A. and Soskic, V. *Curr. Topics Med. Chem.* **2006**, *6*, 663–686.
76 Kim, J. J., Patton, W. C., Corselli, J., Jacobson, J. D., King, A., and Chan, P. J. *J. Reprod. Med.* **2005**, *50*, 533–538.
77 Huuskonen, H. *Toxicol. Appl. Pharmacol.* **2005**, *207*, 495–500.
78 Spielmann, H. *Toxicol. Appl. Pharmacol.* **2005**, *207*, 375–380.
79 Whitebread, S., Hamon, J., Bojanic, D., and Urban, L. *Drug Discov. Today* **2005**, *10*, 1421–1433.
80 Shou, M. *Curr. Opin. Drug Discov. Devel.* **2005**, *8*, 66–77.
81 Lennernas, H. and Abrahamsson, B. *J. Pharm. Pharmacol.* **2005**, *57*, 273–285.
82 Pellizzer, C., Bremer, S., and Hartung, T. *ALTEX* **2005**, *22*, 47–57.
83 Balls, M., Goldberg, A. M., Fentem, J. H., Broadhead, C. L., Burch, R. L., Festing, M. F., Frazier, J. M., Hendriksen, C. F., Jennings, M., van der Kamp, D., Morton, D. B., Rowan, A. N., Russell, C., Russell, W. M., Spielmann, H., Stephens, M. L., Stokes, W. S., Straughan, D. W., Yager, J. D., Zurlo, J., and van Zutphen, B. F. *Altern. Lab. Anim.* **1995**, *23*, 838–866.
84 Li, Y., Powell, S., Brunette, E., Lebkowski, J., and Mandalam, R. *Biotechnol. Bioeng.* **2005**, *91*, 688–698.
85 Moon, S. Y., Park, Y. B., Kim, D. S., Oh, S. K., and Kim, D. W. *Mol. Ther.* **2006**, *13*, 5–14.
86 Plaia, T. W., Josephson, R., Liu, Y., Zeng, X., Ording, C., Toumadje, A., Brimble, S. N., Sherrer, E. S., Uhl, E. W., Freed, W. J., Schulz, T. C., Maitra, A., Rao, M. S., and Auerbach, J. M. *Stem Cells* **2006**, *24*, 531–546.

87 Greenlee, A. R., Kronenwetter-Koepel, T. A., Kaiser, S. J., and Liu, K. *Toxicol. In Vitro* **2005**, *19*, 389–397.

88 Seiler, A., Visan, A., Buesen, R., Genschow, E., and Spielmann, H. *Reprod. Toxicol.* **2004**, *18*, 231–240.

89 Cregan, K. *Intern. Med. J.* **2005**, *35*, 126–127.

90 Daley, G. Q., Sandel, M. J., and Moreno, J. D. *Med. Ethics (Burlington, Mass)* **2005**, *12*, 5.

91 Magnus, D. and Cho, M. K. *Science* **2005**, *308*, 1747–1748.

92 Manzoor, S. and Elahi, M. *J. Coll. Physicians Surg. Pak.* **2005**, *15*, 517.

93 Okie, S. *N. Engl. J. Med.* **2005**, *353*, 1–5.

94 Scolding, N. *Lancet* **2005**, *365*, 2073–2075.

95 Jones, D. G. and Towns, C. R. *Hum. Reprod.* **2006**, *21*, 1113–1116.

96 Yamane, T., Hayashi, S., Mizoguchi, M., Yamazaki, H., and Kunisada, T. *Dev. Dyn.* **1999**, *216*, 450–458.

97 Bagutti, C., Wobus, A. M., Fassler, R., and Watt, F. M. *Dev. Biol.* **1996**, *179*, 184–196.

98 Bjorklund, L. M., Sanchez-Pernaute, R., Chung, S., Andersson, T., Chen, I. Y., McNaught, K. S., Brownell, A. L., Jenkins, B. G., Wahlestedt, C., Kim, K. S., and Isacson, O. *Proc. Natl. Acad. Sci. USA* **2002**, *99*, 2344–2349.

99 Chung, S., Sonntag, K. C., Andersson, T., Bjorklund, L. M., Park, J. J., Kim, D. W., Kang, U. J., Isacson, O., and Kim, K. S. *Eur. J. Neurosci.* **2002**, *16*, 1829–1838.

100 Kim, J. H., Auerbach, J. M., Rodriguez-Gomez, J. A., Velasco, I., Gavin, D., Lumelsky, N., Lee, S. H., Nguyen, J., Sanchez-Pernaute, R., Bankiewicz, K., and McKay, R. *Nature* **2002**, *418*, 50–56.

101 Kawasaki, H., Mizuseki, K., Nishikawa, S., Kaneko, S., Kuwana, Y., Nakanishi, S., Nishikawa, S. I., and Sasai, Y. *Neuron* **2000**, *28*, 31–40.

102 Lee, S. H., Lumelsky, N., Studer, L., Auerbach, J. M., and McKay, R. D. *Nat. Biotechnol.* **2000**, *18*, 675–679.

103 Strubing, C., Ahnert-Hilger, G., Shan, J., Wiedenmann, B., Hescheler, J., and Wobus, A. M. *Mech. Dev.* **1995**, *53*, 275–287.

104 Gottlieb, D. I. and Huettner, J. E. *Cells Tissues Organs* **1999**, *165*, 165–172.

105 Fraichard, A., Chassande, O., Bilbaut, G., Dehay, C., Savatier, P., and Samarut, J. *J. Cell Sci.* **1995**, *108* (Pt 10), 3181–3188.

106 Bain, G., Kitchens, D., Yao, M., Huettner, J. E., and Gottlieb, D. I. *Dev. Biol.* **1995**, *168*, 342–357.

107 Liu, S., Qu, Y., Stewart, T. J., Howard, M. J., Chakrabortty, S., Holekamp, T. F., and McDonald, J. W. *Proc. Natl. Acad. Sci USA* **2000**, *97*, 6126–6131.

108 Brustle, O., Jones, K. N., Learish, R. D., Karram, K., Choudhary, K., Wiestler, O. D., Duncan, I. D., and McKay, R. D. *Science* **1999**, *285*, 754–756.

109 Dang, S. M., Kyba, M., Perlingeiro, R., Daley, G. Q., and Zandstra, P. W. *Biotechnol. Bioeng.* **2002**, *78*, 442–453.

110 Wiles, M. V. and Keller, G. *Development* **1991**, *111*, 259–267.

111 Tsai, M., Tam, S. Y., Wedemeyer, J., and Galli, S. J. *Int. J. Hematol.* **2002**, *75*, 345–349.
112 Fairchild, P. J., Brook, F. A., Gardner, R. L., Graca, L., Strong, V., Tone, Y., Tone, M., Nolan, K. F., and Waldmann, H. *Curr. Biol.* **2000**, *10*, 1515–1518.
113 Lieschke, G. J. and Dunn, A. R. *Exp. Hematol.* **1995**, *23*, 328–334.
114 Potocnik, A. J., Nielsen, P. J., and Eichmann, K. *EMBO J.* **1994**, *13*, 5274–5283.
115 Dani, C., Smith, A. G., Dessolin, S., Leroy, P., Staccini, L., Villageois, P., Darimont, C., and Ailhaud, G. *J. Cell Sci.* **1997**, *110* (Pt 11), 1279–1285.
116 Yamashita, J., Itoh, H., Hirashima, M., Ogawa, M., Nishikawa, S., Yurugi, T., Naito, M., Nakao, K., and Nishikawa, S. *Nature* **2000**, *408*, 92–96.
117 Hirashima, M., Kataoka, H., Nishikawa, S., Matsuyoshi, N., and Nishikawa, S. *Blood* **1999**, *93*, 1253–1263.
118 Bader, A., Al-Dubai, H., and Weitzer, G. *Circ. Res.* **2000**, *86*, 787–794.
119 Muller, M., Fleischmann, B. K., Selbert, S., Ji, G. J., Endl, E., Middeler, G., Muller, O. J., Schlenke, P., Frese, S., Wobus, A. M., Hescheler, J., Katus, H. A., and Franz, W. M. *FASEB J.* **2000**, *14*, 2540–2548.
120 Westfall, M. V., Pasyk, K. A., Yule, D. I., Samuelson, L. C., and Metzger, J. M. *Cell Motil. Cytoskeleton* **1997**, *36*, 43–54.
121 Klug, M. G., Soonpaa, M. H., Koh, G. Y., and Field, L. J. *J. Clin. Invest.* **1996**, *98*, 216–224.
122 Wobus, A. M., Rohwedel, J., Maltsev, V., and Hescheler, J. *Ann. N. Y. Acad. Sci.* **1995**, *752*, 460–469.
123 Prelle, K., Wobus, A. M., Krebs, O., Blum, W. F., and Wolf, E. *Biochem. Biophys. Res. Commun.* **2000**, *277*, 631–638.
124 Rohwedel, J., Maltsev, V., Bober, E., Arnold, H. H., Hescheler, J., and Wobus, A. M. *Dev. Biol.* **1994**, *164*, 87–101.
125 Drab, M., Haller, H., Bychkov, R., Erdmann, B., Lindschau, C., Haase, H., Morano, I., Luft, F. C., and Wobus, A. M. *FASEB J.* **1997**, *11*, 905–915.
126 Kramer, J., Hegert, C., Guan, K., Wobus, A. M., Muller, P. K., and Rohwedel, J. *Mech. Dev.* **2000**, *92*, 193–205.
127 Bourne, S., Polak, J. M.,, S. P., and Buttery, L. D. *Tissue Eng.* **2004**, *10*, 796–806.
128 Kahan, B. W., Jacobson, L. M., Hullett, D. A., Ochoada, J. M., Oberley, T. D., Lang, K. M., and Odorico, J. S. *Diabetes* **2003**, *52*, 2016–2024.
129 Lumelsky, N., Blondel, O., Laeng, P., Velasco, I., Ravin, R., and McKay, R. *Science* **2001**, *292*, 1389–1394.
130 Roche, E., Enseat-Wase, R., Reig, J. A., Jones, J., Leon-Quinto, T., and Soria, B. *Handbook Exp. Pharmacol.* **2006**, 147–167.
131 Kubo, A., Shinozaki, K., Shannon, J. M., Kouskoff, V., Kennedy, M., Woo, S., Fehling, H. J., and Keller, G. *Development* **2004**, *131*, 1651–1662.
132 Heng, B. C., Cao, T., Liu, H., and Phan, T. T. *Exp. Dermatol.* **2005**, *14*, 1–16.
133 Schuldiner, M., Yanuka, O., Itskovitz-Eldor, J., Melton, D. A., and Benvenisty, N. *Proc. Natl. Acad. Sci. USA* **2000**, *97*, 11307–11312.

134 Li, X. J., Du, Z. W., Zarnowska, E. D., Pankratz, M., Hansen, L. O., Pearce, R. A., and Zhang, S. C. *Nat. Biotechnol.* **2005**, *23,* 215–221.
135 Shin, S., Dalton, S., and Stice, S. L. *Stem Cells Dev.* **2005**, *14,* 266–269.
136 Schuldiner, M., Eiges, R., Eden, A., Yanuka, O., Itskovitz-Eldor, J., Goldstein, R. S., and Benvenisty, N. *Brain Res.* **2001**, *913,* 201–205.
137 Carpenter, M. K., Inokuma, M. S., Denham, J., Mujtaba, T., Chiu, C. P., and Rao, M. S. *Exp. Neurol.* **2001**, *172,* 383–397.
138 Zhang, S. C., Wernig, M., Duncan, I. D., Brustle, O., and Thomson, J. A. *Nat. Biotechnol.* **2001**, *19,* 1129–1133.
139 Vodyanik, M. A., Bork, J. A., Thomson, J. A., and Slukvin, I. I. *Blood* **2005**, *105,* 617–626.
140 Tian, X. and Kaufman, D. S. *Methods Mol. Med.* **2005**, *105,* 425–436.
141 Kaufman, D. S. and Thomson, J. A. *J. Anat.* **2002**, *200,* 243–248.
142 Dolnikov, K., Shilkrut, M., Zeevi-Levin, N., Gerecht-Nir, S., Amit, M., Danon, A., Itskovitz-Eldor, J., and Binah, O. *Stem Cells* **2006**, *24,* 236–245.
143 He, J. Q., Ma, Y., Lee, Y., Thomson, J. A., and Kamp, T. J. *Circ. Res.* **2003**, *93,* 32–39.
144 Xu, C., Police, S., Rao, N., and Carpenter, M. K. *Circ. Res.* **2002**, *91,* 501–508.
145 Thomson, J. A., Itskovitz-Eldor, J., Shapiro, S. S., Waknitz, M. A., Swiergiel, J. J., Marshall, V. S., and Jones, J. M. *Science* **1998**, *282,* 1145–1147.
146 Levenberg, S., Golub, J. S., Amit, M., Itskovitz-Eldor, J., and Langer, R. *Proc. Natl. Acad. Sci. USA* **2002**, *99,* 4391–4396.
147 Correia, A. S., Anisimov, S. V., Li, J. Y., and Brundin, P. *Ann. Med.* **2005**, *37,* 487–498.
148 D'Amour, K. A., Agulnick, A. D., Eliazer, S., Kelly, O. G., Kroon, E., and Baetge, E. E. *Nat. Biotechnol.* **2005**, *23,* 1534–1541.
149 Assady, S., Maor, G., Amit, M., Itskovitz-Eldor, J., Skorecki, K. L., and Tzukerman, M. *Diabetes* **2001**, *50,* 1691–1697.
150 Rambhatla, L., Chiu, C. P., Kundu, P., Peng, Y., and Carpenter, M. K. *Cell Transplant.* **2003**, *12,* 1–11.
151 Poznanovic, S., Wozny, W., Schwall, G. P., Sastri, C., Hunzinger, C., Stegmann, W., Schrattenholz, A., Buchner, A., Gangnus, R., Burgemeister, R., and Cahill, M. A. *J. Proteome Res.* **2005**, *4,* 2117–2125.
152 Dorsett, Y. and Tuschl, T. *Nat. Rev. Drug Discov.* **2004**, *3,* 318–329.
153 Meister, G., Landthaler, M., Dorsett, Y., and Tuschl, T. *RNA* **2004**, *10,* 544–550.

Part III
The Use of Human Tissues in Drug Discovery: Scientific, Ethical, Legal, and Regulatory Environments

Drug Testing In Vitro: Breakthroughs and Trends in Cell Culture Technology
Edited by Uwe Marx and Volker Sandig

ISBN: 978-3-527-31488-1

9
Availability, Standardization and Safety of Human Cells and Tissues for Drug Screening and Testing

Glyn N. Stacey and Thomas Hartung

9.1
Introduction

A very broad range of approaches is available for the assay of the effects of drugs and chemicals. Historically, studies in the fields of toxicology and pharmacology have relied on the use of animals to provide safety data on potential toxicants and indications of drug efficacy and safety. The "3Rs" movement to refine, reduce and replace the use of animals in experimentation [1] has gathered momentum over a number of decades with strong international support [2, 3], and today the focus is on the development of alternative *in-vitro* techniques. However, the requirement for functional cell-based assays has maintained the need for tissue preparations which are used in a number of forms, including tissue pieces, organ slices, primary cell cultures, and primary cell reaggregates. Where such preparations remain the only scientific option to achieve necessary results, it is still possible to refine and reduce the use of animal and human sources of the tissue, for example by cryopreservation of primary cells for later use [4, 5]. Increasingly, cell-based assays that utilize human continuous cell lines are becoming established, and these provide significant improvements from ethical and standardization perspectives. Some of the key issues in using human cells and cell lines for laboratory testing are outlined in this chapter.

9.2
Availability of Human Cells and Tissues for *In-Vitro* Testing

9.2.1
Selecting a Cell-Based System

9.2.1.1 Considering the Options for Human Cell-Based Testing

Securing the donation of fresh human tissue for experimental investigations is a challenging process, and it is important to consider for any testing procedure if an

Drug Testing In Vitro: Breakthroughs and Trends in Cell Culture Technology
Edited by Uwe Marx and Volker Sandig

ISBN: 978-3-527-31488-1

appropriate human cell culture is available that will either provide an alternative to the use of human tissue, or will enable preliminary testing that will narrow down the number and types of testing required to be performed on tissues. Even if a method based on a particular cell line is not available, it may be worthwhile developing such an assay or incorporating the derivation of new human cell lines as part of the program for the use of fresh donated tissue.

9.2.1.2 Establishing a Method Based on an Existing Human Cell Line

If a cell line-based assay for the tests to be undertaken is already available, then it is important to assess the level of validation that has been achieved in line with international requirements [6]. If proceeding with the selected method it is important that the cell line should be obtained from the original or otherwise qualified source of cells, as a cell line transferred through different laboratories may undergo significant changes [7]. Preferably, the source of cells will be a *bona fide* biological resource center or culture collection which has performed basic tests to help assure authenticity and an absence of microbiological contamination (e.g., ATCC, www.atcc.org; Coriell Institute, http://locus.umdnj.edu/ccr/; ECACC, www.ecacc.org.uk; DSMZ, www.dsmz.de; ICLC, www.iclc.org.it; JCRB, http://www.jhsf.or.jp/). However, whilst such organizations operate professionally, they cannot be expected to perform qualification of the cells for all applications of each cell line, and this should therefore be an important part of the testing work-program. If an original and quality-controlled source of the cell line is not available, it is wise to obtain cells from more than one source in order to compare their performance and authenticity (see below), as cross-contamination of cell lines is all too common [7–9]. Likewise, cells with different histories of laboratory use have been shown to display different characteristics.

In addition to considering the cells it is also important to establish critical control materials. These may include preparations of particular drugs or chemicals of known activity, or standardized biological reagents used for the culture of cells (see below). Where these are chemically defined they may be available as commercial preparations which meet Pharmacopoeia standards. The European Centre for the Validation of Alternative Methods recently held a workshop on reference standards in the context of test standardization and validation (www.ecvam.jrc.it). Aspects of stability, availability, homogeneity, purity, adequate potency within the response range of the test and scientific soundness of choice must all be considered (in addition to cost, safety and practicability) when choosing positive/negative controls and calibrants of test systems. Where the control materials are of biological origin (e.g., antibodies, growth factors, cytokines, interleukins), a local in-house control may be very difficult to standardize with outside groups. However, the World Health Organization manages a program generating biological reference materials of medical value, which are available as International References Materials (see www.nibsc.ac.uk/aboutus).

9.2.1.3 Developing New or Improved Cell Line-Based Techniques

When starting from scratch, the selection of an authentic and effective cell line becomes even more crucial. Ideally, the cells should accurately replicate *in vitro* the characteristics of interest in the target human tissue. A first step is to draw up a list of candidate cell lines from such tissue sources. However, it should be borne in mind that the process of generating new cell lines often involves numerous deleterious changes or transformation of the cells. So-called cell "immortalization" techniques usually cause genetic damage or intervention in normal cellular function, and include viral or carcinogenic transformation, genetic damage by irradiation or chemical carcinogens, and transfection with recombinant DNA vectors. Any one of these could elicit abnormal cell biology. Thus, in many cases cell lines cannot precisely replicate the particular functions of a certain natural cell or tissue type and it is therefore vital, when starting out to develop a new *in-vitro* cell-based assay, to start with a number of candidate cell lines which can be compared and selected to provide the best and most representative response compared to the target cell or tissue *in vivo*.

As mentioned above (see Section 9.2.1.2), it is also vital to establish key reference materials to evaluate cell responses.

9.2.2 Using Donated Human Tissue

Obtaining human tissue for testing requires great care to ensure that the material is obtained with fully informed consent and ethical approval for the proposed studies. The general requirements on those in charge of utilizing human material have been enshrined in international guidance such as Good Cell Culture Practice [10] and accords such as the Hinxton Statement (http://www.hopkinsmedicine.org/bioethics/finalsc.doc), which emphasize the need for attention to central generic criteria, including compliance with national laws and "norms", the avoidance of any coercion to donate tissue, and the maintenance of donor anonymity.

The 32nd Workshop of the European Centre for the Validation of Alternative Methods addressed the issue of the availability of human tissue [11]. The report from this workshop made 23 recommendations (see Table 9.1) which, in respect of tissue availability for testing purposes, included:

- the need for governments to facilitate appropriate legislation to permit use of human material for research and testing;
- the important role of tissue banks in provision of tissue; and
- the importance of accreditation for users of human tissue to ensure safe and ethical use.

Making human tissue available for research, with all appropriate approvals in place, can take some considerable time and effort, and it is therefore important to optimize and maximize its use. An important step to achieve this is to put in place a protocol for obtaining tissues that is agreed between user and providers [12]. This will help to ensure that tissues are optimally harvested, stored and

Table 9.1 Recommendations made by the ECVAM Workshop on Availability of Human Tissue [12].

General	Recommendations
	1. Governments and legislators should be kept fully informed about the important contributions to healthcare that the use of human tissue for research purposes can make, as well as contributing to the replacement of animal experiments.
	2. Where necessary, changes to the legal frameworks operating in European countries should be sought in conjunction with the relevant authorities, in order to facilitate the procurement of surgically removed human material surplus to medical requirements, and its supply in a suitable form to end-users.
Legal and ethical issues	3. The consent form used by Queen Mary's Hospital, Roehampton, UK, is recommended as a suitable example for obtaining general consent from living donors for the provision of human tissue for research.
	4. The use of tissue banks should be recognized as the most legally and ethically acceptable approach to the procurement and distribution of donated non-transplantable human tissue for research.
	5. A tissue-tracking system should be established to guarantee the anonymity of the donor, the donor's family, and health professionals involved in a donation. This should also involve routes of communication between the researcher and the surgeon, via the tissue bank that originally supplied the tissue in question, when appropriate.
	6. Users of human tissue should undergo a process of accreditation, to ensure that a minimum set of criteria are fulfilled. These criteria will include safety issues, scientific reputation, ethical credibility, proper training, suitable facilities for safe handling and disposal of human tissue, and confidentiality. Wherever possible, this accreditation should be conducted via an appropriately qualified and independent expert group.
	7. Informative articles on the establishment and role of tissue banks, and on the need for a wide variety of human tissue for research, for publication in the medical, scientific and general literature, should be prepared by those interested and involved in this area of work.
	8. The availability for research purposes of human tissue, surplus to medical requirements, should be facilitated by all available means; for example, by distributing newsletters, short articles and instructional videos, to improve communication between end-users and health professionals.
Safety	9. All human tissue should be regarded as potentially biohazardous and should be handled accordingly by appropriately trained personnel. Specific and appropriate precautions should be taken against known hazards, and general precautions against unknown hazards.
	10. Tissue banks should ensure that human tissues have been screened for specific biohazards, where possible and as appropriate. Such information should be made available to the end-user as quickly as possible. Ideally, this should be before any exposure of the end-user to the tissue. It is recognized, however, that this might not always be possible. Safety information should be used to modify any risk assessment already undertaken, with subsequent adjustment of handling procedures, or immediate disposal of the material, as appropriate.

Table 9.1 (continued)

General	Recommendations
	11. All those potentially exposed to human tissue should be vaccinated against hepatitis B virus, and should be shown to exhibit the nationally accepted level of immunization before any possible exposure; they should also be regularly checked for level of immunity.
	12. All human material, and disposable equipment used in conjunction with human tissue, should be sterilized according to a recommended procedure before transport for incineration.
	13. All research facilities involved in the use of human tissue should have relevant, standard protocols for ensuring its safe handling and disposal. Guidelines to be used as a basis for these protocols should be made available.
	14. Safety courses should include information specific to the handling of human tissue.
	15. National governments and appropriate international agencies should be encouraged by the European Network of Research Tissue Banks to officially recognize and regulate a non-profit research tissue-banking service to academia and industry.
	16. The nature and content of such courses, and records of staff attendance, should be documented, and maintained within the relevant establishments.
	17. There should be standardization of containers and labeling with respect to the movement of surgical and non-transplantable human tissue for research, so that they are easily recognizable.
	18. Information regarding regulations for packaging and national and international transport of human tissue should be collated into a database and made generally available.
Logistics	19. Tissue banks should establish procedures for providing advice to end-users who find themselves with human tissue surplus to their requirements which could be useful to other end-users.
	20. Tissue banks and others, including end-users, should undertake research which will enable, as far as possible, the regular and reliable supply of cells from all human tissue of sufficiently high quality for research. Issues requiring further research include isolation, preservation and storage techniques, and suitable conditions for distribution.
End-user requirements	21. Tissue banks should be transparent about the procedures involved in the isolation, storage and sterilization of human tissue. This would be facilitated if staff at tissue banks and end-users visited each others' establishments to view techniques and procedures.
	22. A list of suitable quality control criteria for each type of tissue and use should be drawn up by tissue banks, in conjunction with end-users. This is particularly important where a use is for regulatory purposes, when specific criteria will have to be met to satisfy a particular regulatory guideline. Flexibility in structural and performance criteria for using human tissue for regulatory purposes should be increased where this is possible without compromising scientific output, by dialogue among tissue banks, end-users and regulatory bodies.
	23. Attempts should be made to satisfy end-user requirements by standardizing the preparation and quality of specific cell types, and standardizing the transport conditions for their distribution.

transported to the user's laboratory, and provide good quality tissues for testing purposes. This protocol can specify a number of technical details to promote successful delivery of tissues, as illustrated in Table 9.2. Where tissues are transported across national borders it is important to ensure that the ethical approvals and any other requirements such as provision of viral marker screening are completed appropriately. Any laboratory receiving such material should be at least able to demonstrate traceability to a system of ethically approved tissue procurement for each sample.

Table 9.2 Key technical details to be agreed for procurement of tissue samples.

Preliminary arrangements:

- Consult local and national rules for use of patient cells.
- Establish a specification for the tissues samples required, e.g. type of tissues to be included, time between harvesting and reception at the laboratory, type of labeling.
- Coordination meetings between laboratory and clinical staff to agree protocol for provision of tissue including: tissue specification, description of necessary equipment and consumables, communications and labeling.
- Ensure ethical approval and appropriate consenting procedures are in place.
- Awareness/training for those who are involved in the storage and transfer of tissue.
- Ensure adequate storage facilities are available at appropriate sites in the tissue supply chain.

Routine operation during supply of tissue:

- Points of contact at clinical and laboratory sites.
- Forewarning of sample delivery.
- Assessment process for tissue received to ensure it is appropriate for use.
- Documentation of tissue receipt, evaluation and use.

Monitoring procedures:

- Review of consents for donated tissue.
- Notification of any changes in local practice which may impact on supply.
- Review any proposed change of use or supplementary use for donated tissues to ensure ongoing compliance with ethical approvals in place or activation of new applications for ethical approval.

Closure of protocol:

- All key staff informed.
- All remaining sample tubes and dedicated storage facilities recalled.
- Arrangements made for dealing with residual unused tissues.
- If appropriate, communication with local ethics committees.

9.3
Standardization of Cells and Tissues for Testing Purposes

9.3.1
Standardization of Primary Cells and Tissues

Where tissues are used for assays, comparison of data from various time points may be difficult, as the tissue will be derived from different individuals and may therefore respond differently [13]. It is important to establish that each tissue preparation used meets a certain minimum specification for the testing schedule, and where primary cell cultures are derived from the tissue and cryopreserved it may be possible to provide identical cell preparations for a series of experiments included in the associated application for ethical approval [5].

Differentiation, in its broadest sense, describes the development of a cell from embryonic to adult status as well as the maintenance of specific cell functions *in vivo* or in culture. Generally speaking, the genome of the cell in a culture holds the capacity to deliver a broad repertoire of differentiated states. The differentiation state is dynamic, depending on endogenous processes and exogenous factors, and will vary for individual cells within a culture. There are certain prerequisites to evaluate the differentiation state and validity of a cell culture, and these are characteristics that include: (1) species; (2) the stage of development; (3) a correlation of features with the original tissue ("lineage"); (4) the position of the cell within the identified cell lineage; (5) cellular transformation; (6) stability/variability of the cell phenotype; (7) a correct provenance and identity of the cells; and (8) the individual characters of a sub-cloned cell line or product of hybridization.

The choice of culture conditions is aimed first at sustaining the vitality of cells and, second, at achieving a certain a differentiation status representative of the *in-vivo* situation. Any change in culture conditions will potentially change the differentiation state as a result of adaptation of the living material to its environment. Thus, any change in culture conditions must be evaluated for its effect on the endpoints under study, as well as the overall differentiation status of the culture. It is important to recognize that cells cultured *in vitro* are rarely cultured under homeostatic conditions, and the frequency of changes of culture media will influence availability of nutrients and accumulation of metabolites, which will influence differentiation over the course of culture.

Accordingly, it is important to have indicators of the differentiated state of a culture that may include:

- morphology, histochemistry;
- enzyme content, isoenzyme pattern, gene expression;
- capacity to synthesize, secrete, and eliminate or accumulate substances;
- growth characteristics (e.g., growth curve, population doubling time);
- vitality and sensitivity towards toxins;
- cell functions which can be stimulated (e.g., by signal molecules);
- surface markers or antigens;

- nature of cell–cell interactions;
- nature of adherence to extracellular matrix.

The differentiation state is a combination of these functions, some of which are vital for the cell and will be expressed to some degree, or in some other cases may be absent. Notably, the increase in one parameter can influence negatively, or be accompanied by, changes in other parameters. An important example in this context is that growth rate and differentiation of cells are often opposing parameters, with cells that show a high degree of proliferation generally having a lesser degree of differentiation. Changes in one parameter often impact on others, and are part of complex feedback and control mechanisms of the cell.

Before starting a study, suitable parameters to monitor the differentiation status should be defined. In general, the endpoint of the experiment itself should not serve to control differentiation; rather, this parameter should be controlled over the course of the study in a series of experiments.

The desired differentiation state will usually – though not exclusively – mirror the *in-vivo* situation. To achieve this, culture conditions should approximate the *in-vivo* situation. Optimally, organotypic culture conditions ("state of the art") are employed, if they can be achieved with reasonable effort. It is desirable to establish reference tissue culture banks with materials for comparison of known differentiation state.

The most important means of standardization is control of the culture conditions, to assure a comparable differentiation state in all parts of the study. It is advisable to restrict the number of different potential conditions by choosing well-defined medium supplements, extracellular matrices, etc. The adequate documentation of all factors affecting cell function is crucial, as is the choice of monitoring culture parameters, the time point of their measurement, and maintenance of culture systems within acceptable limits.

In order to better understand this complex area of cell differentiation for *in-vitro* processes, the reader is directed to review by Coriell [14], Freshney [15, 16], Harris [17], Hill [18], Oshima [19], and Watt [20].

9.3.2 Standardization of Cell Lines

9.3.2.1 Challenges for Standardization of Cell Lines

Despite the fact that cell lines usually appear to be quite monomorphic when examined microscopically, they are inherently prone to variation and contamination and can be difficult to control. In some cases they may contain a variety of subpopulations representing different states of differentiation. Thus, standardization is a critical activity to obtain reliable data or products from tissue and cell-based systems [7, 10, 21].

Microbial contamination can have dramatic effects on cell cultures. In most cases, a bacterial or fungal contamination will cause death of the cell culture. Contamination with organisms such as *Mycoplasma* and viruses can become established as persistent infections of the cells, without causing overt cytotoxic effects. However, these persistent infections may substantially alter the characteristics of the cell line. In the case of *Mycoplasma* there may be irreversible changes including genetic effects, physiological changes, and transformation [22]. Effects can also be quite subtle, including increased chromosomal abnormality [23], induction of cytokines in the infected cells [24, 25], and interference with the selection of hybrids [26]. The eradication of viral infection is not generally feasible, while treatment with antibiotics for *Mycoplasma* infections is not routinely successful and may lead to deleterious effects on the cells themselves.

Genetic stability is also a key issue, and cell lines may show evidence of genetic change through a shift in karyotype (Table 9.3) which may be reflected in a change of culture characteristics. Furthermore, other subtle genetic changes that do not affect the gross chromosomal structure of the cells may yet have dramatic effects on cell function.

Cells will also vary in their response to environmental factors and treatments depending on their particular phase of growth (i.e., lag phase, exponential growth phase versus stationary or plateau phase) or the proportion of cells in a particular part of the cell cycle (i.e., G_0, G_1, S G_2, M). Standardizing the status of cells prior to use in a cell-based assay is central to assuring the reproducibility and quality of data from cell-based assays [7, 11], and will require investigation for each assay and cell type.

9.3.2.2 **Achieving Standardization of Cell Lines**

The above challenges to the stability and standardization of cells may be tackled through a combination of attention to appropriate cell banking, quality control, and well-defined conditions and procedures for routine passage and preparation of cells.

An important first step is the establishment of a master bank, and its characterization and quality control. Subsequently, individual vials of this stock are taken to generate large "working" cell banks that are retested for critical characteristics prior to provision for routine use. This tiered system is central to assuring the long-term provision of good-quality cells (both prokaryotic and eukaryotic), and should be considered best practice for any cell culture laboratory (Fig. 9.1).

The quality control applied to each cell bank will vary depending on the application of the cells, but the core testing for all cell lines should include some means of determining viability, *Mycoplasma* testing, and sterility testing [7]. The suitability of the viability test method used should be established for each cell line, and the specificity and sensitivity of the *Mycoplasma* and sterility tests should be carefully considered, as should the use of appropriate controls and reference strains and materials. It should be borne in mind that some fastidious organisms may not be detected in standard protocols, and sterility tests may need to be extended to include microbiological growth media that will support the growth of organisms that are dependent on carbon dioxide, anaerobic or microaerophilic conditions and serum. A typical generic core testing regime for a cell bank is provided in Table 9.4.

Table 9.3 Typical examples of variation in karyotype data for Vero cells (A) and HeLa cells (B) over a period of *in-vitro* culture (L. Young, unpublished data; NIBSC).

Passage level of Vero cell culture	**Numbers of cells with respective chromosome count[a] (200 cell metaphases)**														**Modal no.[b]**	**Range**
(a) Vero cells	*51*	*52*	*53*	*54*	*55*	*56*	*57*	*58*	*59*	*60*	*61*	*62*	*63*	*64*		
148				2	11	16	26	38	35	27	17	17	9	2	58	54–64
165				4	12	15	14	22	19	18	13	13	6	4	58	54–64
179				5	14	18	17	28	25	22	21	16	10	4	58	54–64
199				5	16	14	19	30	27	25	18	13	10	3	58	54–64
(b) HeLa cells	*53/54*	*55/56*	*57/58*	*59/60*	*61/62*	*63*	*64*	*65*	*66*	*67*	*68*	*69/70*	*71/72*	*73/74*		
$n + 14$	0/0	0/0	0/1	1/2	5/12	17	19	23	26	24	22	20/13	6/4	3/2	66	58–74
$n + 33$	0/0	0/0	0/1	1/8	9/12	15	22	27	26	30	23	13/7	3/2	1/0	67	58–73
$n + 76$	1/5	6/8	7/9	12/14	16/17	23	24	21	16	11	6	3/1	0/0	0/0	64	53–70
$n + 85$	0/1	1/4	5/7	8/15	19/23	26	25	22	17	11	6	5/3	2/0	0/0	63	54–71

a) Column headers indicate the number of chromosomes per cell metaphase.
b) Note that the Vero cell line has a stable modal number, and the HeLa cells a variable modal number.

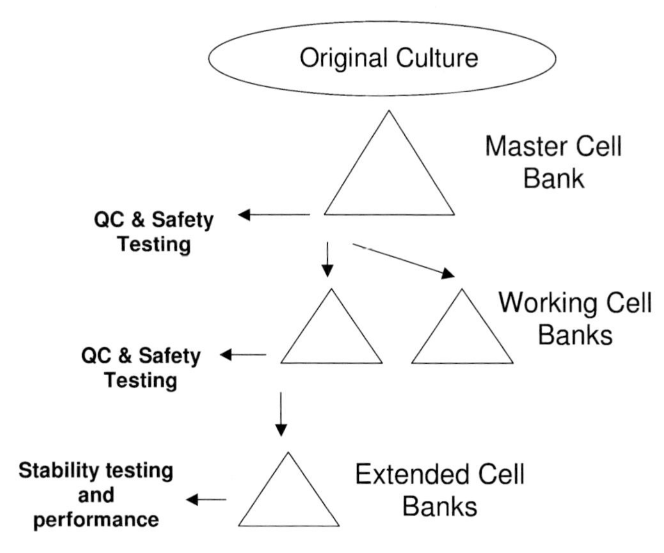

Fig. 9.1 A cell banking scheme for cell lines.

Table 9.4 Typical quality control regimes for cell banks.

Characteristic	Types of test
Viability	Trypan blue dye exclusion [57]
Growth	Recovery of cultures from frozen ampoules Plating efficiency [58]
Morphology	Light microscopy
Species of origin	Isoenzyme analysis Karyology
Genetic identity (i.e., authenticity)	STR profiling and DNA fingerprinting
Sterility	Bacteriological broth and agar culture [59]
Mycoplasma contamination	Culture [59] Hoechst stain PCR [60]
Performance in respective bioassay	As required with use of appropriate reference materials

A Master Cell Bank provides an important reference point for any project in which cell cultures are a critical component. However, it is also important to recognize that cells are prone to variation. When individual bank vials are recovered, careful specification of the culture media, growth conditions and subculture regime are important to assist standardized results in different laboratories, and over time. In addition, for critical applications it is good practice to passage cells beyond the expected limit of use so that their continued stability of performance in the respective assays can be determined. The response of cells to pharmacological or toxic agents may vary, depending upon factors such as the cell concentration at seeding of cultures and the degree of confluency or cell density at the point of use. Accordingly, care should be taken to control and standardize the state of cells used in bioassays.

9.4
Safety Issues

Provided that certain precautions are taken in the handling of human cells and tissues, laboratory workers will be unlikely to become infected by contamination from the cells and tissues they are using. However, there have been notable cases where laboratory staff have suffered serious infection (and even death) from infected animal tissue [27, 28]. Accordingly, it is wise to treat all cells and tissues as potentially infectious. The means of evaluating and prioritizing this risk are described in the following sections.

9.4.1
Hazards Associated with Human Cells and Tissues

The primary safety concern regarding the use of human tissues is the risk of infection, and this will depend upon a number of basic criteria including:

- The quantity of each tissue sample received.
- The number of samples to be received.
- The background of the donor cohort: risk groups include individuals not screened for viral infection, and lifestyle risks such as foreign travel or a history of *intravenous* drug abuse.
- The degree of manipulation and dissection required for each specimen.
- The type of test system (i.e., closed or open to the laboratory atmosphere).

Where available, virological test data and lifestyle questionnaires from donors can be helpful to identify higher-risk patient tissue, thus enhancing safety for laboratory staff. However, it should be borne in mind that a negative antiviral antibody test on a tissue donor may miss infections where antibody has not yet been produced in the infected donor. Other important factors in containment of infectious risks for both tissues and cell lines are: (1) the careful use of aseptic techniques for cell processing; (2) storing new untested material in quarantine storage vessels; and (3) the careful control of waste disposal procedures. All of these measures are in turn critically dependent on effective staff training and an awareness of risks in the laboratory environment.

9.4.2 Risks from Cell Lines

A key risk factor for cell lines, other than the contamination of the original tissue, is the potential of any *in-vitro* cell culture to be contaminated with (and possibly support) the growth of fastidious and potentially pathogenic microorganisms (e.g., viruses, prions, *Mycoplasma*). As discussed earlier, a variety of factors can influence infectious risk, and in particular the species and tissue of origin will influence the level of risk associated with a particular cell culture. The relative risks from different species/tissues has already been reviewed [29, 30], and will enable some critical viral and other risks to be identified. However, other factors may also influence this risk, an important example being the geographical origin of the cells or tissue, as disease type and incidence in human and animal populations will clearly vary between different countries and regions

For most routine cell culture laboratories, the major concerns will relate to the potential for cell contamination with serious human pathogens that would put laboratory workers at risk (e.g., HIV1 and 2, HTLV I and II, hepatitis B). Testing cell banks for such viruses can provide confidence in the safety of work with cell lines, though this is dependent on the availability and quality of the testing performed. A negative test which is not adequately sensitive or not appropriate (e.g., it detects antibody response to viral infection only found in blood samples, not qualified with cell lines) may in fact be dangerous in that it could give false confidence to laboratory staff by giving false-negative results. Such testing should be considered carefully and carried out by an accredited laboratory.

9.5 The Validation of Cell- and Tissue-Based Assays

The validation of cell- and tissue-based tests is the process by which the reliability and relevance of a test is established for a particular purpose [31]. Over the past 12 years, the European Centre for the Validation of Alternative Methods (ECVAM) has set up guidelines for validation, in cooperation with international experts. These guidelines have identified five stages in the evolution of new tests, namely *development, prevalidation, validation, peer-review,* and *regulatory acceptance.* These stages reflect the sequence of activities to be performed for a prospective validation exercise and are documented in detail in several publications [32–35]. This validation process has proved effective in establishing regulatory acceptance of several alternative tests [36]. Furthermore, these principles are generic and can be applied to any type of test, including *in-vivo* tests.

A typical validation programme lasts three years, and the subsequent regulatory implementation from two to seven years. At the time of writing, the cost for validation of a single test is approximately 300 000 Euro. The primary constraints in the validation process are the availability of reference substances and animal test data, the need for further optimization of test methods, the duration of financial/administrative procedures, and the ongoing consensus process of regulatory implementation.

Nine alternative methods for chemicals/cosmetics have been endorsed by ECVAM, including tests for skin corrosivity, skin sensitization, phototoxicity and embryotoxicity, as well as the percutaneous absorption of chemicals. In addition, seven methods reached scientific acceptance for the safety evaluation of biologicals such as vaccines. Currently, nine alternative methods for myelotoxicity, pyrogen testing and chronic toxicity in dogs are undergoing peer-review by the ECVAM Scientific Advisory Committee.

A number of the methods cited above have been accepted by both the European Commission (Annex V to Council Directive on dangerous substances 67/548/EEC), as well as by the Test Guideline Programme of the Organisation of Economic Cooperation and Development (OECD, press release JRC April 2004), and three potency tests of biologicals have been accepted by the European Directorate for the Quality of Medicines (European Pharmacopoeia).

A number of organizations are engaged primarily in the validation of alternative methods to the use of animals. These include ECVAM, The Interagency Co-ordinating Committee on the Validation of Alternative Methods (ICCVAM) and specific actions by OECD. In the future, the recently established Japanese Centre for Evaluation of Alternative Methods (JaCVAM) will hopefully also play a valuable role in this area. ECVAM operates as a co-coordinator of international validation studies, a focal point of information, a central database on alternative methods, a center of public dialogue, and a pre-normative research facility of the European Commission's Joint Research Centre. Participation in validation studies requires a running infrastructure, while active research maintains a practical and realistic view on science and technology. Furthermore, high-quality research ensures credibility in the scientific community.

An alternative, but complimentary, approach is used in establishing standards for the assay of biological medicines such as vaccines, insulin, growth factors, and monoclonal antibodies [37, 38]. In this model, batches of different control materials are prepared and distributed to a range of expert laboratories using a range of different assay methods. A nominal value for biological activity is assigned to the distributed materials, which are characterized in great detail through international collaborative studies in which different techniques and assays are used [39]. A number of organizations are engaged in standardization and the provision of reference materials (e.g., Institute for Reference Materials and Methods (Belgium), National Institute for Biological Standards and Control (UK), National Institute for Standards and Testing (USA), national pharmacopoeia institutes, WHO Expert Committee on Biological Standardisation). For more information on biological reference materials, see www.nibsc.ac.uk or http://www.who.int/bloodproducts/ref_materials/en/.

9.6 Conclusions and Future Prospects

The use of human tissue is increasingly difficult due to increasing regulations. In addition, the variability of such materials limits their value for drug development and toxicology. It is to be hoped that cell-based systems will increasingly use human cell lines, many of which have qualities yet to be fully realized through the application of cell differentiation protocols and three-dimensional (3-D) cell culture methods. A number of cell differentiation protocols are available that can be used to enhance the performance of cell lines as models of tissue, and these may include the treatment of cultures with bioactive molecules [40] or 3D culture systems [41].

Nanotechnology and systems for single cell analysis may well provide new approaches to the rapid generation of data on the response of cells to drugs and toxicants [42–44]. Another area currently generating great interest is that of embryonic stem cell lines to provide a diverse range of differentiated cell types as models of human tissue for drug testing and toxicology [45]. Although stem cell testing has been under development for *in-vitro* assays for some years [46], mouse embryonic cell lines have been used more recently [47–49]. Test methods using human embryonic stem cell (hESC) lines are also being developed, and hESCs are beginning to be investigated as tools for developmental toxicology [49, 50]. Potentially, the hESC offer systems that in many countries are fundamentally more acceptable when compared to other technologically advanced sources of embryonic tissue such as zebra fish embryos [51, 52].

A number of research groups are currently generating hESC lines from cells used in the prenatal diagnosis of inherited disorders and which carry genetic defects predisposing to known diseases [53]. Such lines have significant potential in generating *in-vitro* tissue models that could significantly enhance our ability to investigate therapeutic drugs for inherited genetic diseases such as cystic fibrosis.

At present, however, the culture of hESC lines remains very difficult, and the standardization of data is hampered by the significant levels of differentiated cells that arise in each new culture (see Table 9.4). Nonetheless, when the basic methods of culture and control of differentiation are mastered, these cultures can be expected to provide valuable methods for drug screening and toxicology. Significantly, a number of initiatives have already been launched to develop the standardization of hESC cultures [54, 55]. Hopefully these developments, combined with the biological potential of hESC, will deliver new and improved systems for *in-vitro* testing and drug development.

References

1 Russell, W. M. S., and Burch, R. L. (1959). *The Principles of Humane Experimental Techniques*. Methuen, London, UK.

2 Milstein, J., Grachev, V., Padilla, A., and Griffiths, E. (1996). WHO activities towards the three Rs in the development and control of biological medicines. *Dev. Biol. Stand.* 86, 31–39.

3 Hartung, T. (2001). Three Rs potential in the development and quality control of pharmaceuticals. *ALTEX* 18, 3–11.

4 Chesne, C., Guyomard, C., Fautrel, A., Poullain, M. G., Fremond, B., De Jong, H., and Guillouzo, A. (1993). Viability and function in primary culture of adult hepatocytes from various animal species and human beings after cryopreservation. *Hepatology* 18, 406–414.

5 Day, J. G. and Stacey, G. N. (Eds.), *Cryopreservation and Freezedrying Methods*. Humana Press, Totowa, USA (in press).

6 Hartung, T., Bremer, S., Casati, S., Coecke, S., Corvi, R., Fortaner, S., Gribaldo, L., Halder, M., Hoffmann, S., Roi, A. J., Prieto, P., Sabbioni, E., Scott, L., Worth, A., and Zuang V. A. (2004). Modular approach to the ECVAM principles on test validity. *Altern. Lab. Anim.* 32, 467–472.

7 Stacey, G. N. (2002). Standardisation of cell lines. *Dev. Biol.* 111, 259–272.

8 MacLeod, R. A. F., Dirks, W. G., Matsuo, Y., Kaufman, M., Milch, H., and Drexler, H. G. (1999). Widespread intra-species cross-contamination of human tumour cell lines arising at source. *Int. J. Cancer* 83, 555–563.

9 Masters, J. R. W., Thompson, J., Daly-Burns, B., Reid, Y. A., Dirks, W. G., Packer, P. I., Toji, L. H., Ohno, T., Tanabe, T. H., Arlett, C. F., Kelland, L. R., Harrison, M., Virmani, A., Ward, T. H., Ayres, K. L., and Debenham, P. G. (2001). Short tandem repeat profiling provides international reference standards for human cell lines. *Proc. Natl. Acad. Sci. USA* 98, 8012–8017.

10 Coecke, S., Balls, M., Bowe, G., Davis, J., Gstraunthaler, G., Hartung, T., Hay, R., Merten, O.-W., Price, A., Shechtman, L., Stacey, G., and Stokes, W. (2005). Guidance on Good cell culture practice: a report of the second ECVAM task force on good cell culture practice. *Altern. Lab. Anim.* 33, 261–287.

11 Stacey, G. N., Doyle, A., and Tyrrell, D. (1998). Source materials. In: Stacey, G. N., Doyle, A., and Hambleton, P. (Eds.), *Safety in Cell and Tissue Culture*. Kluwer Academic Publishers, Dordrecht, Netherlands, pp. 1–25.

12 Anderson, R., O'Hare, M., Balls, M., Brady, M., Brahams, D., Burt, A., Chesné, C., Combes, R., Dennison, A., Garthoff, B., Hawksworth, G., Kalter, E., Lechat, A., Mayer, D., Rogiers, V., Sladowski, D., Southee, J., Trafford, J., van der Valk, J., and van Zeller, A.-M. (1998). The Availability of Human Tissue for Biomedical Research – The Report and Recommendations of ECVAM Workshop. *Altern. Lab. Anim.* 26, 763–777.

13 Hartung, T., Balls, M., Bardouille, C., Blanck, O., Coecke, S., Gstraunthaler, G., and Lewis, D. (2002). Report of ECVAM task force on good cell culture practice (GCCP). *Altern. Lab. Anim.* 30, 407–414.

14 Coriell, L. L. (1984). Establishing and characterizing cells in culture. In: Acton, R. T. and Lynn, J. D. (Eds.), *Eukaryotic Cell Cultures*. Plenum Press, New York, pp. 1–31.

15 Freshney, R. I. (1990). Charakterisierung von Zellinien. In: *Tierische Zellkulturen*. Walter de Gruyter, Berlin, pp. 171–187.

16 Freshney, R. I. (1990). Induktion der Differenzierung. In: *Tierische Zellkulturen*. Walter de Gruyter, Berlin, pp. 188–197.

17 Harris, M. (1989). Phenotypic changes in cell cultures. *Dev. Biol. (New York)* 6, 79–85.

18 Hill, D. J. (1989). Cell multiplication and differentiation. *Acta Paediatr. Scand. [Suppl.]* 349, 13–20.

19 Oshima, J. and Campisi, J. (1991). Symposium: mammary cell proliferation and morphogenesis. *J. Dairy Sci.* 74, 2778–2787.

20 Watt, F. M. (1991). Cell culture models of differentiation. *FASEB J.* 5, 287–294.

21 The Standard Metabolic Reporting Structures Working Group (2005) Summary recommendations for standardisation and reporting of metabolic analyses. *Nat. Biotechnol.* 23, 833–838.

22 Rottem, S. and Naot, Y. (1998). Subversion and exploitation of host cells by *Mycoplasma*. *Trends Microbiol.* 6, 436–440.

23 Polyanskaia, G. G. and Samokisch, V. A. (1999). The study of quantitative karyotypic variability by induction of chromosomal instability in cultured cells of the Indian muntjac skin fibroblasts. *Tsitologiia* 41, 752–757.

24 Zurita-Salinas, C. S., Palacios-Boix, A., Yanez, A., Gonzalez, F., and Alcocer-Varela, J. (1996) Contamination with *Mycoplasma* spp. induces interleukin-13 expression by human skin fibroblasts in culture. *FEMS Immunol. Med. Microbiol.* 15, 123–128.

25 Fabisiak, J. P., Weiss, R. D., Powell, G. A., and Dauber, J. H. (1993). Enhanced secretion of immune-modulating cytokines by human lung fibroblasts during in vitro infection with *Mycoplasma fermentans*. *Am. J. Respir. Cell Mol. Biol.* 18, 358–364.

26 Boyle, J. M., Hopkins, J., Fox, M., Allen, T. D., and Leach, R. H. (1981). Interference in hybrid clone selection caused by *Mycoplasma hyorhinis* infection. *Exp. Cell Res.* 132, 67–72.

27 Mahy, B. W., Dykewicz, C., Fischer-Hoch S., et al., (1991). Virus zoonoeses and their potential for contamination of cell cultures. *Dev. Biol. Stand.* 75, 183–189.

28 Jones, N. and Lloyd, G. (1984). Infection of laboratory workers with hantavirus acquired from immunocytomas propagated in laboratory rats. *J. Infect.* 12, 117–125.

29 Frommer, W., Archer, L., Boon, B., Brunuis, G., Collins, C. H., Crooy, P., et al., (1993). Safe Biotechnology (5). Recommendations for safe work with animal and human cell cultures concerning potential human pathogens. *Appl. Microbiol. Biotechnol.* 39, 141–147.

30 Stacey, G. N. Risk assessment for cell culture processes. In: Stacey, G. N. and Davis, D. (Eds.), *Medicines from Animal Cells.* Wiley, Chichester, UK (in press).

31 Balls, M., Blaauboer, B., Brusick, D., Frazier, J., Lamb, D., Pemberton, M., Reinhardt, C., Roberfroid, M., Rosenkranz, H., Schmid, B., Spielmann, H., Stammati, A. L., and Walum, E. (1990). Report and recommendations of the CAAT/ERGATT workshop on validation of toxicity test procedures. *Altern. Lab. Anim.* 18, 303–337.

32 Balls, M., Blaauboer, B. J., Fentem, J. H., Bruner, L., Combes, R. D., Ekwall, B., Fielder, R. J., Guillouzo, A., Lewis, R. W., Lovell, D. P., Reinhardt, C. A., Repetto, G., Sladowski, D., Spielmann, H., and Zucco, F. (1995). Practical aspects of the validation of toxicity test procedures. The report and recommendations of ECVAM workshop 5. *Altern. Lab. Anim.* 23, 129–147.

33 Curren, R. D., Southee, J. A., Spielmann, H., Liebsch, M., Fentem, J. H., and Balls, M. (1995). The role of prevalidation in the development, validation and acceptance of alternative methods. *Altern. Lab. Anim.* 23, 211–217.

34 Worth, A. P. and Balls, M. (2001). The role of ECVAM in promoting the regulatory acceptance of alternative methods in the European Union. *Altern. Lab. Anim.* 29, 525–536.

35 Balls, M. and Karcher, W. (1995). The validation of alternative methods. *Altern. Lab. Anim.* 23, 884–886.

36 Hartung, T., Bremer, S., Casati, S., Coecke, S., Corvi, R., Fortaner, S., Gribaldo, L., Halder, M., Janusch Roi, A., Prieto, P., Sabbioni, E., Worth, A., and Zuang, V. (2003). ECVAM's response to the changing political environment for alternatives: Consequences of the European Union Chemicals and Cosmetics Policies. *Altern. Lab. Anim.* 31, 473–481.

37 Saldanha, J. (1999). Standardisation: a progress report. *Biologicals* 27, 285–289.

38 Seagroatt, V. and Kirkwood, T. B. (1986). Principles of biological standardisation and the conduct of collaborative studies to establish international standards. *Lymphokine Res.* 5 (suppl.), S7–S11.

39 World Health Organization Expert Committee on Biological Standardization (1990) Guidelines for the preparation characterisation and establishment of international and other standards and reference reagents for biological substances. WHO Technical Report Series, No. 800, World Health Organization, Geneva, Switzerland. N. B. This document was reviewed in 2004; details may be obtained through WHO and at www.who.int/).

40 Fleck, R. A., Athwal, H., Bygraves, J. A., Hockley, D., Feavers, I. M., and Stacey, G. N. (2003). Optimisation of NB-4 and HL-60 differentiation for use in oposonophagocytosis assays. *In Vitro Cell Dev. Biol. Anim.* 39, 235–242.

41 Battle, T., Maguire, T., Moulsdale, H., and Doyle, A. (1999). Progressive maturation resistance to microcystin-LR cytotoxicity in two different hepatospheroidal models. *Cell Biol. Toxicol.* 15, 3–12.

42 Li, N., Tourovskaia, A., and Folch, A. (2003). Biology on a chip: microfabrication for studying the behavior of cultured cells. *Crit. Rev. Biomed. Eng.* 31, 423–488.

43 Park, T. H. and Shuler, M. L. (2003). Integration of cell culture and microfabrication technology. *Biotechnol. Prog.* 19, 243–253.

44 Sniadecki, N. J., Desai, R. A., Ruiz, S. A., and Chen, C. S. (2006). Nanotechnology for cell-substrate interactions. *Ann. Biomed. Eng.*, March 9, 2006; [E-pub ahead of print]

45 Spielmann, H., Pohl, I., Doering, B., Liebsch, M., and Moldenhauer, F. (1997). The embryonic stem cell test (EST), an in vitro embryotoxicity test using two permanent cell lines: 3T3 fibroblasts and embryonic stem cells. *In Vitro Toxicol.* 10, 119–127.

46 Buesen, R., Visan, A., Genschow, E., Slawik, B., Spielmann, H., and Seiler, A. (2004). Trends in improving the embryonic stem cell test (EST): an overview. *ALTEX* 21, 15–22.

47 Bremer, S. and Bigot, K.(1997). Comparison of three different toxicological end points in the establishment of an *in vitro* embryotoxicity screening system based on embryonic stem cells. *Eur. J. Cell Biol.* 74, 26.

48 Scholz, G., Pohl, I., Genschow, E., Klemm, M., and Spielmann, H. (1999). Embryotoxicity screening using embryonic stem cells in vitro: correlation to in vivo teratogenicity. *Cells Tissues Organs* 165, 203–211.

49 Rolletschek, A., Blyszczuk, P., and Wobus, A. M. (2005). Embryonic stem cell-derived cardiac, neuronal and pancreatic cells as model systems to study toxicological effects. *Physiol. Rev.* 85, 635–678.

50 Bremer, S. and Hartung, T. (2004). The use of embryonic stem cells for regulatory developmental toxicity testing *in vitro* – the current status of test development. *Curr. Pharm. Design* 10, 2733–2747.

51 Van Leeuwen, C. J., Grootelaar, E. M., and Niebeek, G. (1990). Fish embryos as teratogenicity screens: a comparison of embryotoxicity between fish and birds. *Ecotoxicol. Environ. Safety* 20, 42–52.

52 Nagel, R. (2002). DarT: The embryo test with the zebrafish *Danio rerio* – a general model in ecotoxicology and toxicology. *ALTEX* 19, 38–48.

53 Pickering, S. J., Minger, S. L., Patel, M., Taylor, H., Black, C., Burns, C. J., Ekonomou, A., and Braude, P. R. (2005). Generation of a human embryonic stem cell line encoding the cystic fibrosis mutation deltaF508, using preimplantation genetic diagnosis. *Reprod. Biomed. Online* 10, 390–397.

54 Andrews, P. W., Benvenisty, N., McKay, R., Pera, M. F., Rossant, J., Semb, H., and Stacey, G. N. (2005). The International Stem Cell Initiative: toward bench marks for human embryonic stem cell research. *Nat. Biotechnol.* 23, 795–797.

55 Loring, J. F. and Rao, M. S. (2006). Establishing standards for the characterization of human embryonic stem cell lines. *Stem Cells* 24, 145–150.

56 Anderson, R., Balls, M., Burke, D., Cummins, M., Fehily, D., Gray, N., de Groot, M. G., Helin, H., Hunt, C., Jones, D., Price, D., Richert, L., Ravid, R., Shute, D., Sladowski, D., Stone, H., Thasler, W., Trafford, J., van der Valk, J., Weiss, T., Womack, C., and Ylikomi, T. (1998). The Establishment of Human Research Tissue Banking in the UK and Several Western European Countries; The Report and Recommendations of ECVAM Workshop 44. *Alt. Lab. Anim.* 29, 125–134.

57 Patterson, M. K. (1979). Measurement of growth and viability of cells in culture. *Methods Enzymol.* 58, 141–152.

58 Freshney J. (1994). *A Culture of Animal Cells.* 3rd edition, Wiley Liss Inc., New York.

59 Stacey, A. and Stacey, G. N. (2000). Routine quality control testing of cell cultures: Dectection of *Mycoplasma.* In: Kinchington, D. and Schinazi, R. F. (Eds.), *Antiviral Methods and Protocols.* Humana Press, Totowa, IL, pp. 27–40.

60 Stacey, G. N. (2001). Detection of *Mycoplasma* by DNA amplification (PCR). In: Doyle, A., Griffiths, J. B. (Eds.), *Cell and Tissue Culture for Medical Research.* Wiley, Chichester, Chap. 2, pp. 58–61.

10
Ethical Environment and Scientific Rationale Towards *In-Vitro* Alternatives to Animal Testing: Where Are We Going?

Horst Spielmann

> *"If we are to use a criterion for choosing experiments to perform, the criterion of humanity is the best we could possibly invent. The greatest scientific experiments have always been the most humane and the most aesthetically attractive, conveying that sense of beauty and elegance which is the essence of science at its most successful".*
>
> The humanity criterion proposed by William Russell and Rex Burch in *The Principles of Humane Experimental Technique* (Methuen, London, 1959).

10.1
Introduction

In 1959, William Russell and Rex Burch published their book, *The Principles of Humane Experimental Technique* [1], in which they proposed the concept of the three Rs (3Rs) *reduction, refinement,* and *replacement* – in relation to the humane treatment of experimental animals. Their concept was little recognized outside the United Kingdom for about 20 years, until the animal welfare movement, the general public, some committed politicians and, finally, the international scientific community, raised concerns about the suffering of experimental animals. The 3Rs concept has become the generally accepted scientific basis of institutions serving the development of alternatives to animal experiments. In 1989, ZEBET (Zentralstelle zur Erfassung und Bewertung von Ersatz- und Ergänzungsmethoden zum Tierversuch = National Centre for Documentation and Evaluation of Alternative Methods to Animal Experiments) was established at the Federal Health Institute (BGA) as the first government agency promoting the development and validation of non-animal testing procedures. In 1993, the European Union (EU) followed this example by establishing the European Centre for the Validation of Alternative Methods (ECVAM) at the Joint Research Centre (JRC) in Ispra (Italy); in 1997, the federal government agencies of the USA formed ICCVAM (Interagency Coordinating Center for the Validation of Alternative Methods); and

Drug Testing In Vitro: Breakthroughs and Trends in Cell Culture Technology
Edited by Uwe Marx and Volker Sandig

ISBN: 978-3-527-31488-1

in 2005, the Japanese government established JaCVAM (Japanese Centre for the Validation of Alternative Methods) at the National Institute of Health in Tokyo.

The activities of these institutions focuses on replacing regulatory animal tests that must be conducted in order to identify the toxic properties of chemicals to which humans or the environment are exposed, when the chemicals are used in a specific product or for a specific purpose. These tests have been criticized, since the exposure of test animals to hazardous chemicals may lead to considerable suffering and even death of the animals. In Europe, these activities were driven by the cosmetics and chemicals policies which are aimed at replacing the animal tests within the next decade by advanced non-animal testing procedures. The German Ministry for Education and Research (BMBF, Bundesminsiterium für Bildung und Forschung) has, for the past 20 years, provided substantial funding to research aimed at replacing testing in animals for regulatory purposes. Since 1990 the European Commission has sponsored research according to the 3Rs in several of its framework programs. Within the program of policy supporting research, the European Commission has more recently assigned a substantial amount of funding in the life sciences to research on the development and validation of alternative methods for regulatory safety testing of drugs, cosmetics, pesticides and other chemicals.

The 3Rs concept has proven successful in reducing the suffering of laboratory animals used in regulatory safety testing. At the international level, the harmonization of test guidelines has proven successful in reducing safety testing in animals for regulatory purposes. However, it is still a scientific challenge to replace a given regulatory animal test by a non-animal test that must provide sufficient information for the safety assessment of chemicals. Meanwhile, there are a few examples which prove that regulatory animal tests can indeed be replaced when the mechanistic basis of the specific area of toxicology is well understood and an appropriate *in-vitro* model is available. Using the validation of the 3T3 NRU *in-vitro* phototoxicity test as an example, it will be illustrated how the first non-animal toxicity tests were experimentally validated for several years and accepted for regulatory purposes by EU Member-States, and in the year 2004, even at a worldwide level, by the OECD. Examples will also be provided as to how animal tests in the pharmacopoeia were replaced by advanced non-animal methods.

10.2
Legal Framework in Europe for Developing Alternatives to Experimental Animals

According to Article 7.2 of EU Directive 86/609/EEC (EU Commission, 1986) on the use of experimental animals *"... an experiment shall not be performed if another scientifically satisfactory method of obtaining the result sought, not entailing the use of an animal, is reasonably and practicably available"*. Moreover, in the same Directive it is proposed in article 23 that *"... the Commission and Member States should encourage research into the development and validation of alternative techniques which could provide the same level of information as that obtained in experiments using animals, but which involve fewer animals or which entail less painful procedures, and shall take such other steps as they consider appropriate to encourage research in this field. The commission and Member States shall monitor trends in experimental methods".* The German animal welfare legislation (Tierschutzgesetz) is taking into account the principles of EU Directive 86/609/EEC as far as the protection of laboratory animals is concerned.

To promote the implementation of the EU Directive 86/609/EEC on the use of experimental animals, the European Commission and several Member States have established centers for the validation of alternative methods. For example, in 1989 ZEBET, the German Centre for the Documentation and Evaluation of Alternative Methods, was established at the Federal Health Institute BGA in Berlin, and in 1993 ECVAM was established at the JRC in Ispra, Italy. In the UK, FRAME had already been established before EU Directive 86/609/EEC had been adopted, and in the Netherlands, the National Centre for Alternatives (NCA) was established in 1992 in the National Institute of Health (RIVM) in Bilthoven. However, since ECVAM was established in 1993, none of the other EU Member States has established a national center for the promotion of alternatives to testing in animals; while more recently, in 2001, in Poland the NCA "Vitryna" started its activity at the Nofer Institute for Occupational Health in Lodz.

In the USA, several institutions serving the development of alternative methods have been established during the past two decades. These included, in 1981, the Johns Hopkins University alternatives center CAAT (Centre for Alternatives to Animal Testing), in Baltimore; in 1997, the Institute for *In Vitro* Sciences in Gaithersburg (MD); and in 1998, the validation center of the federal government ICCVAM (Interagency Co-ordinating Centre for the Validation of Alternative Methods) in Washington DC. Finally, in 2005, the government Japanese Validation Centre JaCVAM was established at the NIH in Tokyo.

To provide an example of the duties assigned to validation centers, the national German center ZEBET has served the following mission:

- to establish a database and information service on alternatives at the national and international level;
- to develop alternatives according to the 3Rs principle of Russell and Burch;
- to fund research on alternatives;
- to co-ordinate validation studies;
- to co-operate with national and international funding agencies and validation centers; and
- to provide a forum for information on alternatives to animal testing.

At the Federal Institute for Risk Assessment BfR, ZEBET's main activity is the reduction of animal tests conducted for regulatory purposes. The EU validation center ECVAM is also focusing its activity on reducing regulatory safety testing in animals, and the US validation agency ICCVAM is serving a similar mission. ECVAM and ZEBET have funds to support the development and validation of alternative methods, while ICCVAM has so far limited its activity to evaluating the results of validation studies conducted elsewhere.

10.3 Cell and Tissue Culture Systems used in Pharmacology and Toxicology

Culturing cells is the most widely used *in-vitro* method in pharmacology and toxicology, with cells being used either as permanent cell lines or as freshly obtained "primary cultures". *Co-cultures* of two or more cell types express organ-specific functions even better, for example, human keratinocytes and fibroblasts in bioengineered human skin models. Today, human skin models are commercially available and have been used successfully to investigate the pharmacology and toxicology of new drugs and cosmetics. This example illustrates that human tissues and tissue models are the most promising tools in drug development. However, it must be borne in mind that each cell and tissue culture model has its inherent limitations, which are usually identified in validation studies.

In *primary cultures* the organ-specific properties are well conserved, and primary liver cells both from humans and test animals are widely used in drug metabolism studies. *Permanent cell lines* are easy to handle and can be obtained from cell banks in high quality.

Molecular genetics allow the development of permanent cell lines expressing molecules of interest, thus opening a new perspective for cell culture methods in pharmacology and toxicology. *Transgenic cells* can express genes coding for receptor molecules on their surface or genes coding for drug-metabolizing enzymes within the cytosol. The genes introduced may even be of human origin.

The *advantages* of *in-vitro* systems include the following:

- controlled testing conditions;
- reduction of systemic effects;
- reduction of variability between experiments;
- the same dose range can be tested in a variety of test systems (cells and tissues);
- time-dependent studies can be performed and samples taken;
- testing can be conducted rapidly and at low cost;
- very small amounts of test material are required
- a limited amount of toxic waste is produced;
- human cells and tissues can be used;
- transgenic cells carrying human genes can be used; and
- there is a reduction of testing in animals.

The *limitations* of *in-vitro* systems include the following:

- systemic effects and side effects cannot be evaluated;
- interactions between tissues and organs cannot be tested;
- metabolism is limited;
- pharmacokinetic effects cannot be evaluated;
- specific organ sensitivity cannot be assessed; and
- chronic effects cannot be tested.

Today, *in vitro* systems serve as the basis for studying the effects of drugs and other chemicals (Fig. 10.1), whereby they are used initially for screening purposes

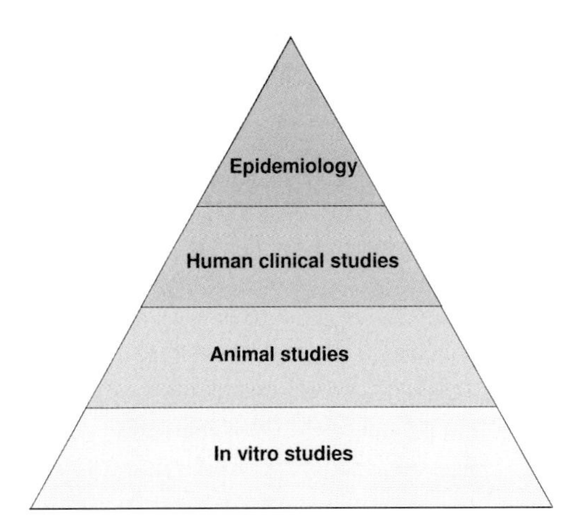

Fig. 10.1 The role of *in-vitro* studies in pharmacology and toxicology. Today, both in drug development and in toxicology, experimental investigations are starting with *in-vitro* studies. Studies in animals are conducted as the next step. If the results are still promising, clinical studies on a limited number of patients must be conducted before a new drug can be released onto the market. Epidemiological studies on a representative population will provide the final proof of the efficacy and effects of a new drug.

and in mechanistic studies. Although animal studies take considerably longer to conduct and are more expensive, until today only studies in animals have allowed the investigation of the systemic effects of exposure to chemicals. It must also be taken into account that studies in animals are used as surrogates for studies in humans. Thus, only clinical studies in humans can provide the final proof that a new drug is effective in man. It must also be considered that clinical studies in patients are even more expensive to conduct than studies in animals, and are more time-consuming.

At present, the assessment of adverse effects of chemicals can only be monitored by using *in-vitro* systems to a limited extent. Although some *in-vitro* systems are available that allow prediction of the local effects of chemicals when applied to the skin (e.g., irritation and phototoxicity), even the most sophisticated tests or test batteries cannot yet be used to predict systemic effects.

10.4 Drug-Metabolizing Systems

One general disadvantage of cell and tissue culture systems is their limited drug-metabolizing capacity, although drug-metabolizing cells or subcellular systems can be added to cultures in order to overcome this problem. An example is when hepatocytes rapidly lose their metabolic capacity. The following approaches have been used successfully used to improve the drug-metabolizing capacity of *in-vitro* systems:

- Active liver microsomal fractions from the rat, such as the standard "S-9 mix" used in the Ames' test.
- Freshly isolated hepatocytes from humans and other mammalian species.
- Co-cultures of hepatocytes and supporting liver cells essential to metabolize xenobiotics, such as bile duct epithelia.
- Liver slices.
- Hepatoma cell lines.
- Transgenic permanent cell lines expressing specific isoenzymes of human cytochrome P-450.

In addition, *ex-vivo* metabolites from exposed patients or animals can be added to cultures. Since transgenic cells do not contain the same set of regulating genes as, for example, human liver cells, the expression of human cytochrome P-450 in transgenic cells may be regulated differently. A more general disadvantage of the use of metabolizing systems is the lack of standardization among laboratories.

Although *in-vitro* systems can only be used to a limited extent in toxicity testing, they clearly offer a range of benefits (see Section 10.3).

Two aspects are especially important: first, due to their limitations, *in-vitro* systems allow attention to be focused on specific mechanisms, while second, studies can be conducted on human cells and tissues. In this way the toxicity of highly toxic chemicals (e.g., TCDD and dioxins) can be conducted on human cells, tissues and organs, without any ethical considerations.

10.5 Reductions in Experimental Animal Numbers During the Past Decade in Europe: The Situation in Germany

In 1989, EU Directive 86/609/EEC, for the first time, allowed government authorities in EU Member States to collect test animal numbers according to a standardized procedure [2]. Since that time, the annual numbers of experimental animals in Germany have decreased from 2.7 million in 1989 to 2 million in 2004. A closer analysis shows that the decrease was predominantly due to a reduction in animal numbers used for the development of drugs, which fell by more than 50% in the same 15-year period, from 1.4 million to 0.5 million. The decrease was even more impressive when taking into account earlier data provided by the German association of drug manufacturers, which showed that in 1977 a total of 4.4 million experimental animals was used for the development and safety testing of new drugs in Germany (Fig. 10.2) [3].

This dramatic fall in laboratory animal numbers has not been due to the high priority which EU Directive 86/609 gives to the development of *in-vitro* alternatives, but rather to general changes of methodology in the life sciences, from animal models to molecular biology and genetics, including cell and tissue culture models. In the field of drug development the new technology allows the high-throughput screening (HTS) of thousands of new drug candidates. This approach is, of course,

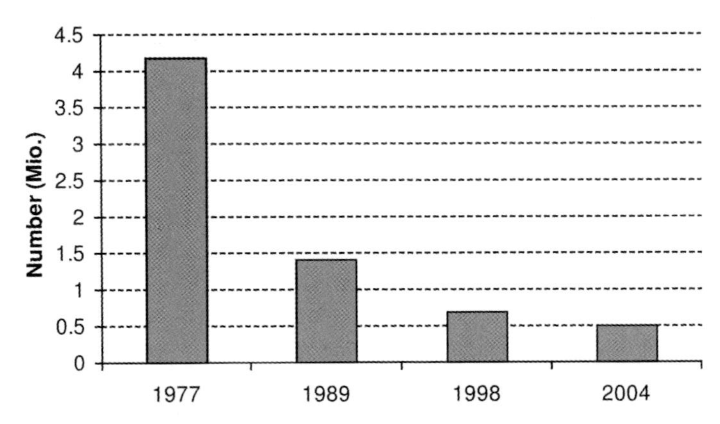

Fig. 10.2 Test animal numbers used by the German drug industry between 1977 and 2004. The data for 1977 were collected by the German Drug Manufacturers' Association (Bundesverband der Pharmazeutischen Industrie, BPI); the data for 1989 and 2004 were officially compiled by the German Federal Minister of Agriculture according to EU Directive 86/609/EEC.

more predictive, faster and cheaper than animal models, which in the past have been used successfully for drug development. However, at the same time, it has proven extremely difficult to reduce animal numbers in regulatory safety testing by non-animal methods, as the established endpoints in toxicity tests in laboratory animals are usually organ-specific and cover endpoints that are quite similar to the situation in humans.

10.6 Reducing Animal Numbers in Regulatory Testing by International Harmonization of Test Guidelines

For the past 30 to 40 years, toxicity testing has been developed empirically in many laboratories worldwide. For a given specific area of toxicology – such as eye and skin irritation or embryo toxicity – the standard animal procedures have differed considerably between countries, in terms of species, regimen of treatment, and numbers of animals per treatment group. In addition, there were different ways in which information from specific animal tests for classification and labeling (e.g., the Draize eye test) was used by regulatory agencies in various countries, and even by different agencies of one country.

10.7
Harmonization of OECD Guidelines for the Testing of Chemicals

The international harmonization of toxicity tests by the OECD in 1982 [4] was the first, and – so far – has been the most effective step in reducing the duplication of testing in animals for regulatory purposes. This is because a toxicity test conducted according to the OECD guidelines will be accepted by regulatory agencies in all OECD Member States. The 30 Member States of the OECD are the world's major industrial nations. A similar approach has thereafter been used for the safety and efficacy testing of drugs by the International Conference on Harmonisation (ICH) [5], which represents the three major economic regions, namely Europe, Japan, and the USA. Since 1990, the ICH has accepted harmonized guidelines for efficacy and safety testing of drugs and medicines, including animal tests. Again, the harmonization of test guidelines has led to significant reductions of testing in animals, as regulatory agencies worldwide now accept the results of a test conducted according to ICH guidelines.

The most important areas which require safety testing in animals, and in which the test guidelines have been harmonized at the international level, include:

- Industrial chemicals, pesticides, cosmetics, food additives, etc. (OECD-Guidelines for the Testing of Chemicals).
- Drugs and medical devices (International Conferences on Harmonization; ICH).
- Safety and efficacy of hormones and biologicals (Pharmacopoeias; European and US Pharmacopoeia).
- Vaccines and other immunologicals (WHO recommendations, European and US Pharmacopoeia).

This categorization shows that, in addition to drugs, industrial chemicals and pesticides, international test guidelines have also been harmonized for hormones and biologicals by the pharmacopoeias, and for vaccines by the World Health Organization (WHO). To date, the harmonization of international test guidelines for toxicity and safety testing has been the most successful approach to reducing animal testing for regulatory purposes.

10.8
Principles of Scientific Validation: The Amden Validation Workshops

Regulators will only accept alternatives to animal tests in toxicology, if the new tests will allow them to classify and label chemicals in the same way as the current animal tests. The OECD has therefore decided that *in-vitro* toxicity tests can be accepted for regulatory purposes only after a successful experimental validation study. In order to approach this problem scientifically, European and American scientists agreed in 1990 in Amden, Switzerland, on a definition of experimental validation and the essential steps in this process. At this workshop, validation was defined as "... the process by which reproducibility and relevance of a toxicity testing procedure are established for a particular purpose ..." [6], regardless of whether the method is an *in-vitro* or *in-vivo* test.

The four essential steps of the experimental validation process were defined in the following manner:

1. Test development in a single laboratory.
2. Experimental validation under blind conditions in several laboratories.
3. Independent assessment of the results of the validation trial.
4. Regulatory acceptance.

Steps 2 and 3 serve as the essential parts of a formal validation study conducted for regulatory purposes. The report of the First Amden workshop on validation [6] encouraged scientists to initiate several international validation studies. Since the Draize eye test has been the most widely criticized toxicity test, a worldwide validation study on nine alternatives to this procedure was coordinated by the EU and the British Home Office. However, this and other extensive international validation attempts failed [7].

Therefore, the leading scientists involved met for the Second Amden validation workshop in 1994, to improve the concept of the validation procedure. The workshop recommended the inclusion of new elements into the validation process [8], which had not been sufficiently identified in the First Amden validation workshop. The following three essential elements were recommended:

- The definition of a biostatistically based *prediction model.*
- The inclusion of a *prevalidation stage* between test development and formal validation under blind conditions.
- A well-defined management structure.

With regard to *in-vitro* tests, a *prediction model* should allow the prediction of *in-vivo* endpoints in animals or humans from the endpoints determined. The prediction model must be defined mathematically in the standard operation procedure of the test that will undergo experimental validation under blind conditions with coded chemicals [8]. In order to assess the limitations of a new test before it is evaluated in a validation study, the test should be standardized in a *prevalidation study* with a few test chemicals in a several laboratories [9]. This will ensure that the *in-vitro* test method, including the prediction model, is robust and that the formal validation study with coded chemicals is likely to be successful. Finally, the goal of a validation study has to be clearly defined, and the *management structure* must ensure that, within the study, the scientists who are responsible for essential tasks can conduct their duties independently of the sponsors and the managers of the study, for example, biostatistical analysis, and the selection, coding and shipment of the test chemicals.

The improved concept of experimental validation for regulatory purposes defined in the Second Amden workshop was accepted by ECVAM in 1995, and in 1996 by US regulatory agencies [10] and also by the OECD [11]. Since this agreement was made at the international level, scientists have tried to follow the ECVAM/

US/OECD principles for new validation trials. The improved validation concept was immediately introduced into ongoing validation studies, for example, the ECVAM/COLIPA validation study on *in-vitro* phototoxicity tests and the ECVAM validation study of *in-vitro* skin corrosivity tests. When these validation studies had been finished successfully, the two new methods were accepted as the first *in-vitro* toxicity tests at the international level for regulatory purposes by the EU Commission in 2000 [12, 13] and by the OECD in 2004 [14, 15].

10.9
Regulatory Acceptance of the Successfully Validated 3T3 NRU *In-Vitro* Phototoxicity Test

As no standard guidelines for testing photoirritation potential, either *in vivo* or *in vitro*, had been accepted for regulatory purposes at the international level by the OECD, in 1991 the European Commission (EC) and the European Cosmetics, Toiletry and Perfumery Association (COLIPA) established a joint program to develop and validate *in-vitro* photoirritation tests. In the first phase of the study, which was funded by DG XI of the EC and co-ordinated by ZEBET, *in-vitro* phototoxicity tests established in laboratories of the cosmetics industry, and a new assay – the 3T3 NRU PT test, which is a photocytotoxicity test using the mouse fibroblast cell line 3T3 with neutral red uptake (NRU) as the endpoint for cytotoxicity – were evaluated.

In the prevalidation study conducted with 20 test chemicals, the 3T3 NRU PT *in-vitro* phototoxicity test was unexpectedly the only such test in which all 20 chemicals were correctly identified as phototoxic or non-phototoxic [16, 17]. Quite independently, a laboratory in Japan subsequently obtained the same correct results in the 3T3 NRU PT, when testing the same set of 20 test chemicals.

In the second phase of the study, which was funded by ECVAM and co-ordinated by ZEBET, the 3T3 NRU PT test was validated with 30 carefully selected test chemicals in 11 laboratories in a blind trial. A representative set of test chemicals covering all major classes of phototoxins was selected according to results from standardized photopatch testing in humans [18]. The results obtained for this *in-vitro* test under blind conditions were reproducible, and the correlation between *in vitro* and *in vivo* data was almost perfect. Therefore, the ECVAM Scientific Advisory Committee [19] concluded in 1998 that the 3T3 NRU PT is a scientifically validated test which is ready to be considered for regulatory acceptance. However, the EU expert committee on the safety of cosmetics, the Scientific Committee on Cosmetology and Non-Food-Products (SCCNFP), criticized that an insufficient number of UV-filter chemicals (widely used as sunblockers) were tested in the formal validation study. In a subsequent blind trial on UV filter chemicals, which was again funded by ECVAM and co-ordinated by ZEBET, the phototoxic potential of all test chemicals was predicted correctly in the 3T3 NRU PT *in-vitro* phototoxicity test [20]. Therefore, in 1998, the EU – having accepted the 3T3 NRU PT test as the first experimentally validated *in-vitro* toxicity test for regulatory purposes – officially applied to the OECD for worldwide acceptance of this *in-vitro* toxicity test. In 2000, the European Commission officially accepted and published the 3T3 NRU PT phototoxicity test in *Annex V* of *Directive 67/548 EEC on the Classification, Packaging and Labelling of Dangerous Substances* [21]. Thus, this *in-vitro* test is the first formally validated *in-vitro* toxicity test to be accepted into Annex V, and is the only phototoxicity test accepted for regulatory purposes in Europe. Meanwhile, in 2004 the OECD accepted the 3T3 NRU PT phototoxicity test at the worldwide level as the first *in-vitro* toxicity test to be introduced into the OECD Guidelines for the testing of chemicals [15].

10.10 Use of QSAR and Physico-Chemical Exclusion Rules to Predict Skin Irritation Potential

During the past 20 years, the EU legislation for the notification of chemicals has focused on new chemicals and, at the same time, has failed to cover the evaluation of existing chemicals in Europe. Therefore, in a new EU chemicals policy (REACH; Registration, Evaluation and Authorisation of Chemicals), the European Commission proposes to evaluate 30 000 existing chemicals within a period of 15 years. Data on the hazardous properties of chemicals must be provided by industry, which must also cover the costs if testing is required. There is agreement within the EU Commission that additional toxicity testing should, in the first place, rely on non-animal *in-vitro* tests, both for financial and animal welfare reasons. A realistic scenario, based on an in-depth discussion of potential toxicological developments and an optimized "tailor-made" testing strategy, shows that in order to meet the goals of the REACH policy, animal numbers may be significantly reduced, if industry were to use in-house data from toxicity testing (which is confidential), and if non-animal tests were used as well as information from quantitative structure–activity relationships (QSARs) applied to substance-tailored testing schemes.

An evaluation of a set of QSAR rules for predicting the absence of skin irritation and/or corrosion was recently conducted by Ingrid Gerner and colleagues from the BfR [22–24]. Scientists from the National Institute of Health in the Netherlands (RIVM) have critically evaluated the BfR approach on behalf of the European Chemicals Bureau (ECB) of the European Commission [25]. The evaluation of the rule-base on irritation and corrosion was developed by the BfR, the German Federal Institute for Risk Assessment [23, 24]. This rule-base predicts non-irritation and non-corrosion using physico-chemical cut-off values, defining general rules applicable to all substances and separate rules for special chemical classes of substances.

The evaluation on behalf of the ECB includes first, the compliance of the rule-base with the OECD principles on (Q)SARs; second the derivation of the (Q)SAR rules; and finally, the external validation of these rules, including an assessment of the suitability of the dataset used for validation.

The distribution of the training set data over the domains of the physico-chemical parameters used in the rule-base is visualized and analyzed. Recommendations are given for setting the cut-off values of the rules at a consequently "safe" level (not allowing for any exception to the rule in the training set), and for including a consistently calculated safety margin. Specific results of the analysis were that:

- the rule-base fulfils the OECD principles on (Q)SARs for the largest part;
- most rules cover all irritant/corrosive substances in the training set; however:
- some physico-chemical parameters have a limited predictive value:
  - lipid solubility, as this is not a generally available parameter,
  - vapor pressure, as the experimental data used for derivation of the rules is not conclusive, and
  - melting point, as the cut-off values for this parameter were not set at a "safe" level, making predictions based on melting point less reliable.

An external validation of the set of rules using 201 new substances not present in the training set, showed 99.3% correct predictions of non-corrosivity, and 96.6% correct predictions of non-irritancy. These predictions would allow declassification as R34/R35 – corrosive for 28.4% of the chemicals, and R38 – irritant for 42.3% of the chemicals. These results would thus allow the waiving of skin irritation tests for at least 42.3% of the EU's new substance notifications. Four predictions were incorrect, however: for three of these, reasons could be given why the set of rules failed (two substances were misclassified based on melting point rules shown to be unreliable in our analysis), and/or how these incorrect predictions can be avoided in the future. The performance of the rules then increases to 100% correct predictions of non-corrosivity and 98.8% correct predictions of non-irritancy.

The results show that the evaluated rule-base is highly useful for regulatory purposes, as almost all OECD principles on QSARs are met, and the good predictivity could lead to the waiving of skin irritation tests for at least 42.3% of EU new substance notifications [25].

10.11 Alternative Methods Used in the Development and Safety Testing of Drugs, Biologicals, and Medical Devices (Table 10.1)

As outlined above, the number of experimental animals used by the drug industry in Germany has decreased during the past 25 years by more than 90% (see Fig. 10.2). This development is typical for the international drug industry, and is due to a general change of models that are used in the life sciences. Research in pharmacology is no longer centered around animal models of disease but rather is focused on molecular models. Biochemical pharmacology has been replaced by molecular pharmacology, which is shifting towards molecular genetics. The new molecular methods allow HTS of several thousand new drug candidates each month such that, today, animal models are only to monitor the efficacy and safety of the most promising lead chemicals before they enter clinical trials in humans.

Table 10.1 *In-vitro* tests which have replaced test animals in the quality control of biological drugs and in toxicology.

Purpose of testing			
In-vitro method	**Animal test**	**Degree of replacement**	**Accepted by regulatory authorities (country)**
Quality control: biological drugs			
Pregnancy test (immune assay)	Frog test	Complete replacement	Worldwide
Pyrogenicity test (limulus-(LAL)-test) (human cytokine TNF-α; ELISA)	Rabbit test	Replacement for protein-free solutions Replacement for all applications	US-, EU- and Japan-Pharmacopoeia EU-Pharmacopoeia submitted
Vitamins and hormones (vitamins, oxytocin, calcitonin, parathormone, sexual hormones)	Bioassay in chicken, rats, and mice	Complete replacement	EU-Pharmacopoeia
HPLC, immune assays			
Insulin-determination (HPLC)	Convulsion test in mouse	Complete replacement	US- and EU-Pharmacopoeia
Insulin-determination (HPLC)	Blood glucose determination in mouse and rabbit	Complete replacement	US- and EU-Pharmacopoeia
Toxicity testing			
Eye irritation (HET-CAM test; BCOP test; isolated chicken eye; isolated rabbit eye	Draize test in rabbit's eye	Replacement for severely irritating materials	EU according to OECD Test Guideline 405 Germany, Belgium Netherlands, U.K.
Phototoxicity (3T3 NRU phototoxicity test)	Phototoxicity tests in rabbit, rat, and guinea pig	Complete replacement	OECD Test Guideline 432
Skin corrosion (human skin constructs)	Corrosivity testing on rabbit skin	Complete replacement for corrosive materials	OECD Test Guideline 431
Skin penetration (human skin)	Skin penetration test on the skin of rats	Complete replacement	OECD Test Guideline 428
Delayed neurotoxicity of organophosphates (NTE-esterase determination in neuroblastoma cells)	OECD: neurotoxicity test of organophosphates in chicken	Partial replacement for esterase inhibitors	Worldwide according to OECD Guidelines 418 and 419

Due to progress in molecular genetics and biotechnology during the past 20 years, most hormones and other biologicals are no longer extracted from human and animal tissues but rather are produced by bacteria and cell lines. Insulin is an excellent example of this process as, in the past, it had to be extracted from the pancreas of cattle and pigs, and the potency of the extracts standardized in tests on rats and rabbits. In addition, safety tests had to be conducted on animals to detect impurities resulting from the extraction and purification procedures. For reasons of quality control, both potency and safety tests had to be conducted on each batch of insulin, and consequently, companies producing insulin conducted animal tests continuously in order to ensure the quality and safety of the drug.

Today, in the treatment of diabetes, bovine and porcine insulin have been replaced by pure human insulin produced by bacteria or genetically engineered mammalian cell lines. Therefore, extraction procedures are no longer needed, there is no contamination from the tissues of the donor species and, consequently, animal tests needed for quality control are no longer necessary. In contrast, the production of human insulin today is conducted according to the high standards of good manufacturing practice (GMP), which includes quality control of the final product by sensitive analytical methods. Experience has proven that this type of testing is sufficient for providing both doctors and patients with the highest quality of human insulin, which is free of contamination. For the reasons described, animal tests are today no longer conducted to ensure the safe use of insulin in diabetic patients.

Likewise, many other hormones and biologicals which previously required extraction from human or animal tissues are today produced by applying advanced technologies at the highest level of purity. Hence, most of the animal tests required to standardize the quality of these materials need no longer be conducted. Thus, the humanity criterion proposed by Russell and Burch has indeed become true in the production of hormones and other biologicals.

10.12
The Way Forward

When the 3Rs concept of reducing animal experiments was first developed by Russell and Burch more than 40 years ago, the challenge was made from the ethical point of view, and was not appreciated by the scientific community in the life sciences. In their book *The Principles of Humane Experimental Technique,* Russell and Burch defined the humanity criterion in the following manner: "*The greatest scientific experiments have always been the most humane and the most aesthetically attractive, conveying that sense of beauty and elegance which is the essence of science at its most successful.*" [1]. Due to the rapid technical progress in cell and tissue culture, as well as in molecular biology and genetics, the majority of experimental studies in the life sciences today are conducted *in vitro,* and the number of studies in experimental animals is decreasing. In fact, *in-vitro* studies form the basis of the tiered experimental approach to drug development and toxicity testing, before any experiments are conducted on animals, or clinical studies on patients (see Fig. 10.1). The examples of the toxicity testing and production of hormones and other biologicals show that Russell and Burch's humanity criterion is in fact scientifically superior to the animal experiments conducted for the same purposes in the past.

Acknowledgments

The author is indebted to his collaborators at ZEBET in the BfR; Dr. Barbara Grune, Helena Kandarova, Dr. Manfred Liebsch, Dr. Andrea Seiler, and Dr. Richard Vogel, without whom ZEBET would not have reached its international reputation as one of the leading centers devoted to the 3Rs, as proposed by William Russell and Rex Burch. Our studies at ZEBET have depended on generous funding from ECVAM, the EU Centre for Validation, and the BMBF, the National Department for Research and Technology of the Federal Government of Germany.

References

1 Russell, W. M. S. and Burch, R. L. (1959). *The Principles of Humane Experimental Technique*. Methuen, London, UK.

2 European Commission (1986). *EU Directive 86/609/EEC on protection of animals used for experimental and other scientific purposes*. EU DG Environment, Brussels, Belgium.

3 Bundesverband der Pharmazeutischen Industrie e.V. (BPI) (1981). *Tiere in der Arzneimittelforschung. Nutzen und Grenzen von Tierversuchen und anderen experimentellen Modellen*. BPI Abt. Presse und Öffentlichkeitsarbeit, Frankfurt, Germany, 32 pp.

4 OECD (Organisation for Economic Co-operation and Development) (1982). OECD Guidelines for Testing of Chemicals. OECD Publication Office, Paris, France.

5 D'Arcy, P. F. and Harron, D. W. G. (Eds.) (1995). *Proceedings of the Third International Conference on Harmonisation (ICH) Yokohama 1995*. The Queens University of Belfast, Belfast, UK.

6 Balls, M., Blaauboer, B., Brusik, D., Frazier, J., Lamp, D., Pemberton, M., Reinhardt, C., Robertfroid, M., Rosenkranz, H., Schmid, B., Spielmann, H., Stammati, A. L., and Walum, E. (1990). Report and recommendations of the CAAT/ERGATT workshop on the validation of toxicity test procedures. *Alt. Lab. Anim.* 18, 313–337.

7 Balls, M., Botham, P. A., Bruner, L. H., and Spielmann, H. (1995). The EC/HO international validation study on alternatives to the Draize eye irritation test. *Toxicol. In Vitro* 9, 871–929.

8 Balls, M., Blaauboer, B. J., Fentem, J., Bruner, L., Combes, R. D., Ekwal, B., Fiedler, R. J., Guillouzo, A., Lewis, R. W., Lovell, D. P., Reinhardt, C. A.,

Repetto, G., Sladowski, D., Spielmann, H., and Zucco, F. (1995). Practical aspects of the validation of toxicity test procedures. The report and recommendations of ECVAM Workshop 5. *Alt. Lab. Anim.* 23, 129–147.

9 Curren, R. D., Southee, J. A., Spielmann, H., Liebsch, M., Fentem, J., and Balls, M. (1995). The role of prevalidation in the development, validation and acceptance of alternative methods. *Alt. Lab. Anim.* 23, 211–217.

10 National Institute of Environmental Health (NIEHS) (1997). *Validation and Regulatory Acceptance of Toxicological test Methods: a report of the Ad hoc Interagency Co-ordinating Committee on the Validation of Alternative Methods.* NIH Publication No. 97-3981. NIH, Research Triangle Park, USA.

11 OECD (1996). *Final Report of the OECD Workshop on Harmonisation of Validation and Acceptance Criteria for Alternative Toxicological Tests Methods.* OECD Publication Office, Paris, France.

12 European Commission (2000). EU Directive 2000/33/EU for the 21st Amendment of Annex V of the EU Directive 86/906/EEC for classification and labelling of hazardous chemicals: Test guideline B-40 "skin corrosivity – *in vitro* method". O. J. European Communities, 8th June 2000, L136, 85–97.

13 European Commission (2000). EU Directive 2000/33/EU for the 21st Amendment of Annex V of the EU Directive86/906/EEC for classification and labelling of hazardous chemicals: Test guideline B-41 "phototoxicity – *in vitro* 3T3 NRU phototoxicity test". O. J. European Communities, 8th June 2000, L136, 98–107.

14 OECD (2004). *OECD guidelines for the testing of chemicals: Test Guideline 431 "In Vitro corrosivity test using human skin models"*. OECD Publication Office, Paris, France.

15 OECD (2004). *OECD guidelines for the testing of chemicals: Test Guideline 432 "In vitro 3T3 NRU phototoxicity test"*. OECD Publication Office, Paris, France.

16 Spielmann, H., Balls, M., Brand, M., Döring, B., Holzhütter, H. G., Kalweit, S., Klecak, G., L'Epattenier, H., Liebsch, M., Lovell, W. W., Maurer, T., Moldenhauer, F., Moore, L., Pape, W. J. W., Pfannenbecker, U., Potthast, J., De Silva, O., Steiling, W., and Willshaw, A. (1994). EC/COLIPA project on *in vitro* phototoxicity testing: first results obtained with the Balb/c 3T3 cell phototoxicity assay. *Toxicol. In Vitro* 8, 793–796.

17 Spielmann, H., Lovell, W. W., Hölzle, E., Johnson, B. E., Maurer, T., Miranda, M., Pape, W. J. W. Sapora, O., and Sladowski, D. (1994). *In vitro* phototoxicity testing. The report and recommendations of ECVAM Workshop 2. *Alt. Lab. Anim.* 22, 314–348.

18 Spielmann, H., Balls, M., Dupuis, J., Pape, W. J. W., Pechovitch, G., de Silva, O., Holzhütter, H. G., Clothier, R., Desolle, P., Gerberick, F., Liebsch, M., Lovell, W. W., Maurer, T., Pfannenbecker, U., Potthast, J. M., Csato, M., Sladowski, D., Steiling, W., and Brantom, P. (1998). The international EU/COLIPA *in vitro* phototoxicity validation study: results of Phase II (blind trial); part 1: the 3T3 NRU phototoxicity test. *Toxicol. In Vitro* 12, 305–327.

19 ESAC (ECVAM Scientific Advisory Committee) (1998). Statement on the scientific validity of the 3T3 NRU PT test (an *in vitro* test for phototoxic potential). *Alt. Lab. Anim.* 26, 7–8.

20 Spielmann, H., Balls, M., Dupuis, J., Pape, W. J. W., de Silva, O., Holzhütter, H. G., Gerberick, F., Liebsch, M., Lovell, W. W., and Pfannenbecker U. (1998). A study on UV filter chemicals from Annex VII of European Union Directive 76/768/EEC, in the *in vitro* 3T3 NRU phototoxicity test. *Alt. Lab. Anim.* 26, 679–708.

21 European Commission (1983). Directive 83/467EEC adapting to technical progress for the fifth time Council Directive 67/548/EEC on the approximation of the laws, regulations and administrative provisions relating to the classification, packaging and labelling of dangerous substances. EU DG Environment, Brussels, Belgium.

22 Gerner, I., Graetschel, G., Kahl, J., and Schlede, E. (2000). Development of a decision support system for the introduction of alternative methods into local irritation/corrosion testing strategies: development of a relational data base. *Alt. Lab. Anim.* 28, 11–28.

23 Gerner, I., Zinke, S., Graetschel, G., and Schlede, E. (2000). Development of a decision support system for the introduction of alternative methods into local irritancy/corrosivity testing strategies: creation of fundamental rules for a decision support system. *Alt. Lab. Anim.* 28, 665–698.

24 Gerner, I., Spielmann, H., Hoefer, T., Liebsch, M., and Herzler, M. (2004). Regulatory use of (Q)SARs in toxicological hazard assessment strategies. *SAR QSAR Environ. Res.* 15, 359–366.

25 Rorje, E. and Hulzebos, E. (2005). Evaluation of (Q)SAR for the prediction of skin irritation/corrosion potential – physico-chemical exclusion rules. European Chemicals Bureau www.ecb.jrc.eu.int, 46 pp.

Part IV
Summary and Visions

Drug Testing In Vitro: Breakthroughs and Trends in Cell Culture Technology
Edited by Uwe Marx and Volker Sandig

ISBN: 978-3-527-31488-1

11
How Drug Development of the 21st Century Could Benefit from Human Micro-Organoid *In-Vitro* Technologies

Uwe Marx

11.1
Introduction

Each year, approximately 20 to 25 new drugs and biologics are approved by the US Food and Drug Administration (FDA), with a growing share of biologics. In recent years, the development costs of new drugs and biologics have increased dramatically. Currently, drugs and biologics proceed from preclinical development into the market with a probability of success between 2 and 15%, but with total aggregated costs of US$ 500–800 million for each successful market approval achieved. Despite growing regulatory bureaucracy, capabilities on safety and efficacy assessment are the key cost drivers.

During these early years of the 21st century, with the emergence and maturation of the "-omic" technologies (genomics, proteomics, glycomics, lipomics), unprecedented discoveries are ongoing in the fields of system biology. With these accumulated data, for the first time in human history, rational approaches for the discovery, evaluation and development of new medicines might become reality. This will allow for selection of the correct candidates at very early stages in the drug development timeline, thus substantially reducing the failure rate in clinical trials. The lack of reliable drug-testing technologies able to predict the outcome of individual human exposure turned out to be the most prohibitive bottleneck enforcing these rational, risk-benefit-structure, balanced, cost-effective development approaches. Experience during the past decades has shown clearly that neither animal testing – even on our nearest relatives, the chimpanzees – nor conventional culture screening and testing using cell lines, sufficiently emulate the complex behavior in human individuals which is the basis of proper rational drug development.

Recent achievements in stem cell research, human cell engineering, biomaterial sciences, tissue bioreactor design and micro-biosensor developments are crucial prerequisites for the development of novel platform technologies for human micro-organoid *in-vitro* cultures to overcome these bottlenecks. Despite exciting achievements in each of these areas, predictive procedures for assessing human

Drug Testing In Vitro: Breakthroughs and Trends in Cell Culture Technology
Edited by Uwe Marx and Volker Sandig

ISBN: 978-3-527-31488-1

exposure via drug screening and testing based on such platform technologies remain very rudimentarily available at the research level. The combined efforts of biologists, physicians, engineers and other different disciplines are necessary to derive *in-vitro* equivalents for the most important human organs and systems for adsorption, distribution, metabolism, excretion, toxicity (ADMET) and efficacy testing on an industrial scale and throughput.

11.2
One Hundred Years of *In-Vitro* Culture

Over the past 100 years, scientists have been trying to culture human tissue *in vitro*, both to gain knowledge and to develop new medicines. Surprisingly, as early as 1912 Alexis Carrel from the Rockefeller Institute for Medical Research in New York published in an article "On the permanent life of tissues outside of the organism", that some *in-vitro* "... cultures could be maintained in active life for fifty, fifty-five, and even for sixty days. These results showed that the early death of tissues cultivated *in vitro* was preventable, and, therefore, that their permanent life was not impossible" [1].

During the early 1900s, synthetic cell culture media, antibiotics and disposable tissue culture flasks were far from being invented, and clean benches and bioreactors (if any) were regarded as pure science fiction. In 1929, an avian bone of more than 7 mm length and with clear signs of calcification was generated *in vitro* from embryonic cells for the first time [2]. At this time, scientists had long concentrated on the research of tumor cell lines in suspension or monolayer cultures. Historically, this was triggered by the fact that conventional culture systems during the past 100 years had selectively supported the growth of cells that relied mainly on glycolysis for their energy supply. These included tumor cells, rare types of differentiated oxygen-independent tissues and, eventually, most of the very early progenitor cells. Tremendous achievements have been made along this development line during the past few decades, resulting in cell lines and test systems that were perfectly suited to dedicated areas of drug screening and testing. Chapters 5 to 8 of this book emphasize all aspects of those latest achievements, such as the use of primary cells, cell lines and embryonic cells in different discovery and testing strategies. These studies also demonstrate impressively the high-quality standards of modern tools available to manipulate subcellular and cellular levels within these systems. Today, these suspension or monolayer-based systems are valuable test systems that are capable specifically of answering questions regarding the interference of drugs with ligand–receptor interactions and intracellular pathways. Substantial improvements in the metabolic and genetic engineering of cells, as described notably in Chapter 7, may further improve the meaningfulness of these *in-vitro* systems within the frame of their specific applications.

The low solubility of oxygen in culture media prevents many differentiated primary cells and tissues from behaving physiologically in culture, as most of them prefer to use the biologically most effective energy supply of generating ATP via the oxidative chain. The second historical development line – human histotypic cultures – underwent a revival during the late 1960s, when the crucial role of efficient oxygen supply was fully recognized [3] and innovations and technical systems appeared that improved oxygen distribution in cell culture. Interestingly, some of the early human histotypic cultures, such as Dexter's culture of human hematopoietic stem cells on feeder layers, first showed the importance of the interaction of different cell types for growth and functionality. During the late 1990s, tissue engineers first applied the principles of biology and engineering to develop a functional substitute for damaged human tissue, and this raised tremendous hope for the treatment of as-yet irreparable cell damage. Although the first wave of tissue therapies did not meet expectations and failed commercially, it did provide crucial initial knowledge of how to engineer tissues that in time would emulate their human counterparts. More recently, our instinctive hopes of identifying the ultimate solutions for organ and tissue repair were raised again with the discovery of stem cell technologies, and the prospect of regenerating each and every human organ from embryonic or adult stem cells. For future tissue culture techniques it has become clear that, in addition to efficient oxygen and nutrient supplies, it is also vital to establish local gradients such as growth factors, oxygen tension, and pH. Moreover, other – as yet undiscovered – parameters, as well as structured surfaces for chemotaxis and local settlement (including intercellular cross-talk through tight junctions between cells), are crucial prerequisites for the proper emulation of *in-vivo* environments [4].

These proposals provoked a shift from the development of homogeneous culture systems to heterogeneous systems, and placed an emphasis on controlled, continuously adjustable, long-term culture processes. The basic aims of these cell culture device and process developments are to create an architecture and homeostasis that mimics the relevant human microenvironment for self-organization of a specific tissue. During the past 30 years, the medical and biopharmaceutical manufacturing industries have, in their separate ways, developed bioreactors capable of maintaining functionally viable mammalian cells *in vitro* at tissue densities, over long periods. Examples of these developments include hybrid extracorporeal livers and skin equivalents as medical devices, and perfusion hollow-fiber bioreactors for tissue-density cell culture.

With recent discoveries in human stem cell research, substantial knowledge has been acquired as to how stem cells self-renew and produce differentiated progeny under homeostatic conditions, both during ontogeny [5] and in adults. It is assumed that pluripotent stem cells reside in specific stem cell niches of each organ or tissue with, under physiological conditions, the number of tissue stem cells remaining relatively constant. The driving mechanisms for differentiation in these niches are divisional and environmental asymmetries. Divisional asymmetry is caused by intrinsic cellular factors within the cell division process, whereas the exposure of two identical daughter stem cells to different extrinsic signals may lead to environmentally driven differentiation. Currently, initial attempts are being made to explore these recently identified characteristics in novel cell culture systems and devices supporting divisional and/or environmental asymmetry for

Table 11.1 Examples of human sub-organoid structures with a prominent functionality and highly variable conglomerate geometry.

Organ	Sub-organoid structure	Function	Shape/Size
Lung	Alveolar sacs and alveoli	• Gas exchange • Adsorption of compounds from gaseous phases • Distribution to the blood	Spheroid/hundreds of μm in diameter
Colon	Mucosa and submucosa	• Adsorption and distribution of nutrients or drugs • Secretion of mucosa	Multilayer wall/several dozens to several hundred μm thick
Skin	Epidermis	• Mechanical protection • Adsorption and distribution of compounds	Multilayer barrier/several dozens μm thick
Capillaries	Blood–organ barrier	• Distribution of compounds and cells to organs	Tubular form/tens of μm in cross-section
Liver	Liver lobules	• Metabolism of plasma compounds • Bile production	Hexagonal cross-section/several hundred μm in diameter
Kidney	Renal corpuscle	• Excretion of metabolic compounds	Bladder shape/hundreds of μm in diameter
Lymph node	Germinal centers	• Immune recognition • Antibody affinity maturation	Spheroid/hundreds of μm in diameter
Pancreas	Islets of Langerhans	• Insulin production	Spheroid/a few mm in diameter
Pituitary gland	Adenohypophysis	• Hormone secretion	Bladder shape/a few mm in diameter

tissue differentiation *in vitro*. One impressive human adult stem cell niche which illustrates the complexity and importance of the tissue microenvironment is that of bone marrow hematopoietic stem cells [6].

11.3 A Unique Chance Has Been Created by Nature

A paradigm of stringent correlation between architecture and functionality applies to all levels of biological existence on Earth. These levels of increasing biological complexity appeared step by step during a multi-million year process of evolution. Existence was most likely triggered by slight changes of the external environment which, in turn, created the ability for self-assembly to the next level of complexity. For humans, molecules, cells, sub-organoid tissues, organs, systems, and finally the individual organisms themselves, were thought to represent these levels.

For a long time, the role and function of the sub-organoid structures in man were underestimated, but today it has been proven that almost all organs and systems are built up by multiple, identical, functionally self-reliant, structural units. Interestingly, these sub-organoid units are of very small dimensions, ranging from several cell layers to a few millimeters. A small selection of examples of such human sub-organoid structures, all with a prominent functionality and highly variable conglomerate geometry, are listed in Table 11.1.

Due to distinguished functionality, a high degree of self-reliance and multiplicity of such micro-organoids within the respective organ, their reactivity pattern to drugs and biologics seems representative of the whole organ. Nature created very small, but sophisticated, biological structures to realize most prominent functions of organs and systems. The multiplication of these structures within a given organ is Nature's risk-management tool to prevent the total loss of functionality during partial organ damage. In evolutionary terms, however, this concept has allowed the easy adjustment of organ size and shape to the needs of a given species (e.g., liver in mice and man), while still using almost the same master plan to build up the single functional micro-organoid unit.

A unique and outstanding chance for modern drug discovery and development lies in the establishment of equivalents of functionally relevant to ADMET- and immunogenicity testing of human micro-organoids *in vitro* predictive to human exposure. With such proper *in-vitro* equivalents, a new quality of efficacy data for drugs and biologics can be envisioned prior to clinical trials.

11.4 How Do We Explore This Unique Chance?

The overview in Chapter 1 provides a taste of the extent to which these technologies could be improved today, and suggests how they might be applied to drug screening or testing. In Chapters 2 and 3, the authors illustrate the progress and challenges to meet the exceptionally high specifications for bioreactors and biosensors in this area, while Chapter 4 provides some examples of how to extend the use of cell culture technologies emulating human organ functions to other areas of application. Today, a number of *in-vitro* test systems based on three-dimensional (3-D) skin equivalents are available. Most notably, the cosmetics industry – which at least in Europe is facing a regulatory ban on animal tests within the next few years – is relying increasingly on testing strategies involving 3-D human tissue culture. Generally speaking, due to the great complexity of such cultures and the substantial hurdles in terms of human cell supply, standardization issues, miniaturization and automation, the pharmaceutical industry does not currently capitalize on such technologies. However, a number of pharmaceutical companies have implemented drug discovery programs on stem cell and 3D tissue culture tools into their R&D activities. In summarizing the forefront developments in 3-D histotypic tissue cultures, as well as in conventional suspension and monolayer high-throughput testing on primary cells and cell lines, we have attempted to provide a comprehensive review of the methods and tools which form the basis for the development of new platform technologies, allowing the generation of equivalents of each and every human sub-organoid structure of relevance *in vitro*. As such technologies rely essentially on human cell sources, an effective supply in terms of yield and quality are prerequisites for success. The current status, addressing needs in terms of further expansion of cell and tissue collection and characterization, is outlined in Chapter 9.

11.5
A Roadmap to Enforce New Platform Technologies

Clearly, with recent breakthroughs in these different disciplines, a unique and specialized knowledge can be established. It appears to be a matter of time and correct interdisciplinary cooperation and management to bring this knowledge together, ultimately to enable the development of robust, high-content drug-testing systems, devices and procedures on the basis of long-term human micro-organoid cell culture of each relevant functional human entity.

In practice, the following program should be utilized in order to achieve this final goal:

- The design of bioreactor/cell culture systems supporting architecture and homeostasis.
- The development of processes for the assembly and maintenance of all relevant human micro-organoids in the bioreactor/cell culture systems, of drug application tools and of read-out assays.
- The establishment of a sufficient human cell supply for each type of human micro-organoid and each relevant genotype.

Each of these program points has its challenges and hurdles, and these are discussed briefly in the following sections.

11.5.1 The Design of Cell Culture Systems and Bioreactors

The aim in designing any drug-testing bioreactor should be to adhere as much as possible to the architecture and homeostasis of the respective sub-organoid structure in man. Thus, the development of biomaterials and shapes that emulate the extracellular habitat and architecture are mandatory. Due to the fact that Nature miniaturizes these functionally self-reliant sub-organoids in the body (often to a volume of less than 1 mm^3), bioreactors should employ this possibility with regard to decreased cell demand and ease of high throughput to the same scale. The development of appropriate microsensors to monitor and control such small cell culture spaces is necessary and ongoing. Currently, the major hurdles are: (1) a lack of biomaterials that efficiently emulate the natural extracellular matrix; (2) the biocompatible miniaturized cell culture spaces themselves; and (3) the automation and multiplication of such systems for high-throughput testing.

These hurdles apply to all types of bioreactor, and are not necessarily limited to the final testing devices themselves. In order to succeed in a drug-testing strategy of human micro-organoids *in vitro*, stem cell expansion devices may be essential for cell supply. Within such cell- and tissue-manufacturing devices, both intrinsic symmetrical division as well as intrinsic and environmental asymmetric division for the expansion and differentiation of the necessary cell populations to charge the drug-testing bioreactors are mandatory. In contrast to the drug-testing bioreactors where differentiated micro-organoids are kept alive, these devices must emulate the embryonal environment or somatic stem cell niches. An additional challenge here is to prevent the induction of irreversible molecular pathways by traumatic events in stem cell fate. Thus, it is essential that a proper environment is applied, not only for the formation of differentiated human micro-organoids but also for human stem cells directly after delivery of the primary material from the human donor.

11.5.2 Process Development

The following parameters are crucial to the proper emulation of primary human tissues *in vitro*, but are currently underestimated in bioreactor and tissue culture developments:

- the constancy of the systemic pH, pCO_2, pO2;
- the physiological nature and effective range of the buffer systems (e.g., carbonate, hemoglobin, etc.); and
- the consistency and composition of interstitial fluids (redox potential, osmolality, growth factors, lipids, proteins, etc.).

The reasons for this are either limited knowledge of their importance in human organ and system behavior, or an insufficiency of currently available technical solutions to maintain these parameters within a specific range of action. Such parameters must be designed for the *in-vitro* needs of each and every relevant human micro-organoid, which is thought to be in accordance with their original environment (which might differ dramatically between sub-organoids in the liver, kidney or brain). Nonetheless, generalized ranges of acceptance parameters for the supply of nutrients and the removal of waste products can be drawn up for human tissues. Moreover, the individual layout for each envisioned micro-organoid can differ substantially, and this will impact on both the bioreactor design and process development.

11.5.3
Human Cell Supply

Rather than collecting differentiated tissues from humans to reassemble each and every relevant sub-organoid structure *in vitro*, the expansion and differentiation of the stem cells of a given donor with a known genotype into the relevant cell types for the *in-vitro* assembly of micro-organoids should be the preferred strategy. The expansion and differentiation capabilities of human stem cells for the generation of histotypic micro-organoids of each and every organ is illustrated in Figure 11.1.

With respect to the individual human life span, stem cells can be derived from three phases: (1) prenatal (embryonic and fetal stem cells derived prior to birth); (2) postnatal (derived directly after birth from the umbilical cord or during childhood); and (3) adult stem cells (derived from organ-specific stem cell niches of a fully developed individual). It is generally accepted that the expansion potential decreases along with the increasing life time of an individual, and that the potency to differentiate into each type of tissue is higher at early development stages. Assuming a tissue density of 1×10^9 cells mL^{-1}, approximately 40 divisions (symmetric and asymmetric) are, in theory, necessary to develop 1 kg of differentiated tissue from a single embryonic stem cell. If future drug-testing bioreactors are to provide proper culture conditions in cell culture spaces miniaturized to the smallest biological scale of the respective sub-organoid structure, then for very rough estimates of cell demands one can assume a median volume of 1 mm^3 per single donor-related histotypic micro-organoid culture. Thus, 1 kg of tissue corresponds to approximately one million micro-organoids for high-throughput testing. It remains unclear as to which of the several hundred sub-organoid structures in man are relevant to drug screening and testing within the drug development timelines, but with the latest achievements, the necessary quality and yield of donor-related differentiated histotypic micro-organoids should be achievable in the near future. Genotypic diversity and predisposition can be addressed by using respective donor populations. Large characterized cell and tissue banks, including stem cells and somatic cells and tissues at different stages of their development, need to be stored to ensure a constant and consistent supply of the *in-vitro* testing technologies with cells and tissues. Ideally, those cells and tissues would be generated from only a few initially derived donor stem cells by means of expansion *in vitro*.

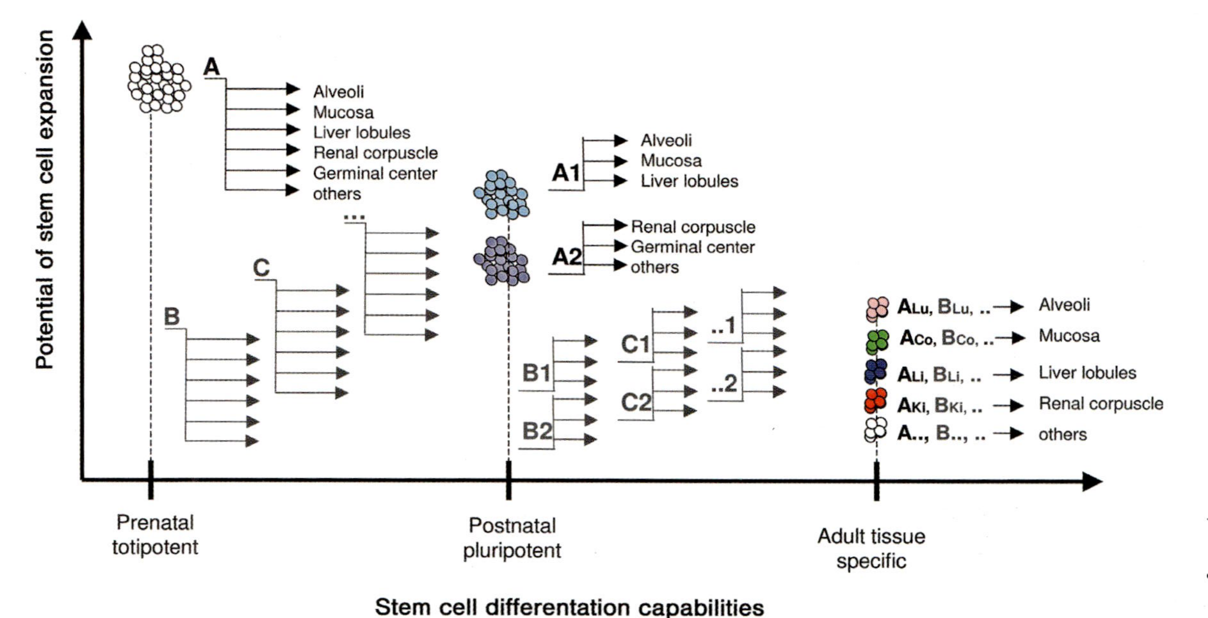

Fig. 11.1 Human stem cell sources for the *in-vitro* generation of organ specific micro-organoids. Prenatal totipotent stem cells are thought to have significant undifferentiated expansion potential if proper expansion devices can be developed. From a resulting totipotent stem cell pool of each relevant genotype (A, B, C ...), equivalents of each relevant sub-organoid human structure can be produced and maintained in testing bioreactors. In contrast, the adult lineage committed tissue-specific stem cells can be expanded only to a much lower degree, providing a somatic stem cell niche environment in respective expansion bioreactors. From each resulting organ-specific adult donor stem cell pool (e.g., A liver) the respective micro-organoid (liver lobules of donor A) should be developed and maintained *in vitro*. *In-vitro*-expanded pools of postnatal pluripotent stem cells (e.g., umbilical cord stem cells) can be differentiated in several, but not all, tissue specimens. Expansion devices for these stem cells need to support the respective pattern of pluripotency (e.g., hematopoietic lineage) at high expansion rates.

11.6
Outlook

In the past, technical systems capable of orchestrating all relevant parameters into functional micro-organoid cultures with balanced homeostasis *in vitro* seemed to be a dream. However, the various chapters of this book have addressed recent advancements in different research areas which, once they can be synchronized into comprehensive interdisciplinary development programs, will overcome the remaining challenges and hurdles. Moreover, when the dream becomes reality, broad applications in modern drug discovery and testing can be envisioned. Meanwhile, preclinical drug testing seems to be the most important application, and its use in early phases of the drug development time line for predictive drug screening, as well as for post-approval extension or restriction of indication to specific genotypic subpopulations, might be of great value (Fig. 11.2). Finally, properties relating to individuals can be tested for individualized therapies. As discussed earlier, an extension of use to non-pharmaceutical areas such as cosmoceuticals and nutraceuticals can also be envisioned.

The envisioned developments are embedded in a sensitive socio-ethical environment. On the one hand, success in this area will not only replace currently used animal tests – as highlighted in Chapter 10 – but also significantly increase predictiveness. On the other hand, however, there are reasonable doubts of whether – and to what extent – human embryonic cell and tissues might be involved in the development and exploitation of such technologies. We strongly believe that within the next few years, by carefully weighing ethical issues against medical needs, programs will lead to revolutionary new robust and reliable technologies emulating human organ function *in vitro*. Details of conference programs and activities related to this topic are summarized in Table 11.2, which provides the reader with access to information on the latest developments in this field.

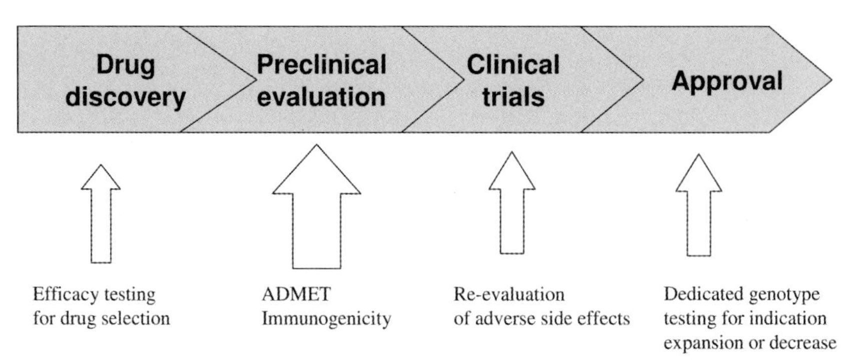

Fig. 11.2 The potential impact of micro-organoid testing technologies on the drug development timeline. ADMET and immunogenicity testing are the primary targets for newly envisioned micro-organoid drug-testing systems. The use of this technology might also be widened towards efficacy testing in drug discovery and the re-evaluation of drug effects in clinical trials, or after-market approval with regard to genetic predisposition.

Table 11.2 Conferences related to advanced drug screening and testing technologies.

	Conference	Organizer	Website
1	Tissue Models for Therapeutic Development	Cambridge Healthtech Institute	http://www.healthtech.com
2	Tissue Technologies for Discovery Research	Cambridge Healthtech Institute	http://www.healthtech.com
3	Drug Discovery Technology® & Development	IBC Life Sciences	http://www.drugdisc.com/section.asp
4	World Congresses on Alternatives & Animal Use in the Life Sciences	Alternatives Congress Trust	http://www.worldcongress.net/
5	Characterization and Comparability for Complex Biological Products	The Williamsburg BioProcessing Foundation	http://www.wilbio.com/
6	Annual Conference of SBS	The Society for Biomolecular Sciences	http://www.sbsonline.org/
7	Systems Biology	International Society for Computational Biology	http://www.iscb.org/events/
8	Annual Meeting of ESACT	European Society for Animal Cell Technology	http://www.esact.org/
9	Congress of the European Society for Artificial Organs	ESAO	http://www.umu.se/phmed/medicin/ESAO/
10	Cell & Gene Therapy Forum	Phacilitate	http://www.phacilitate.co.uk/index.html
11	European Stem Cells & Regenerative Medicine Congress	Terrapinn	http://www.lifescienceworld.com

References

1. Carrel, A. On the permanent life of tissue outside of the organism. *J. Exp. Med.* **1912**, *15*, 516–528.
2. Fell, H. B. and Robison, R. The growth, development and phosphatase activity of embryonic avian femora and limb buds cultivated *in vitro*. *Biochem. J.* **1929**, *23*, 767–784.
3. McLimans, W. F., et al. Kinetics of gas diffusion in mammalian cell culture systems. *Biotech Bioeng.* **1968**, *10*, 725–740.
4. Griffith, L. G. and Swartz, M. A. Capturing complex 3D tissue physiology *in vitro*. *Nat. Rev. Mol. Cell Biol.* **2006**, *7*, 211–224.
5. Okazaki, K. M. and Maltepe, E. Oxygen, epigenetics and stem cell fate. *Regen. Med.* **2006**, *1*, 71–83.
6. Wilson, A. and Trumpp, A. Bone-marrow haematopoeitic-stem-cell niches. *Nat. Rev. Immunol.* **2006**, *6*, 93–106.

Subject Index

Drug Testing In Vitro: Breakthroughs and Trends in Cell Culture Technology
Edited by Uwe Marx and Volker Sandig

ISBN: 978-3-527-31488-1

d

e

f

g

i

Related Titles

G. Vunjak-Novakovic, R. I. Freshney (Eds.)

Culture of Cells for Tissue Engineering

2005

ISBN-13: 978-0-471-62935-1
ISBN-10: 0-471-62935-9

R. I. Freshney

Culture of Animal Cells

A Manual of Basic Technique

2005

ISBN-13: 978-0-471-45329-1
ISBN-10: 0-471-45329-3

W. W. Minuth, R. Strehl, K. Schumacher

Tissue Engineering

Essentials for Daily Laboratory Work

2005

ISBN-13: 978-3-527-31186-6
ISBN-10: 3-527-31186-6

R. Pfragner, R. I. Freshney (Eds.)

Culture of Human Tumor Cells

2004

ISBN-13: 978-0-471-43853-3
ISBN-10: 0-471-43853-7

R. I. Freshney, M. G. Freshney (Eds.)

Culture of Epithelial Cells

2002

ISBN-13: 978-0-471-40121-6
ISBN-10: 0-471-40121-8

A. Doyle, J. B. Griffiths (Eds.)

Cell and Tissue Culture for Medical Research

2000

ISBN-13: 978-0-471-85213-1
ISBN-10: 0-471-85213-9